Chemical Reactions:
Principles and Processes

Chemical Reactions: Principles and Processes

Contributors

Dennis Görlich, Peter Dittrich et al.

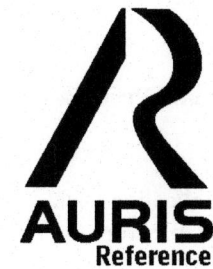

AURIS
Reference

www.aurisreference.com

Chemical Reactions: Principles and Processes

Contributors: Dennis Görlich, Peter Dittrich et al.

Published by Auris Reference Limited
www.aurisreference.com

United Kingdom

Chemical Reactions: Principles and Processes

ISBN: 978-1-78154-860-8

British Library Cataloguing in Publication Data
A CIP record for this book is available from the British Library

Printed in the United Kingdom

Exclusively distributed by CBS Publishers & Distributors Pvt. Ltd.

Sales & Distribution Rights only for India, Pakistan, Bangladesh, Sri Lanka, Nepal and Bhutan. This book is not to be sold outside these territories.

Contents

List of Abbreviations

aaRS	Amino acyl tRNA synthetases
BMC	Binary molecular code
CRS	Catalytic reaction system
CRNs	Chemical Reaction Networks
DAG	Directed acyclic graph
FHP-II	Frisch–Hasslacher–Pomeau
GRN	Gene regulatory network
IDH	Isocitrate dehydrogenase
IDHKP	Isocitrate dehydrogenase kinase phosphatase
LBM	Lattice Boltzmann methods
MD	Molecular dynamics
NCL	Native chemical ligation
PFD	Perfluorodecalin
pNP	P-nitrophenyl
PDMS	Polydimethylsiloxane
QSGS	Quartet structure generation set
RAF	Reflexively autocatalytic and F-generated
SEM	Scanning Electron Microscopy
SCL Graph	Species-Complex-Linkage Graph
SR Graph	Species-Reaction Graph
SRM	Spectral Relaxation Method
Boc	Tert-butyloxycarbonyl
TGA	Thermo-gravimetric analysis
TF	Transcription factor
TEM	Transmission electron microscopy
TCA	Tricarboxylic acid
W/O	Water-in-oil

List of Contributors

Dennis Görlich
Bio Systems Analysis Group, Institute of Computer Science, Jena Centre for Bioinformatics and Friedrich Schiller University Jena, Jena, Germany

Peter Dittrich
Institute of Biostatistics and Clinical Research, University of Muenster, Muenster, Germany

Diogo A. R. S. Latino
CQFB, REQUIMTE, Departamento de Química, Faculdade de Ciências e Tecnologia, Universidade Nova de Lisboa, Caparica, Portugal, CCMM, Departamento de Química e Bioquímica, Faculdade de Ciências, Universidade de Lisboa, Lisboa, Portugal

João Aires-de-Sousa
CQFB, REQUIMTE, Departamento de Química, Faculdade de Ciências e Tecnologia, Universidade Nova de Lisboa, Caparica, Portugal

Hongsheng Chen
School of Materials Science and Engineering, Chongqing University, Chongqing 400044, China

Zhong Zheng
School of Materials Science and Engineering, Chongqing University, Chongqing 400044, China

Zhiwei Chen
Fluidization Research Center, Department of Chemical and Biological Engineering, University of British Columbia, Vancouver V6T 1Z3, Canada

Xiaotao T. Bi
Fluidization Research Center, Department of Chemical and Biological Engineering, University of British Columbia, Vancouver V6T 1Z3, Canada

Dnyaneshwar B. Rasale
Department of Chemistry, Indian Institute of Technology Indore, Khandwa Road, Indore 452017, India

Apurba K. Das
Department of Chemistry, Indian Institute of Technology Indore, Khandwa Road, Indore 452017, India

Hassan H. Abdallah
School of Chemical Sciences, University Sains Malaysia, Penang 11800, Malaysia

Janez Mavri
National Institute of Chemistry, Hajdrihova 19, SI-1001 Ljubljana, Slovenia
EN-FIST Centre of Excellence, Dunajska 156, SI-1000 Ljubljana, Slovenia
Department of Chemistry, Faculty of Science, University of Malaya, Kuala Lumpur 50603, Malaysia
School of Pharmaceutical Sciences, University Sains Malaysia, Penang 11800, Malaysia

Matej Repič
National Institute of Chemistry, Hajdrihova 19, SI-1001 Ljubljana, Slovenia

Vannajan Sanghiran Lee
Department of Chemistry, Faculty of Science, University of Malaya, Kuala Lumpur 50603, Malaysia

Habibah A. Wahab
School of Pharmaceutical Sciences, University Sains Malaysia, Penang 11800, Malaysia

Naren Ramakrishnan
Department of Computer Science, Virginia Tech, Blacksburg, Virginia, United States of America
Institute of Bioinformatics and Applied Biotechnology, Bangalore, India

Upinder S. Bhalla
National Centre for Biological Sciences, Tata Institute of Fundamental Research, Bangalore, India

Ernesto Suárez
Departamento de Química Física y Analítica, Universidad de Oviedo, Julián Clavería 8, 33006, Oviedo, Spain; Tel.: +34-985103492; Fax: +34-985103125

Guy Shinar
InBrain Therapeutics Ltd., 12 Metzada St., Ramat Gan 52235, Israel

Martin Feinberg
The William G. Lowrie Department of Chemical & Biomolecular Engineering and Department of Mathematics, Ohio State University, 140 W. 19th Avenue, Columbus, OH, USA 43210

Wim Hordijk
Department of Ecology and Evolution, University of Lausanne, 1015 Lausanne, Switzerland

Stuart A. Kauffman
Tampere University of Technology, Finland
University of Vermont, 85 South Prospect Street Burlington, VT 05405, USA
Santa Fe Institute, 1399 Hyde Park Road Santa Fe, NM 87501, USA

Mike Steel
Department of Mathematics and Statistics, University of Canterbury, Private Bag 4800, Christchurch, New Zealand

Katalin M. Hangos
Process Control Research Group, Computer and Automation Research Institute, Kende u. 13-17, 1111 Budapest, Hungary; Phone:

Guy Shinar
InBrain Therapeutics Ltd., 12 Metzada St., Ramat Gan 52235, Israel
Martin Feinberg
The William G. Lowrie Department of Chemical & Biomolecular Engineering and Department of
Mathematics, Ohio State University, 140 W. 19th Avenue, Columbus, OH, USA 43210

Faiz G. Awad
Department of Pure & Applied Mathematics, University of Johannesburg, Auckland Park, Johannesburg, South Africa

Sandile Motsa
School of Mathematics, Statistics and Computer Science, University of KwaZulu-Natal, Scottsville, Pietermaritzburg, South Africa

Melusi Khumalo
Department of Pure & Applied Mathematics, University of Johannesburg, Auckland Park, Johannesburg, South Africa

Venkatachalam Chokkalingam
Experimental Physics, Saarland University, Saarbrücken

Ralf Seemann
Max Planck Institute for Dynamics and Self-Organisation (MPIDS), Göttingen
Technical Chemistry, Saarland University, Saarbrücken Germany

Boris Weidenhof
Technical Chemistry, Saarland University, Saarbrücken Germany

Wilhelm F. Maier
Technical Chemistry, Saarland University, Saarbrücken Germany

Preface

A chemical reaction is a process that leads to the transformation of one set of chemical substances to another. Classically, chemical reactions encompass changes that only involve the positions of electrons in the forming and breaking of chemical bonds between atoms, with no change to the nuclei, and can often be described by a chemical equation. The text *Chemical Reactions Principles and Processes* provides complete coverage of the fundamentals, including in-depth coverage of chemical reactions. First chapter focuses on molecular codes in biological and chemical reaction networks. Second chapter demonstrates the possibility of applying machine learning methods to automatically identify types of co-occurring chemical reactions from NMR data. In third chapter, we explore the application of a lattice gas automata (LGA) method to the heat transfer and chemical reaction of fluid flow around and through a porous circular cylinder in a channel. The aim of fourth chapter is to describe various mild chemical reactions in the development of peptide self-assembly. The purpose of fifth chapter is to critically examine the reactivity toward N7 position of guanine of four flavonoids and, for each of them, the both reactive forms are considered. Sixth chapter focuses on memory switches in chemical reaction space. Chemical reactions, conformational, and residual entropy have been discussed in seventh chapter. Concordant chemical reaction networks and the species-reaction graph have been outlined in eighth chapter. In ninth chapter, we investigate the levels of catalysis required for a self-sustaining autocatalytic network to form. In tenth chapter, the structural properties of chemical reaction systems obeying the mass action law have been investigated and related to the physical and chemical properties of the system. Concordant chemical reaction networks have been presented in last chapter.

Chapter 1

MOLECULAR CODES IN BIOLOGICAL AND CHEMICAL REACTION NETWORKS

Dennis Go¨ rlich[1,2], Peter Dittrich[1]

[1] Bio Systems Analysis Group, Institute of Computer Science, Jena Centre for Bioinformatics and Friedrich Schiller University Jena, Jena, Germany,

[2] Institute of Biostatistics and Clinical Research, University of Muenster, Muenster, Germany

ABSTRACT

Shannon's theory of communication has been very successfully applied for the analysis of biological information. However, the theory neglects semantic and pragmatic aspects and thus cannot directly be applied to distinguish between (bio-) chemical systems able to process "meaningful" information from those that do not. Here, we present a formal method to assess a system's semantic capacity by analyzing a reaction network's capability to implement molecular codes. We analyzed models of chemical systems (martian atmosphere chemistry and various combustion chemistries), biochemical systems (gene expression, gene translation, and phosphorylation signaling cascades), an artificial chemistry, and random reaction networks. Our study suggests that different chemical systems posses different semantic capacities. No semantic capacity was found in the model of the martian atmosphere chemistry, the studied combustion chemistries, and highly connected random networks, i.e. with these chemistries molecular codes cannot be implemented. High semantic capacity was found in the studied biochemical systems and in random reaction networks where the number of second order reactions is twice the number of species. We conclude that our approach can be applied to evaluate the information processing capabilities of a chemical system and may thus be a useful tool to understand the origin and evolution of meaningful information, e.g. in the context of the origin of life.

INTRODUCTION

In recent years great advances have been made in understanding the biochemical basis of biological information processing. For theoretical analysis of biological information Shannon's theory of communication [1] has been applied very successfully in various domains, like gene regulatory networks [2], bacterial quorum sensing [3], or signaling in molecular systems [4],[5]. The mathematical theory of communication focusses on uncertainty of events and intentionally neglects semantic aspects of information, because *"they are irrelevant for the engineering problem"* (Shannon [1], p. 1). However, in order to obtain a full understanding of biological information, studying also semantic as well as pragmatic aspects would be important, if not necessary [6], [7]. Although syntax, semantics, and pragmatics are interdependent [8], we focus here only on the semantic aspects of molecular networks in order to keep our formalism and analysis clear and concise.

In general, semantics refers to the relation between a sign and its meaning. This relation can be characterized by a code, which is a mapping from the signs to their meanings [9]. For example, the genetic code is a mapping between codons and amino acids [10], which is realized in cells by a complex translation machinery. An important property of a code is its contingency. This means that the relation between signs and meanings could be different, thus the relation is not determined by the signs and meanings alone [6], [9]. In particular, this implies that natural laws allow to derive the relation only by knowing the context under which the signs are interpreted.

Furthermore, it implies the existence of another context under which the signs are interpreted differently. This is why we say that the relation between signs and meanings, i.e. the code, cannot be explained by physical laws [11], like the natural laws do not help in understanding the written law or the grammar of a language. However, this notion of independence from natural laws sometimes causes confusion [11].

In order to properly use semiotic concepts in biology we should provide a link to the realm of physics by (1) selecting an experimentally grounded and reliable formal description of the targeted biological system, by (2) providing precise, not necessarily formal, definitions of the semiotic concepts that shall be applied to the system, and by (3) interpreting these definitions by linking them to the formal description of the biological system. (1) We use reaction networks as a formal description, (2) link it to the notion of organic codes as reviewed by Barbieri [9] and (3) develop a formal definition of a molecular code with respect to reaction networks [12].

With this approach, the semiotic concept of code gets – at least partially – operationalized by means of physical experiments. In particular, it allows to incorporate contingency in a formal model of molecular codes.

To illustrate the basic idea we will briefly discuss an example reaction network that contains a contingency. Fig. 1A shows a reaction network containing eight molecular species and four reactions. We assume that the network contains all possible reactions that can appear when mixing these molecules. The network then is assumed to be a complete model of the world, i.e. no species and reactions are missing that are physically possible. A reaction network can implement a *mapping* among molecular species. Here, for example, $\{A\}$ can be mapped to $\{C\}$ by reaction $A + E \rightarrow E + C$. $\{E\}$ is necessary for the reaction to happen and thus we call it a *molecular context*. The network can implement a *molecular code*, if there exists a set of molecular species that can be mapped on a second set of molecular species in at least two different ways. In this example network the sets $S = \{A, B\}$ and $M = \{C, D\}$ fulfill this property. **S** (*domain*) maps to **M** (*codomain*) by applying the context $\{E, H\}$. No two elements of the domain **S** map to the same element in the codomain **M**. There exist an alternative molecular context $\{F, G\}$, which realizes a different mapping between domain and codomain, so the mappings qualify as molecular codes.

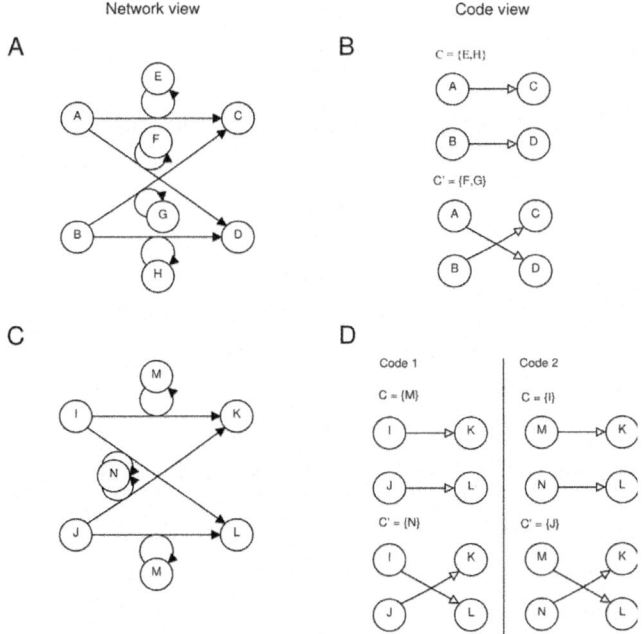

Figure 1: Two exemplary reaction networks containing molecular codes. Panel A: Chemical reaction network $\langle \mathcal{M}, \mathcal{R} \rangle$ with species $\mathcal{M} = \{A, B, C, D, E, F, G, H\}$ and

reaction rules $\mathcal{R}=\{A+E\rightarrow C+E, A+F\rightarrow F+D, B+G\rightarrow G+C, B+H\rightarrow H+D\}$; panel B: Code pair that can be realized by the network in panel A. The binary molecular codes are characterized by $\mathbf{S}=\{A,B\}$, $\mathbf{M}=\{C,D\}$, and the two molecular contexts $\mathbf{C}=\{E,H\}$, and $\mathbf{C'}=\{F,G\}$; panel C: Chemical reaction network with species $\mathcal{M}=\{I,J,K,L,M,N\}$ and the reactions $\mathcal{R}=\{I+M\rightarrow K+M, I+N\rightarrow N+L, J+M\rightarrow M+L, J+N\rightarrow N+K\}$; panel D: Two molecular code pairs can be realized by the network in panel C. Note that our code analysis does not depend on catalysis. Replacing a reaction like $A+E\rightarrow C+E$ by $A+E\rightarrow C$ would lead to the same molecular codes. doi:10.1371/journal.pone.0054694.g001

METHODS

In this section we provide a formal definition of a molecular code as a contingent mapping with respect to a reaction network. Then we formally define a reaction network's semantic capacity based on the number of molecular codes it can realize, and finally describe two algorithms for identifying molecular codes in a reaction network.

Molecular Codes are Contingent Molecular Mappings

A *reaction network* $N=\langle\mathcal{M},\mathcal{R}\rangle$ is defined by a set of molecular species $\mathcal{M}$ and a set of reactions $\mathcal{R}$ occurring among the molecular species $\mathcal{M}$. See Fig. 1A for an example. For each reaction $\rho\in\mathcal{R}$, let $\mathrm{LHS}(\rho)$ and $\mathrm{RHS}(\rho)$ denote the set of reacting and produced species of reaction ρ, respectively.

A subset of molecular species $\mathbf{C}\subseteq\mathcal{M}$ is *closed*, iff the application of all possible reactions from $\mathcal{R}$ on $\mathbf{C}$ does only produce species from $\mathbf{C}$, i.e. for all $\rho\in\mathcal{R}$ with $\mathrm{LHS}(\rho)\subseteq\mathbf{C}$: $\mathrm{RHS}(\rho)\subseteq\mathbf{C}$ [13]. For every set of species $\mathbf{A}\subseteq\mathcal{M}$ there exists a smallest closed set $G_{CL}(\mathbf{A})$ containing $\mathbf{A}$ [14]. We say that $G_{CL}(\mathbf{A})$ is the *closure* of $\mathbf{A}$. Intuitively, the closure of a set of species contains all those species that can be reached by an arbitrary long reaction path among the species of $\mathbf{A}$.

Given a reaction network $N=\langle\mathcal{M},\mathcal{R}\rangle$ and two sets of molecular species $\mathbf{S},\mathbf{M}\subseteq\mathcal{M}$, we say that $f:\mathbf{S}\rightarrow\mathbf{M}$ is a *molecular mapping* with respect to N, iff there exist a set of species $\mathbf{C}\subseteq\mathcal{M}$ (called context), such that for each pair $s,s'\in\mathbf{S}$ with $s\neq s'$: $f(s)\in G_{CL}(\mathbf{C}\cup\{s\})$ and $f(s')\notin G_{CL}(\mathbf{C}\cup\{s\})$. If there exists a molecular mapping f with respect to N, we also say that N can *realize* the molecular mapping f.

Note that in a reaction network there is usually more than one molecular context $\mathbf{C}$ that realizes a particular molecular mapping f. Intuitively, in order to "compute" $f(s)$ with the reaction network N, we put all molecules from the context $\mathbf{C}$ together with s in a reaction vessel. Then we repeatedly apply all applicable reaction rules and add the products to the reaction vessel until no novel molecular species can be added anymore. Then we check which molecular species from $\mathbf{M}$ is present, which must be – according to our definition – only one species and the result of $f(s)$.

Given a reaction network $N = \langle \mathcal{M}, \mathcal{R} \rangle$ and a non-constant (A mapping $f : \mathbf{S} \to \mathbf{M}$ is called non-constant, iff there exists $s, s' \in \mathbf{S}$ such that $f(s) \neq f(s')$) molecular mapping $f : \mathbf{S} \to \mathbf{M}$, with $\mathbf{S}, \mathbf{M}, \mathbf{C} \subseteq \mathcal{M}$ we call the mapping f a *molecular code* with respect to N, if all other mappings $g : \mathbf{S} \to \mathbf{M}$ with the same domain $\mathbf{S}$ and codomain $\mathbf{M}$ can also be realized by the reaction network N, i.e. there exist alternative molecular contexts to map $\mathbf{S}$ to $\mathbf{M}$.

The definition catches the notion of contingency as described above, i.e. the elements of the domain can be mapped to the elements of the codomain in a contingent way by changing the molecular context. In a semiotic interpretation we can also say domain and codomain contain the signs and meanings, respectively. The molecular context thus becomes the "codemaker", i.e. it is necessary to realize the code. In general, the definition given above allows for codes of arbitrary size. In order to keep our study tractable, we will focus on molecular codes that are binary, i.e. where $\mathbf{S}$ as well as $\mathbf{M}$ contain exactly two molecular species [12]. We will also not study molecular mappings that are only partially contingent. For binary molecular codes our definition can be reformulated as follows:

Given a reaction network $N = \langle \mathcal{M}, \mathcal{R} \rangle$ and two binary sets of molecular species $\mathbf{S} = \{s_1, s_2\} \subseteq \mathcal{M}$ and $\mathbf{M} = \{m_1, m_2\} \subseteq \mathcal{M}$. The mapping $f : \mathbf{S} \to \mathbf{M}$ is called *binary molecular code* (BMC), iff there exist two sets $\mathbf{C}, \mathbf{C}' \subseteq \mathcal{M}$, such that the following conditions hold : $f(s_1) \in G_{CL}(\{s_1\} \cup \mathbf{C})$, and $f(s_2) \notin G_{CL}(\{s_1\} \cup \mathbf{C})$, and $f(s_2) \in G_{CL}(\{s_2\} \cup \mathbf{C})$, and $f(s_1) \notin G_{CL}(\{s_2\} \cup \mathbf{C})$, and $f(s_2) \in G_{CL}(\{s_1\} \cup \mathbf{C}')$, and $f(s_1) \notin G_{CL}(\{s_1\} \cup \mathbf{C}')$, and $f(s_1) \in G_{CL}(\{s_2\} \cup \mathbf{C}')$, and $f(s_2) \notin G_{CL}(\{s_2\} \cup \mathbf{C}')$.

Each binary molecular code comes with a second code implementing a different mapping. The alternative code g is determined by $g(s_1) = f(s_2)$ and $g(s_2) = f(s_1)$. $\langle f, g \rangle$ is called *code pair*. Two simple example networks are shown in Fig. 1A and 1C (cf. Dataset S1 and Dataset S2 for the network description). Both networks appear to be very similar in their structure, but contain different numbers of code pairs. While the former network is capable to realize one code pair, the latter network – though being smaller – can realize two code pairs.

A Network's Semantic Capacity can be Measured by Molecular Codes

A system's *semantic capacity* $\mathcal{SC}$ is its ability to realize contingent molecular mappings, i.e. the number of code pairs CP_N that can be identified, $\mathcal{SC}(N) = CP_N$.

To compare different semantic capacities we can also use the *logarithmic semantic capacity* $\mathcal{SC}_{\log}(N) = \log_2(1 + \mathcal{SC}(N)) = \log_2(1 + CP_N)$ especially with very high values of $\mathcal{SC}$. We apply the transformation $1 + x$ to guarantee that $\mathcal{SC}_{\log}(N)$ is well defined and its smallest value is zero, in case the network cannot realize any molecular code.

In future studies, the semantic capacity can be integrated with measures of the code's quality, fitness, or cost [15], [16]. e.g. two networks with the same number of code pairs could be differentiated with respect to the costs to implement those codes.

Molecular Codes can be Identified Algorithmically

The formal definition of binary molecular codes allows to develop code-identifying algorithms. In general, the algorithms search for a combination of molecular species and reactions fulfilling the BMC conditions. Different approaches can be used to implement the BMC conditions, i.e. via closed sets, or via paths.

The closure-based algorithm calculates all closed sets and checks combinations of six closed sets for the BMC conditions. In particular, for the two elements of the domain, and the two elements of the codomain the single molecular closed sets, i.e. the closed sets that are generated by a single molecular species alone $(G_{CL}(\{m\}), m \in \mathcal{M})$, are used. There exist at most $|\mathcal{M}|$ single molecular closed sets. The closure-based algorithm has a worst-case running time complexity of $O(|\mathcal{M}|^4 n_c^2)$ with n_c as number of all closed sets contained in the system.

Domain and codomain are connected by reactions such that an alternative algorithm can be formulated using the network's paths. For the identification of BMCs the paths for all pairs of species are identified. Every combination of four paths is checked for the BMC condition. The running time complexity of this *path-based* algorithm depends on the number of paths the network contains, which can grow enormously with the network's density. Therefore, we apply a parameterized algorithm that uses only the k-shortest paths [17] between every pair of species. The worst case running time of the parameterized algorithm is bounded by $O(|\mathcal{M}|^4 k^4)$. If k is chosen too small the algorithm is not able to find all codes in the system, but gives an approximate measure. Large values of k resemble the non-parameterized path algorithm, since all paths are considered for the analysis. Pseudocode for the parameterized path algorithm, the closure-based algorithm and subroutines is given in Text S1. The different running time complexities suggests a conditional application of the algorithms. The path-based algorithm can be efficiently applied on networks that have a high number of closed sets and a low number of paths, while the closure-based algorithm can be applied in the other case, where the number of paths is high and the number of closed sets in the network is low. Interestingly, systems with high semantic capacity tend to have both, high number of closed sets and many paths, such that an algorithmic challenge remains for analyzing such systems.

RESULTS

We survey different kinds of systems for their semantic capacity by the application of the algorithms described above. In particular we analyze the gene translation chemistry, gene regulatory networks, phosphorylation cascades, combustion chemistries, the martian atmosphere photochemistry, and random reaction networks. As a result of the analysis we can assign semiotic roles to the molecular species. Table 1 summarizes the semiotic structure of the analyzed biological systems. For details on all analyzed networks see Table S1.

Table 1: Overview of semiotic interpretation of the biological systems surveyed. doi:10.1371/journal.pone.0054694.t001

Role	Gene regulatory codes	Genetic codes	Phosphorylation cascade codes
Signs	transcription factor s	DNA codons and/or unloaded tRNAs	high concentration of kinases and/or phosphatases
Meanings	gene product s	amino acid s	high/low concentration of target molecules
Molecular contexts	DNA with promoter and coding region	loaded tRNAs or a combination of loaded tRNAs, aaRSs, and codons	kinases and/or phosphatases

The Genetic Code is a Molecular Code

The genetic code, i.e. the mapping describing the translation from nucleotide triplets to amino acids, was the first biological code described as such [18] and is often used as initial example for molecular codes [9], [15], [19].

To check whether the genetic code is a molecular code as defined in this paper we need to identify contingent molecular mappings in the reaction network describing the translation from codons to amino acids. In recent species only one code is realized, thus the reaction network taken from a certain species will not contain any molecular codes. A reasonable approach to overcome this effect is to merge the known genetic codes in one reaction network, such that the merged network contains all (known) alternatives. Note that merging two chemical networks has to be done carefully to avoid unwanted inconsistencies. In particular, the networks to be merged needs to be from the same physicochemical context, which determines the reactions of the network model. This guarantees that no "artificial" contingencies are introduced. The gene translation chemistries studied here can be merged, because they take place in the same environment.

The fact that there exist more than one genetic code is known for a long time [20], [21]. The 17 known genetic codes, as listed at NCBI [22], cover nuclear and non-nuclear codes of different genera, e.g. bacterial, archaeal, and plant plastid codes, the vertebrate, invertebrate and yeast mitochondrial codes, and the alternative yeast nuclear code. The flexibility of the genetic system is also underlined by the possibility to introduce even unnatural amino acids to the genetic codes of various organisms [23]. For our analysis, we merge the 17 codes listed at NCBI by constructing a reaction network containing the 64 codons, 20 amino acids, and the specific tRNAs, which are necessary for the translation. For all mappings between DNA triplets and amino acids occurring in the 17 codes we add a reaction in the network of the form $codon + tRNA \rightarrow amino\ acid$ (see Dataset S3).

The algorithmic analysis of this network identified 16 binary molecular codes (see Text S2 for a complete list), i.e. a logarithmic semantic capacity of $SC_{log} = 4.09$. The binary codes can partly be assigned to larger molecular codes. For instance, the codons CTT,CTG,CTA, and CTC can be mapped on leucin (L) and threonin (T) and give rise to six of the found BMCs. A second group involves the mapping between AGG,AGA and glycin (G), serine (S), arginine (R) and the translation stop. This code can also be decomposed into six BMCs. There does exist four more BMCs that involve the codons TCA, TTA, TAG and TAA and the amino acids leucine (L), glutamine (Q) and the stop signal. The data suggests that it is easier for the cell to change the mapping for the stop signal, than for an amino acid. Table 2 summarizes the identified BMCs. The general existence of alternative mappings in the genetic translation system suggests that the genetic code qualifies as a molecular code. The relatively small semantic capacity of the merge network demonstrates that the genetic code, thus a principally contingent system, is under strong constraints, regarding the assignment between codons and amino acids.

To calculate the system's potential maximum semantic capacity we extend the reaction network model by including all potential mappings between codons and amino acids even if they have not been observed so far. The model includes all possible tRNA molecules, such that each codon could be read for each amino acid. In such a system the number of binary molecular codes can easily be calculated. Each pair of codons forms a code pair with each pair of amino acids. Since there exist $\binom{64}{2}$ pairs of triplets and $\binom{20}{2}$ pairs of amino acids the number of BMCs is

Table 2: Molecular codes in the reaction network model of the 17 known genetic codes. doi:10.1371/journal.pone.0054694.t002

Signs (codons)	Meanings (amino acids)	#BMC	References
CTT, CTG, CTA, CTC	L, T	6	[20,24]
AGG, AGA	G,S,R, Stop	6	[20,25–36]
AGG, TCA	S, Stop	1	[20,27,28,31,33,37]
AGA, TCA	S, Stop	1	[20,27,28,31,33,37]
TTA, TAG	L, Stop	1	[20,22,37–39]
TAA, TAG	Q, Stop	1	[20,40–43]

$$SC(\textit{gene translation}) = \binom{64}{2} \cdot \binom{20}{2} = 383{,}040.$$

(1)

The logarithmic semantic capacity is approximately 18.55. The difference to the merge network (which relies completely on observed variation in the code) suggests that cells use only a small fraction of their semantic capacity.

The analysis of molecular codes relies on the identification of the adapters [9]. In the two models above the tRNAs are the adapters and carry the combinatorial complexity of the system. In the following we analyze a more realistic model of the gene translation machinery by including the loading step of the tRNA. The refined network model $N_{GC} = \langle M_{GC}, R_{GC} \rangle$ contains all possible mappings between the 64 codons and 20 amino acids as described above. Additionally, we model the loading step of the tRNAs by inserting the respective amino acyl tRNA synthetases (aaRS) (cf. Fig. 2). The reaction network N_{GC} describes the core molecular mechanism realizing the standard genetic code and all alternative codes. The set of molecular species M_{GC} contains all DNA strings of length three (Table S2, Eq. 2), representing the codons, the twenty proteinogenic amino acids in their free form (Table S2, Eq. 3), the twenty amino acids bound in a protein (Table S2, Eq. 4), all possible tRNAs in their unloaded (Table S2, Eq. 5) and loaded form (Table S2, Eq. 6) and all possible aaRS (Table S2, Eq. 7), such that the system is able to load all amino acids to all tRNAs.

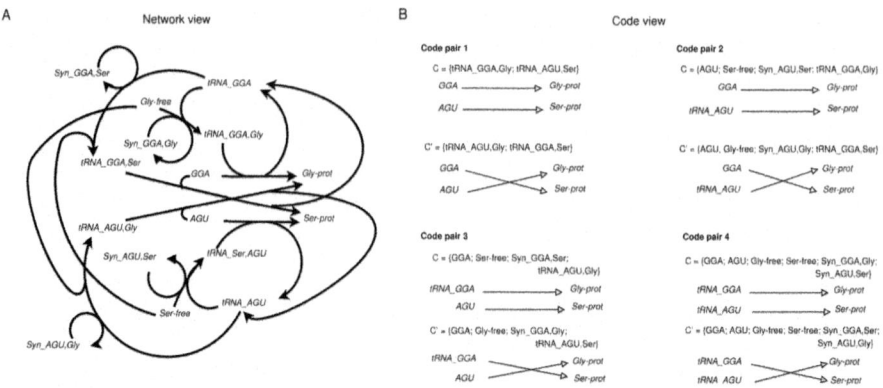

Figure 2: Subnetwork of the full gene translation network model with synthetases ($\mathcal{N}_{GC}$) and the realized molecular codes. The network (panel A) shows a subnetwork of the gene translation network model containing the translation, and loading reactions for two selected codons (GGA, AGU) and amino acids (Gly, Ser). The semantic analysis shows that four code pairs can be implemented by this network (panel B). doi:10.1371/journal.pone.0054694.g002

The set $\mathcal{R}_{GC}$ contains all reactions loading the amino acids onto the tRNAs (Table S2, Eq. 8) and all reactions inserting an amino acid in the peptide sequence (Table S2, Eq. 9). Fig. 2Adisplays a subnetwork (Dataset S4) with two codons (GGA, AGU), two amino acids (Gly, Ser) and the respective other elements of the network (tRNA and synthetases).

Analyzing the subnetwork (Fig. 2, Dataset S4) allows to assess the whole network's semantic capacity. Table 3 shows the four contained molecular code pairs, the respective molecular contexts are listed in Table 4. The identified code pairs (Table 3) show that not only codons can be signs, but also the unloaded tRNAs can function as signs. These additional signs increase the number of code pairs in a combinatoric manner. The "new" codes differ structurally in their molecular context. While, classically, the codons are mapped to the set of amino acids using the loaded tRNAs as context, the new signs, i.e. unloaded tRNAs, are mapped to the set of amino acids by using a molecular context that consists of the free amino acid loaded to the free tRNA, the synthetase performing the loading step, and the codon that needs to be recognized by the tRNA. The number of code pairs in this system can be calculated by

$$CP_{GC} = \left[\binom{n_s}{2} - \frac{n_s}{2} \right] \cdot \binom{n_m}{2},$$

(2)

with n_s as number of signs and n_m as number of meanings (amino acids). For the full gene translation system the number of signs is $n_s = c + t$, with c as number of codons and t as number of unloaded tRNAs. Since there is always one pair of one tRNA and codon belonging together, which therefore can not be combined in an BMC, we have to subtract the number of such pairs $n_s/2$ from the amount of all combinations.

Table 3: Code pairs in the gene translation model. doi:10.1371/journal.pone.0054694. t003

Code pair	Signs	Meanings
1	$\{GGA, AGU\}$	$\{Gly^{prot}, Ser^{prot}\}$
2	$\{GGA, tRNA_{AGU}\}$	$\{Gly^{prot}, Ser^{prot}\}$
3	$\{AGU, tRNA_{GGA}\}$	$\{Gly^{prot}, Ser^{prot}\}$
4	$\{tRNA_{GGA}, tRNA_{AGU}\}$	$\{Gly^{prot}, Ser^{prot}\}$

Table 4: Molecular contexts of the codes in the gene translation model. doi:10.1371/journal.pone.0054694.t004

Code pair	Molecular context	alternative molecular context
1	$\{tRNA_{GGA,Gly}, tRNA_{AGU,Ser}\}$	$\{tRNA_{AGU,Gly}, tRNA_{GGA,Ser}\}$
2	$\{AGU, Ser^{free}, Syn_{AGU,Ser}, tRNA_{GGA,Gly}\}$	$\{AGU, Gly^{free}, Syn_{AGU,Gly}, tRNA_{GGA,Ser}\}$
3	$\{GGA, Ser^{free}, Syn_{GGA,Ser}, tRNA_{AGU,Gly}\}$	$\{GGA, Gly^{free}, Syn_{GGA,Gly}, tRNA_{AGU,Ser}\}$
4	$\{GGA, AGU, Gly^{free}, Ser^{free}, Syn_{GGA,Gly}, Syn_{AGU,Ser}\}$	$\{GGA, AGU, Gly^{free}, Ser^{free}, Syn_{GGA,Ser}, Syn_{AGU,Gly}\}$

Using Eq. (2) the analysis of the whole network (N_{GC}), describing all potential genetic codes with 64 codons and 20 amino acids, results in $1,532,160$ binary code pairs, i.e. $SC_{log}(N_{GC}) \approx 20.55$. This is a different result than for the less detailed model, as calculated by Eq. (1). The extension of the model by aaRS, unloaded tRNAs, and unloaded amino acids increases the semantic capacity. This increase is not only an artifact from increasing the network size, but results from qualitative new code pairs.

The question to what extend a tRNA based code could be employed by the cell is open, but the potential existence of such a code is nevertheless an interesting result.

Gene Regulation by Transcription Factors Allow for Molecular Codes

In general, the gene regulatory network (GRN) of a cell constitutes the regulatory relations between genes. A particular regulatory relation is a fairly complex process involving a gene, the promoter and binding region of that gene, the binding of the transcription factor (TF) plus c ofactors, and the production of a product by the recruitment of the gene expression machinery. We will show here that a cell's GRN is also a highly semantic system.

In order to do so, we model a GRN as a reaction network $N_{GRC} = \langle \mathcal{M}_{GRC}, \mathcal{R}_{GRC} \rangle$ by explicitly inserting the relevant components (Fig. 3). The resulting network is not a generic model to describe all possible gene regulatory networks, but a model that covers the main properties of regulation important for this study. $\mathcal{M}_{GRC}$ contains n transcription factors TF_i, m products P_j, and genes G_{ij}. Each gene G_{ij} represents a combination of a promoter site i and a coding region j, where the promoter site i is specific to TF_i and the coding region j produces P_j. For our model we assume that there exist as many promoter sites and coding regions as transcription factors and products, respectively, such that each promoter-gene combination is possible. In summary $\mathcal{M}_{GRC} = \{ TF_1, TF_2, \ldots, TF_i, \ldots, TF_n, P_1, P_2, \ldots, P_j, \ldots, P_m, G_{11}, G_{12}, \ldots, G_{ij}, \ldots, G_{nm} \}$.

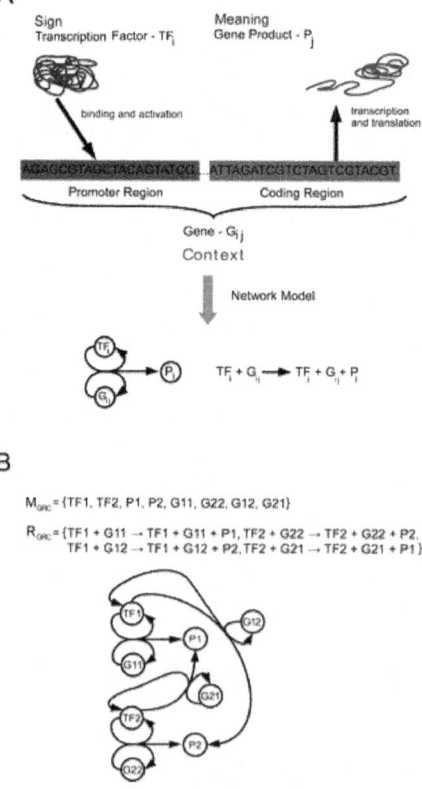

Figure 3: Gene regulatory network model. Panel A: Model of the expression of a gene, and the reaction network formulation of the same process (below). Blue text in panel

A indicates the semantic interpretation, i.e. the transcription factors are the signs, the products are the meanings, and the DNA is the molecular context. Panel B: reaction network constructed according to the formalization of gene regulation shown in (A) containing two transcription factors (TF1, TF2), two gene products (P1, P2) and the according genes (G11, G12, G21, G22). doi:10.1371/journal.pone.0054694.g003

Note that the differences of eukaryotic and prokaryotic gene regulation are abstracted by our model, because only the general mechanism of transcription factor regulated expression that gives rise to a high semantic capacity shall be explored here. Therefore, we consider transcription factors that bind only one promoter and that a promoter is bound by only one transcription factor. Then, the expression of a gene i,j is given by

$$\mathcal{R}_{GRC} = \left\{ TF_i + G_{ij} \rightarrow TF_i + G_{ij} + P_j \right\}, i = 1, 2, \ldots, n,$$
$$j = 1, 2, \ldots, m.$$

The semantic analysis shows that the reaction network can implement molecular codes, but only in one way, i.e. with the transcription factors as signs and the set of products as meanings. The set of genes, i.e. the combination of promoter and coding region, forms the molecular context. So the mapping between transcription factor and gene product can be altered by the exchange of a promoter region of a gene (or vice versa). Such promoter exchanges are also a common tool in molecular biology to allow for the external control of gene expression [44], e.g. to discover the function of silenced gene clusters [45].

Interestingly, in contrast to the model of the gene translation chemistry described above, the DNA is not the sign, but functions as the molecular context. This "role change" suggests an interdependence between different codes. Here the "gene regulatory code" regulates the execution of the "gene translation code", as the former one controls the usage of the latter's signs.

Note that the reaction network model can easily be made more complex by modeling transcription factors as protein complexes and including the respective assembly processes, by modeling different types of transcription factors (activators, repressors, enhancers), or the introduction of several DNA binding sites in the regulatory region to allow a combinatoric regulation by several transcription factors. However, the general conclusion about the semantic capacity of a GRN would not be affected.

Signaling by Phosphorylation Cascades Allows for Molecular Codes Only in a Dynamic Setting

Cells maintain different systems for signal transmission and integration [46]. The most prominent signaling systems rely on reversible phosphorylation of amino acids side-chains for regulation of signaling protein activity. The direct involvement of such systems in signaling suggest that they may be semantic systems. If so, they should be able to realize molecular codes. We have studied phosphorylation cascades, like the mitogen activated kinase regulatory network, as a typical instance of an intra-cellular signaling system. These systems demonstrate the limitation of our static approach. Here, it is necessary not only to distinguish between molecular species, but also between their concentrations. By assigning concentration levels to each species we allow for the dynamic change of these concentrations by the system's reactions. Thus, a molecular species' concentration is decreased if it is used as reactant in a reaction and increased if produced by a reaction. A species can have an effect on another species' concentration through the reactions in the system.

In general, the activation of a kinase by phosphorylation can generate a molecular mapping between the kinase and its target, but this mapping is not necessarily a molecular code (Fig. 4A). In contrast, a two-step cascade is able to implement a molecular code (Fig. 4C).

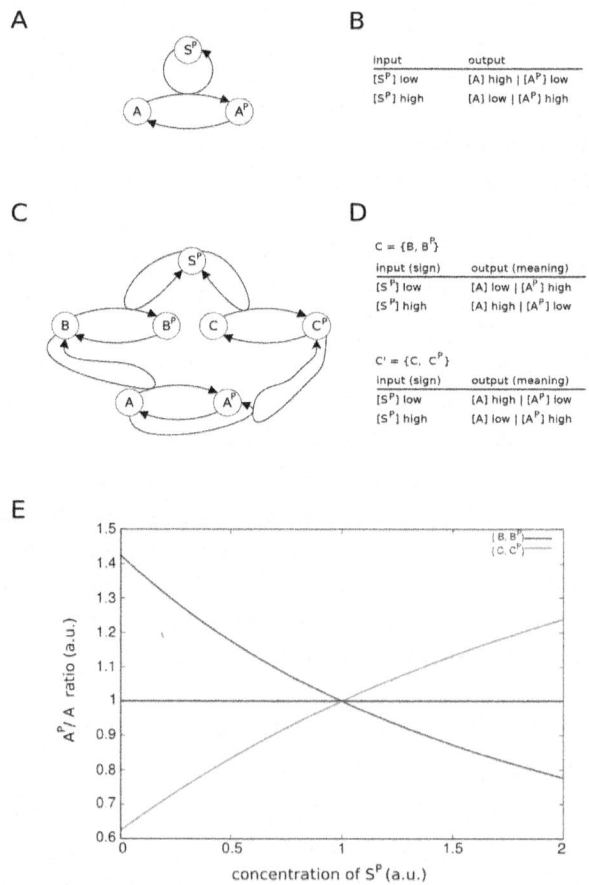

Figure 4: Reaction networks describing phosphorylation motifs. Molecular species in these networks represent kinases that may be activated or inactivated by phosphory-

lation. Activated and non-activated forms of a kinase are modeled as different species (e.g. species A and A^P). Panel A: Reaction network of a simple phosphorylation motif, which can realize a molecular mapping, but not a molecular code. Panel B: Molecular mappings that can be realized by the reaction network from panel A. These mappings do not constitute a molecular code. Panel C: M ore complex reaction network that can realize molecular codes. Panel D: The two binary molecular codes (i.e., one code pair) are realized by either one of the two molecular contexts $\{B, B^P\}$ or $\{C, C^P\}$. In contrast to the other described molecular codes (e.g. the genetic code), here, the code is not only specified by the species also, but also by the species' concentrations. Panel E : Simulation of the second network (panel C) showing the $[A^P]/[A]$ ratio over $[S^P]$ for the two different contexts. The red line shows the system's behavior for the context $\{B, B^P\}$, while the green line shows the system's behavior for the alternative context $\{C, C^P\}$ over varying initial concentrations for S^P. The blue line indicates the (here arbitrary) threshold to separate high and low concentrations. doi:10.1371/journal.pone.0054694.g004

The simple one-step phosphorylation model (Fig. 4A) contains two kinases: an initial kinase (S) and a target kinase (A) which can be phosphorylated by S ($S^P + A \rightarrow A^P$). We also model the dephosphorylation ($A^P \rightarrow A$). For sake of simplicity we do not model the phosphatases, and the phosphate related molecular species (e.g. ATP, ADP, P) involved in the process, but assume a buffered concentration. In the simple one-step model we can identify a molecular mapping between S^P and the two states of kinase A (Fig. 4B). If S^P has a low concentration the system is in a state where the unphosphorylated state A has a high concentration and the phosphorylated state A^P has a low concentration. According to the definition of molecular code given above the system should be able to change the mapping, i.e. be contingent, by the application of a different molecular context to realize a code. Here, no alternative mapping between S and A can be realized, such that the system is not able to realize a molecular code.

If we consider a different system where two kinases are between S^P and A, we obtain a two-step phosphorylation cascade (Fig. 4C). S^P now phosphorylates the inserted species, while these have an effect on A. The system has the possibility to "choose" between two alternative systems, i.e. the inserted species may be "active" in the unphosphorylated state (B), or in the phosphorylated state (C). There exist several mappings in such a system, e.g. between S^P and B, S and C, and S^P and A. The former two mappings behave like the simple model (see above). The mapping between S and A is a molecular code, because the molecular context of the system can be changed, such that the alternative system behavior is generated (Fig. 4D). The molecular context between S and A is either the set $\{B, B^P\}$, or alternatively $\{C, C^P\}$. If we assume two concentration levels denoted by

$[.]high$ and $[.]low$ for high and low concentrations, respectively, we can identify the following codes: Applying the molecular context $\{B,B^P\}$ we get the mappings $[S^P]low \rightarrow [A]low$, $[S^P]low \rightarrow [A^P]high$, $[S^P]high \rightarrow [A]high$, and $[S^P]high \rightarrow [A^P]low$, while the molecular context $\{C,C^P\}$ leads to the mappings $[S^P]low \rightarrow [A]high$, $[S^P]low \rightarrow [A^P]low$, $[S^P]high \rightarrow [A]low$, and $[S^P]high \rightarrow [A^P]high$. We simulated the system and applied both contexts $\{B,B^P\}$ and $\{C,C^P\}$. For the former context a change in $[S^P]$ (x-axis) leads to a decrease in the $[A^P]/[A]$-ratio (y-axis). Applying the alternative context $\{C,C^P\}$ leads to the opposite behavior. Fig. 4 E illustrates these dependencies (for details of the underlying model see Text S3).

The extension of our static approach to a dynamic setting needs more strict definitions, such that the here shown properties are only a first step into this direction.

Random Reaction Networks as Null Model

To check whether the motif describing a BMC can be generated by chance we analyzed random reaction networks of different sizes and densities for their semantic capacity. The networks have been generated by random insertion of reaction rules in an empty network. Each random reaction rule is bimolecular, i.e. contains two reactants, and one product (seeText S1 for pseudocode). The analysis showed that the binary code motif can be generated in random networks (Fig. 5), i.e. contingent mappings can be generated randomly. For a fixed network size and varying densities the average semantic capacity shows a unimodal behavior, which suggests that there exist an optimal range of densities for each network size, leading to maximal semantic capacity. This optimal range shifts to higher densities with increasing size of the network (see Fig. 6). The optimal interval is bounded at lower densities by the low complexity of the network, there are not enough reactions to promote the insertion of molecular codes by chance. On higher densities the network is strongly connected, such that the subsets of the system are hardly closed, therefore it is also harder to implement codes by chance. The optimal interval coincides with two important network properties, i.e. the number of paths, and the number of closed sets. With increasing network density the number of paths grows, while the number of closed sets decreases. High semantic capacity can be found in networks with a high number of pathways and at the same time a high number of closed sets.

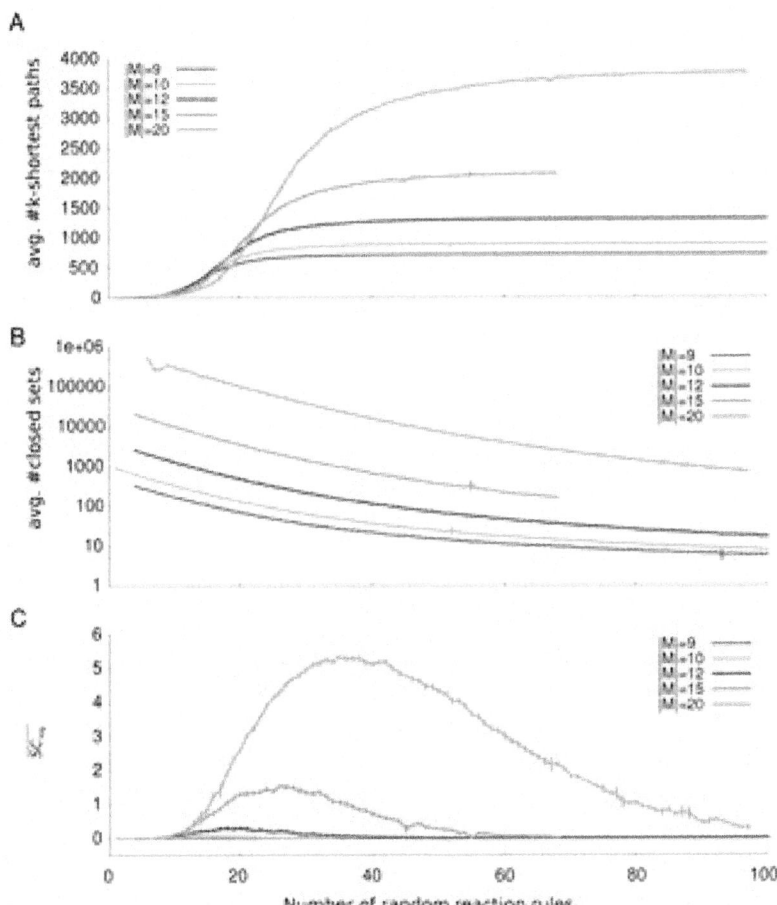

Figure 5: Structural properties of random reaction networks of different size and density. Panels A and B show two important network parameters for five different network sizes over the numbers of reaction rules. The data represents the average values of random replicates. Error bars indicate the standard error of the mean. Panel A shows the average number of paths in the network. Since we applied the path algorithm which only uses the k-shortest paths between each pair of molecular species the curve shows a sigmoidal behavior, which is saturated at the value $|\mathcal{M}| \cdot (|\mathcal{M}| - 1)$, with $k = 10$. Panel B shows the average number of closed sets. With growing density the number of closed sets decreases. Panel C shows the distributions of the average number of code pairs ($< SC_{\log} >$). The semantic capacity follows a unimodal distribution indicating the existence of an optimal interval for the random generation of the BMC motif. If the number of paths is too low no mappings can be implemented because of the missing links between potential signs and meanings. Similarly, if the number of closed sets is too low no mappings can be implemented either. doi:10.1371/journal.pone.0054694. g005

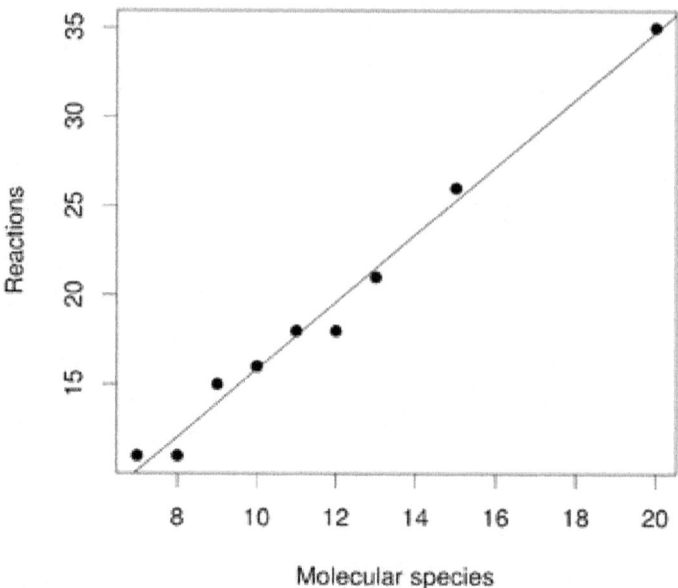

Figure 6: Maximal semantic capacity in random networks. Scatter plot showing the position of the maximal semantic capacity of the random reaction network data (cf. Fig. 5) in a $(|\mathcal{M}| \times |\mathcal{R}|)$-plot. The linear regression of the data shows that the maximal semantic capacity is reached if there are approximately two times more reactions in the system than molecular species: $reactions = -3.06 + 1.89\ species$. doi:10.1371/journal.pone.0054694.g006

Combustion Chemistries and the Martian Atmosphere Photochemistry Show no Semantic Capacity

We analyzed a number of chemical systems, i.e. combustion chemistries of hydrogen [47](Dataset S6), methane [48] (Dataset S7), ethanol [49] (Dataset S8), dimethyl ether [50](Dataset S9). The chemistries are intended to describe all significant processes that can occur in the combustion, i.e. burning, of the respective molecule. The original combustion chemistry data (provided in CHEMKIN format [43]) have been processed to obtain the reaction networks describing the respective chemistry. In the CHEMKIN files reactions are described at equilibrium with additional thermodynamic parameters. Taking these as basis we obtain reaction networks containing the directed reactions depending on the thermodynamic parameters.

The reaction networks cover different sizes (10–79 molecular species) and densities (38–752 reactions). The semantic analysis shows that none of these chemistries is able to realize molecular codes. We can now compare the results with our null model derived from the random reaction networks data (Table 5) to evaluate if both are consistent.

Table 5: Comparison of combustion chemistries and random networks (null model). doi:10.1371/journal.pone.0054694.t005

Combustion chemistry properties						Null model estimate		
	$|M|$	$|R|$	#closedsets	#paths	SC_{log}	est. #closedsets (SEM)	est.#paths	est.SC_{log}
HYD	10	38	16	$7.69 \cdot 10^4$	0	39.84 (0.53)	878.15 (1.27)	0 (0.0)
MET	37	340	4,136	$>10^6$	0	6,423.22 (209.75)	$>1.33 \cdot 10^4$	1.12 (0.08)
ETH	57	752	5,136	$>10^6$	0	82,453.25 (9,545.96)	$>3.19 \cdot 10^4$	3.86 (0.36)
DME	79	708	8	$>10^6$	0	n.a.	n.a.	n.a.

For the hydrogen chemistry the lack of code pairs can be explained by the small number of closed sets compared to the number of paths, such that the molecular species are "too connected" and the network is less structured. In the null model also no molecular codes can be identified. The estimated number of closed sets and paths, although differing from the original chemistry, are also marking that the respective random networks are not in the optimal interval.

In the methane combustion chemistry we see that there exist far more paths than closed sets, such that the network is to some extend "unstructured". The according null model networks also contain a high number of paths, but also a higher number of closed sets. The algorithmic analysis shows that some of the generated null model networks can realize BMCs, such that the average logarithmic semantic capacity is 1.04. Nevertheless, we consider this also as a very low semantic capacity compared to, e.g. the gene translation chemistry. We also analyzed the atmosphere chemistry of Mars [51] (Dataset S5) to check whether other kinds of non-biological systems may contain codes. The model contains 32 molecular species, 104 reactions and 5512 closed sets. In particular, the network describes the reactions happening on the day side of mars. Therefore, light ($h\nu$) is modeled explicitly as inflow reaction $\rightarrow h\nu$. The day side martian photochemistry is not able to realize molecular codes. The comparison of the null model chemistries for ethanol, dimethyl ether, and the martian atmosphere chemistry were not feasible with our current algorithms, due to the large number of paths and closed sets in these networks.

NTOP: An Artificial Chemistry Allowing for Molecular Coding

Recall that with increasing density random networks have a vanishing semantic capacity. In the following we show that even a dense network can have a relatively high semantic capacity. For this purpose we analyze an artificial chemistry with 16-species introduced by Banzhaf [52]called NTOP. For each species there is a 4-bit binary representation and the reaction rules are derived with respect to this representation, which is referred to as a structure-to-function mapping (see Ref. [52] for details and Dataset S10 for the network model).

The algorithmic analysis results in six code pairs (Text S4). Two properties of molecular codes that are of general importance also for biological molecular codes can be observed here. (1) A meaning can take the role of a sign in another code, and (2) molecular species can function as signs (or meanings) in different codes, i.e. they keep their role in different contexts (Fig. S1).

To test the robustness of the network's semantic capacity, we replace 1, 2, 5, 10, 15, 200, and 1000 reaction rules randomly, respectively. In a randomly chosen reaction rule we replace the molecular species, while keeping the number of reactants and products the same. Thus, the type of the reaction stays the same, while the connections are changed. Increased randomization results in a decreased average semantic capacity (Fig. S2). Nevertheless in some cases the randomized network is capable to implement more code pairs. The general trend towards less code pairs can be explained by referring to the analysis of random reaction networks. Random reaction networks with the same number of species and reactions as NTOP show no semantic capacity ($SC_{log} = 0$). Thus the random variation of the NTOP chemistry drives the system towards the mean semantic capacity of random networks.

DISCUSSION

We have introduced a formal criterion for identifying molecular codes in reaction networks and a measure of the semantic capacity of a network, as the number of different code pairs the network can realize. Our notion of contingency, defined as the ability of systems to choose between different mappings, extends and operationalizes the notion of "independence" and "contingency" as discussed by Monod, Barbieri and others.

The structure of molecular codes allows to decompose them into binary molecular codes, which were studied here. Having a list of binary molecular codes it is possible to merge them into larger molecular codes, as has been demonstrated for the genetic code.

Applying the new concepts to different networks, our basic finding demonstrates that the semantic capacity of biological networks tends to be higher than the semantic capacity of the studied non-biological networks. Thus, an important step during the transition from non-life to life must have been the utilization of a chemistry that allows to implement molecular codes. In our opinion it is an open issue how that first coding chemistry has looked like. But, we have now a criterion that can guide us in what we have to look for. Following this line of thought it seems that biological systems "learned" by evolution to make use of chemistries with high semantic capacities by selecting the most appropriate mappings for their purpose. There exist at least three (not necessarily disjoint) evolutionary paths to select a unique mapping from the actual contingency: (1) *compartimentalization*, i.e. spatial separation of the two alternative mappings, (2) separation by *time* of execution, and (3) *fixation*, i.e. separation by deleting one of the alternative mappings. For the genetic code we could argue that at least two paths are used by cells to maintain the uniqueness of the mapping. Different codes are implemented in different species and compartments (compartimentalization) [22] and the genes for the alternative amino acyl tRNA synthetases are not present in the genome (fixation). Time separation can be understood as a regulated switch of mappings, e.g. in mitotic control where the presence of a protein called Cdc20 inhibits the Anaphase-Promoting Complex (APC) during the activated spindle assembly checkpoint (SAC), while in the context of the inactivated checkpoint, Cdc20 activates APC [53], [54].

Moreover, we can now precisely formulate another hypothesis, namely, that during the course of evolution the semantic capacity of the chemistry employed by the biological systems has a tendency to increase, by recruiting new chemistries, though the increase is not necessarily monotonous. One candidate mechanism is the invention and improvement of compositional adaptors, like proteins with exchangeable domains [55] or genes including their promoter- and coding-regions [9]. Note that also the appearance and evolution of neurons and cognitive systems is in line with the hypothesis of increasing semantic capacity.

The analysis of a network model implementing the genetic code showed that not only the codons can be signs, but also tRNA molecules could, in principle, be signs. Apparently, this potential code is not used by the cell. The biomolecular and evolutionary interpretation of this fact has to be left for future studies, because we have to make the notion of code *usage*, that is, the pragmatic aspect of biological information, more precise.

Furthermore, we have shown that DNA not only can function as a sign but also as a molecular context, as the study of gene regulatory networks revealed. The mechanisms in gene regulatory systems and the observation that such systems are highly flexible (i.e. the mapping between transcription factors and gene products can easily be changed) leads to the conclusion that the chemistry of GRNs possesses also a high semantic capacity. This may be the reason why it is the main regulatory subsystem of cells and often is used as typical representant of cellular information processing [56]. From a theoretical point of view it will be interesting to analyse more complex variants (several binding site, different types of transcription factors, transcription factor assembly) of the general GRN network for their influence on the semantic capacity. These extensions can introduce new codes by allowing for additional control and regulation of the system.

Phosphorylation cascades represent a class of biological systems that allow for molecular codes, but requires a quantitative analysis, i.e. the incorporation of concentrations. Thus our qualitative approach is not sufficient here. In the future the molecular code concept needs to be extended to the dynamic interpretation of a system. A molecular code then could be interpreted as a mapping between system states.

The analysis of random networks of different sizes and densities results in a better understanding of the basal rate of code occurrence. We can observe that the distribution of BMCs is unimodal, with high semantic capacity appearing only in sparsely connected random networks, in particular, where the number of second order reactions is approximately twice the number of molecular species. Interestingly, random networks with high semantic capacity show at the same time a high number of closed sets of species (which decreases with increasing network density) and a high number of paths (which increases with increasing network density). The null model estimates the semantic capacity of a reaction network that is generated completely by a random process. For biological and chemical systems this is obviously not true, because of physical constraints like mass conservation on the reactions.

The analysis of the artificial chemistry NTOP suggests that also in dense networks the semantic capacity can be high. We hypothesize that this was caused by the structure-to-function mapping applied in the definition of the chemistry.

There exist certain limitations on the kind of networks that should be analyzed with our approach. The definition of molecular codes requires that, to be applicable, the network model needs to contain all possible reactions among the molecular species. Network data widely available from databases like KEGG, Reactome, BioCyc, or Biomodels DB usually does not fulfill this criteria, yet. The networks found in these databases are becoming now rather complete with respect to the particular organism they belong to. However the network data is rather incomplete with respect to the underlying (bio-) chemistry. That is, with respect to the underlying chemistry many more alternative network species and reactions are possible, which cannot be found in those databases for several practical as well as conceptual reasons. It is the central innovation of our approach that for detecting a molecular code, we need to know the potential reaction network, which in general is not visible in the actual organism. It might sound a bit paradoxical that a network property depends on something that is not part of the network. In our case, however, the link to this "invisible" part is provided by physical laws and chemistry, which determine the alternative network species and reactions.

How to measure the semantic capacity of an actual biochemical system? We suggest a procedure consisting of three major steps: Step 1: Define the system to be studied and its chemistry, Step 2: Obtain the reaction network by physical experiments, Step 3: Compute all molecular codes of the network. In Step 1 we explicate the necessary assumptions: We define the chemical universe we will look at, i.e. the set of potential chemical species and the set of all possible reactions. Note that this depends on the time scale at which our system exists. At a longer timescale more reactions might have to be considered. Further assumptions can include constraints like temperature, pressure, pH, or energy consumption. In Step 2 we construct the reaction network using scientific physical experiments. Methods for this exist in a large variety in Chemistry and the Life Sciences. Note that with proper assumptions (Step 1) we approach with increasing number of experiments a single unique network. In other words, there is a single "true" network, which is defined by the scientific procedure and the assumptions made in Step 1. At least in principle, we can obtain this network with arbitrary precision, provided arbitrary but finite experimental resources. As an open problem remains the question how a measurement error on the network level propagates to the estimation of the semantic capacity. Step 3 is purely formal and in principle deterministic. Practically, however, for large and complex networks (e.g., networks with more than 1000 species) the run time of our deterministic algorithms described here is too long and thus efficient heuristics have to be developed for these networks in the future.

In summary, we conclude that our approach provides a new way to analyze aspects of the information processing capabilities of molecular systems, which might contribute to the understanding of biological information in the context of the origin and evolution of life, cellular signaling, or synthetic molecular computing systems.

ACKNOWLEDGMENTS

We would like to thank the anonymous reviewers for their constructive and helpful comments. We are grateful for many fruitful discussions with Stefan Artmann and his valuable comments and suggestions. We thank Marcel Hieckel for preparing the combustion chemistry data and Bashar Ibrahim for critically reading the manuscript.

AUTHOR CONTRIBUTIONS

Conceived and designed the experiments: DG PD. Performed the experiments: DG PD. Analyzed the data: DG PD. Contributed reagents/materials/analysis tools: DG PD. Wrote the paper: DG PD.

REFERENCES

1.   Shannon CE (1948) A mathematical theory of communication. The Bell Systems Technical Journal 27: 379–423, 623–656.

2.   Tkačik G, Walczak AM (2011) Information transmission in genetic regulatory networks: A review. J Phys Condens Matter 23: 153102. doi: 10.1088/0953-8984/23/15/153102

3.   Mehta P, Goyal S, Long T, Bassler BL, Wingreen NS (2009) Information processing and signal integration in bacterial quorum sensing. Mol Syst Biol 5: 325. doi: 10.1038/msb.2009.79

4.   Lenaerts T, Ferkinghoff-Borg J, Stricher F, Serrano L, Schymkowitz JWH, et al. (2008) Quantifying information transfer by protein domains: Analysis of the Fyn SH2 domain structure. BMC Struct Biol 8: 43. doi: 10.1186/1472-6807-8-43

5.   Waltermann C, Klipp E (2011) Information theory based approaches to cellular signaling. Biochim Biophys Acta General Subjects 1810: 924–932. doi: 10.1016/j.bbagen.2011.07.009

6.   Monod J (1971) *Chance and necessity*. Alfred Knopf, New York/NY. (Originally published 1970).

7.   Küppers BO (1990) *Information and the origin of life*. MIT Press, Cambridge/MA. (Originally published 1986).

8.   Tsuda S, Artmann S, Zauner KP (2009) The Phi-Bot. In Adamatzky A, Komosinski M, eds., Artificial Life models in hardware. Springer, Dordrecht, 213–232.

9.   Barbieri M (2008) Biosemiotics: a new understanding of life. Naturwissenschaften 95: 577–599. doi: 10.1007/s00114-008-0368-x

10. Koonin EV, Novozhilov AS (2009) Origin and evolution of the genetic code: the universal enigma. IUBMB Life 61: 99–111. doi: 10.1002/iub.146

11. Pattee HH (2008) Physical and functional conditions for symbols, codes, and languages. Biosemiotics 1: 147–168. doi: 10.1007/s12304-008-9012-6

12. Görlich D, Dittrich P (2011) Identifying molecular organic codes in reaction networks. In Kampis G, Karsai I, Szathmáry E, eds., Advances in Artificial Life. Darwin Meets von Neumann, vol. 5777 of Lecture Notes in Computer Science. Springer Berlin/Heidelberg, 305–312.

13. Fontana W, Buss L (1994) The arrival of the fittest: Toward a theory of

biological organization. Bull Math Bio 56: 1–64. doi: 10.1016/s0092-8240(05)80205-8

14. Speroni di Fenizio P, Dittrich P, Ziegler J, Banzhaf W (2000) Towards a theory of organizations. In Lange H, et al. (Eds.) German Workshop on Artificial Life (GWAL 2000), in print. Bayreuth, 5.-7. April, 2000, available online: http: //di.ttri.ch/p/SDZB2001gwal.pdf.

15. Tlusty T (2008) Casting polymer nets to optimize noisy molecular codes. Proc Natl Acad Sci U S A 105: 8238–8243. doi: 10.1073/pnas.0710274105

16. Tlusty T (2008) Rate-distortion scenario for the emergence and evolution of noisy molecular codes. Phys Rev Lett 100: 048101. doi: 10.1103/physrevlett.100.048101

17. Martins EQV, Pascoal MMB (2003) A new implementation of yen's ranking loopless paths algorithm. 4OR: A Quarterly Journal of Operations Research 1: 121–133. doi: 10.1007/s10288-002-0010-2

18. Crick FH, Barnett L, Brenner S, Watts-Tobin RJ (1961) General nature of the genetic code for proteins. Nature 192: 1227–1232. doi: 10.1038/1921227a0

19. De Beule J, Hovig E, Benson M (2011) Introducing dynamics into the field of biosemiotics. Biosemiotics 4: 5–24. doi: 10.1007/s12304-010-9101-1

20. Osawa S, Jukes TH, Watanabe K, Muto A (1992) Recent evidence for evolution of the genetic code. Microbiol Rev 56: 229–264.

21. Jukes TH, Osawa S (1993) Evolutionary changes in the genetic code. Comp Biochem Physiol B 106: 489–494. doi: 10.1016/0305-0491(93)90122-l

22. Elzanowski A, Ostell J (2010) The genetic code. Available: http: //www.ncbi.nlm.nih.gov/Taxonomy/Utils/wprintgc.cgi, version 3.9, July 07, 2010. Accessed 2011 February 20.

23. Liu CC, Schultz PG (2010) Adding new chemistries to the genetic code. Annu Rev Biochem 79: 413–444. doi: 10.1146/annurev.biochem.052308.105824

24. Clark-Walker GD, Weiller GF (1994) The structure of the small mitochondrial DNA of Kluyveromyces thermotolerans is likely to reflect the ancestral gene order in fungi. J Mol Evol 38: 593–601. doi: 10.1007/bf00175879

25. Himeno H, Masaki H, Kawai T, Ohta T, Kumagai I, et al. (1987) Unusual genetic codes and a novel gene structure for tRNA(AGYSer) in starfish mitochondrial DNA. Gene 56: 219–230. doi: 10.1016/0378-1119(87)90139-9

26. Jacobs HT, Elliott DJ, Math VB, Farquharson A (1988) Nucleotide sequence and gene organization of sea urchin mitochondrial DNA. J Mol Biol 202: 185–217. doi: 10.1016/0022-2836(88)90452-4

27. Batuecas B, Garesse R, Calleja M, Valverde JR, Marco R (1988) Genome organization of Artemia mitochondrial DNA. Nucleic Acids Res 16: 6515–6529. doi: 10.1093/nar/16.14.6515

28. Osawa S, Ohama T, Jukes TH, Watanabe K (1989) Evolution of the mitochondrial genetic code. I. Origin of AGR serine and stop codons in metazoan mitochondria. J Mol Evol 29: 202–207. doi: 10.1007/bf02100203

29. Garey JR, Wolstenholme DR (1989) Platyhelminth mitochondrial DNA: Evidence for early evolutionary origin of a tRNA(serAGN) that contains a dihydrouridine arm replacement loop, and of serine-specifying AGA and AGG codons. J Mol Evol 28: 374–387. doi: 10.1007/bf02603072

30. Ohama T, Osawa S, Watanabe K, Jukes TH (1990) Evolution of the mitochondrial genetic code. IV. AAA as an asparagine codon in some animal mitochondria. J Mol Evol 30: 329–332. doi: 10.1007/bf02101887

31. Hoffmann RJ, Boore JL, Brown WM (1992) A novel mitochondrial genome organization for the blue mussel, Mytilus edulis. Genetics 131: 397–412.

32. Durrheim GA, Corfield VA, Harley EH, Ricketts MH (1993) Nucleotide sequence of cytochrome oxidase (subunit III) from the mitochondrion of the tunicate Pyura stolonifera: evidence that AGR encodes glycine. Nucleic Acids Res 21: 3587–3588. doi: 10.1093/nar/21.15.3587

33. Boore JL, Brown WM (1994) Complete DNA sequence of the mitochondrial genome of the black chiton, Katharina tunicata. Genetics 138: 423–443.

34. Kondow A, Suzuki T, Yokobori S, Ueda T, Watanabe K (1999) An extra tRNAGly(U*CU) found in ascidian mitochondria responsible for decoding non-universal codons AGA/AGG as glycine. Nucleic Acids Res 27: 2554–9. doi: 10.1093/nar/27.12.2554

35. Telford MJ, Herniou EA, Russell RB, Littlewood DT (2000) Changes in mitochondrial genetic codes as phylogenetic characters: two examples from the flatworms. Proc Natl Acad Sci U S A 97: 11359–11364. doi: 10.1073/pnas.97.21.11359

36. Yokobori S, Watanabe Y, Oshima T (2003) Mitochondrial genome of Ciona savignyi (Urochordata, Ascidiacea, Enterogona): Comparison of gene arrangement and tRNA genes with Halocynthia roretzi mitochondrial genome. J Mol Evol 57: 574–587. doi: 10.1007/s00239-003-2511-9

37. Nedelcu AM, Lee RW, Lemieux C, Gray MW, Burger G (2000) The complete mitochondrial DNA sequence of Scenedesmus obliquus reflects an intermediate stage in the evolution of the green algal mitochondrial genome. Genome Res 10: 819–831. doi: 10.1101/gr.10.6.819

38. Hayashi-Ishimaru Y, Ohama T, Kawatsu Y, Nakamura K, Osawa S (1996) UAG is a sense codon in several chlorophycean mitochondria. Curr Genet 30: 29–33. doi: 10.1007/s002940050096

39. Laforest MJ, Roewer I, Lang BF (1997) Mitochondrial tRNAs in the lower fungus Spizellomyces punctatus: tRNA editing and UAG 'stop' codons recognized as leucine. Nucleic Acids Res 25: 626–632. doi: 10.1093/nar/25.3.626

40. Schneider SU, Leible MB, Yang XP (1989) Strong homology between the small subunit of ribulose-1,5-bisphosphate carboxylase/oxygenase of two species of Acetabularia and the occurrence of unusual codon usage. Mol Gen Genet 218: 445–452. doi: 10.1007/bf00332408

41. Schneider SU, de Groot EJ (1991) Sequences of two rbcS cDNA clones of Batophora oerstedii: structural and evolutionary considerations. Curr Genet 20: 173–175. doi: 10.1007/bf00312782

42. Liang A, Heckmann K (1993) Blepharisma uses UAA as a termination codon. Naturwissenschaften 80: 225–226. doi: 10.1007/bf01175738

43. Keeling PJ, Doolittle WF (1996) A non-canonical genetic code in an early diverging eukaryotic lineage. EMBO J 15: 2285–2290.

44. Kaufmann A, Knop M (2011) Genomic promoter replacement cassettes to alter gene expression in the yeast saccharomyces cerevisiae. Methods Mol Biol 765: 275–294. doi: 10.1007/978-1-61779-197-0_16

45. Brakhage AA, Schroeckh V (2011) Fungal secondary metabolites - strategies to activate silent gene clusters. Fungal Genet Biol 48: 15–22. doi: 10.1016/j.fgb.2010.04.004

46. Krauss G (2008) Biochemistry of Signal Transduction and Regulation. Wiley-VCH, Weinheim, 4 edn.

47. Conaire MO, Curran HJ, Simmie JM, Pitz WJ, Westbrook C (2004) A comprehensive modeling study of hydrogen oxidation. Int J Chem Kinet 36: 603–622. doi: 10.1002/kin.20036

48. Hughes KJ, Turanyi T, Clague AR, Pilling MJ (2001) Development and Testing of a comprehensive chemical mechanism for the oxidation of methane. Int J Chem Kinet 33: 513–538. doi: 10.1002/kin.1048

49. Marinov NM (1999) A detailed chemical kinetic model for high temperature ethanol oxidation. Int J Chem Kinet 31: 183–220. doi: 10.1002/(sici)1097-4601(1999)31:3<183::aid-kin3>3.0.co;2-x

50. Kaiser E, Wallington T, Hurley MD, Platz J, Curran HJ, et al. (2000) Experimental and modeling study of premixed atmospheric-pressure dimethyl ether-air flames. J Phys Chem 104: 8194–8206. doi: 10.1021/jp994074c

51. Nair H, Allen M, Anbar AD, Yung YL (1994) A photochemical model of the martian atmosphere. Icarus 111: 124–150. doi: 10.1006/icar.1994.1137

52. Banzhaf W (1993) Self-replicating sequences of binary numbers. Comput Math Appl 26: 1–8. doi: 10.1016/0898-1221(93)90046-x

53. Musacchio A, Salomon ED (2007) The spindle-assembly checkpoint in space and time. Nat Rev Mol Cell Bio 8: 379–393. doi: 10.1038/nrm2163

54. Ibrahim B, Diekmann S, Schmitt E, Dittrich P (2008) In-silico modeling of the mitotic spindle assembly checkpoint. PLoS One 3(2): e1555. doi: 10.1371/journal.pone.0001555

55. Bornberg-Bauer E, Huylmans AK, Sikosek T (2010) How do new proteins arise? Curr Opin Struct Biol 20: 390–396. doi: 10.1016/j.sbi.2010.02.005

56. Tyson JJ, Novak B (2010) Functional motifs in biochemical reaction networks. Annu Rev Phys Chem 61: 219–240. doi: 10.1146/annurev.physchem.012809.103457

Chapter 2

AUTOMATIC NMR-BASED IDENTIFICATION OF CHEMICAL REACTION TYPES IN MIXTURES OF CO-OCCURRING REACTIONS

Diogo A. R. S. Latino[1], João Aires-de-Sousa[2]

[1] CQFB, REQUIMTE, Departamento de Química, Faculdade de Ciências e Tecnologia, Universidade Nova de Lisboa, Caparica, Portugal, CCMM, Departamento de Química e Bioquímica, Faculdade de Ciências, Universidade de Lisboa, Lisboa, Portugal

[2] CQFB, REQUIMTE, Departamento de Química, Faculdade de Ciências e Tecnologia, Universidade Nova de Lisboa, Caparica, Portugal

ABSTRACT

The combination of chemoinformatics approaches with NMR techniques and the increasing availability of data allow the resolution of problems far beyond the original application of NMR in structure elucidation/verification. The diversity of applications can range from process monitoring, metabolic profiling, authentication of products, to quality control. An application related to the automatic analysis of complex mixtures concerns mixtures of chemical reactions. We encoded mixtures of chemical reactions with the difference between the ^{1}H NMR spectra of the products and the reactants. All the signals arising from all the reactants of the co-occurring reactions were taken together (a simulated spectrum of the mixture of reactants) and the same was done for products. The difference spectrum is taken as the representation of the mixture of chemical reactions. A data set of 181 chemical reactions was used, each reaction manually assigned to one of 6 types. From this dataset, we simulated mixtures where two reactions of different types would occur simultaneously. Automatic learning methods were trained to classify the reactions occurring in a mixture from the ^{1}H NMR-based descriptor of the mixture. Unsupervised learning methods (self-organizing maps) produced a reasonable clustering of the mixtures by reaction type, and allowed the correct classification of 80% and 63% of the mixtures in two independent test sets of different similarity to the training set. With random forests (RF), the percentage of correct classifications was increased to 99% and 80% for the same test sets. The RF probability associated to the predictions yielded a robust indication of their reliability. This study demonstrates the possibility of applying machine learning methods to automatically identify types of co-occurring chemical reactions from NMR data. Using no explicit structural information about the reactions participants, reaction elucidation is performed without structure elucidation of the molecules in the mixtures.

INTRODUCTION

As the chemical composition of complex mixtures change with time, so do their NMR spectra. The interpretation of spectra modifications in terms of chemical reactions taking place has the potential to elucidate underlying chemical phenomena. Machine learning can extract knowledge from complicated databases of experimental observations, to recognize patterns in new situations. Automatic reaction identification can be useful in many different applications, e.g. to study chemical stabilities and aging of consumer/industrial products, to monitor biotechnological processes, or to assess the function of new enzymes in a pool of possible substrates.

Patterns of NMR changes are expected to be associated with types of reactions, because the atoms near the reaction center have their environment modified – and their NMR properties altered –, whereas the substructures of the reactants that are far from the reaction center will be mostly unchanged. Additionally, NMR spectra are sensitive to changes in the 3D environment of atoms, which may be observed even in substructures topologically distant from the reaction center and can be typical of certain reactions. Machine learning techniques should be able to recognize types of chemical reactions from NMR changes, even when more than one reaction occur simultaneously.

The processing of NMR spectra with chemometric and machine learning techniques is extensively used in the analysis of complex mixtures [1], notably in metabonomics [2]. Examples include the classification of lung carcinoma cell lines [3], the classification of human saliva according to treatment with an oral rinse formulation, or donor [4], [5], the analysis of human plasma to study metabolic changes caused by diet [6], the assessment of how concentration patterns of hydrophilic and lipophilic tissue metabolites describe different stages of breast tumor malignancy [7], or the identification of lipoprotein subclasses in plasma samples[8].

Other areas where NMR-based machine learning methods have been applied are the authentication of products [9]–[10], monitoring of enzymatic reactions [11], or assessment of drug toxicity [12]. Alonso-Salces et. al. used pattern recognition techniques (LDA, PLS-DA, SIMCA, and CART) for the geographical characterization of virgin olive oils based on the ^{1}H NMR fingerprint of the unsaponifiable matter [9]. Aursand and co-workers have reported the ability of Kohonen Self-Organizing Maps (Kohonen SOMs) and generative topographic mapping to discriminate ^{13}C NMR spectra of different commercial fish oil-related health food products concerning the nature, composition, refinement, and/or adulteration [10].

NMR techniques are well established for monitoring chemical and enzymatic reactions, industrial processes, and for the elucidation of reaction mechanisms. A few examples are mentioned next that also illustrate the current development of new instruments specifically designed for reaction analysis. Ballard et al. [13] used quantitative NMR to measure the concentration of carbamates over time, in order to study the chemical reaction of CO_2 with mixtures of amines. Shey et al. [14] monitored polymer-supported reactions with conventional[1]H NMR spectroscopy during a liquid-phase combinatorial synthesis. Kalelkar and co-workers applied SOMs to analyse NMR spectra from combinatorial parallel synthesis [15]. Bernstein et al. [16] described an apparatus consisting in a reactor coupled with an NMR flow cell; more recently an NMR flow cell based on a standard 5 mm NMR tube was presented that can be used for homogeneous and heterogeneous reactions [17]. Gomez et al. [18] presented a nanolitre NMR spectroscopy microfluidic chip hyphenated to a continuous flow microlitre-microwave irradiation set-up, for on-line monitoring. The method was also applied for rapid optimization of reaction conditions. Mix et al. [19] developed a double-chamber NMR tube – differently of others, this apparatus provides the full control of the temperature over the range from −80 to 130°C. Foley et al. [20] developed the ReactNMR method for reaction monitoring and *in situ* characterization of reaction intermediates, assisting in mechanism elucidation and in the characterization of complex reaction mixtures.

Our lab has previously shown that Kohonen Self-Organizing Maps (Kohonen SOMs) and random forests can classify individual reactions from the difference between the ^{1}H NMR spectrum of the products and the reactants [21]. The obtained models can then be applied in new situations, even if the structures of the reactants and products are unknown, but their ^{1}H NMR spectra are available.

Here we present an extension of this approach to mixtures of reactions. As before, machine learning methods received as input the difference between the ^{1}H NMR spectra of the products and the reactants – but now the products of two reactions of different classes are taken together, as well as the reactants. This simulates a situation in which two reactions occur simultaneously. The SPINUS program [22]–[24] was used to estimate ^{1}H NMR chemical shifts from the molecular structure, and the chemical shifts were fuzzified to tolerate small variations. Three machine learning methods were explored that differ in the type of learning. Kohonen SOMs are trained with unsupervised learning (competitive learning), counter-propagation neural networks (CPNN) use semi-supervised learning, and random forests (RF) use supervised learning. A dataset of 181 photochemical cycloadditions manually assigned into six types was used to simulate 12,421 mixtures of two reactions. The machine learning algorithms were given the task of predicting the types of reactions, in a simulated situation where two reactions of different types occur simultaneously, from the simulated ^{1}H NMR spectra of the reactants and products.

METHODS

The experiments here described involve three main steps: a) the generation of a reaction descriptor from the simulated ^{1}H NMR spectra of the products and reactants; b) the generation of the simulated mixtures of two reactions from the NMR reaction descriptors; c) the development of classification models for mixtures of reactions taking as input ^{1}H NMR data.

Data Sets of Reactions

A data set of 181 photochemical reactions, involving two reactants and one product (bearing at least one hydrogen atom covalently bonded to a carbon atom) was extracted from the SPRESI database (InfoChem GmbH, Munich, Germany). The reactions were manually assigned into six types (Figure 1): [3+2] photocycloaddition of azirines to C=C (20 reactions), [2+2] photocycloaddition of C=C to C=O (31 reactions), [4+2] and [4+4] photocycloaddition of olefins to carbon-only aromatic rings (20 reactions), [2+2] photocycloaddition of C=C to C=C (73 reactions), [3+2] photocycloaddition of s-triazolo[4,3-b] pyridazine to C=C (10 reactions), and [2+2] photocycloaddition of C=C to C=S (27 reactions). [21]

We simulated all possible mixtures of two reactions (belonging to different types) from the data set of 181 reactions. For example, the 20 *[3+2] photocycloaddition of azirines to C=C* and the 31 *[2+2] photocycloaddition of C=C to C=O* yield 620 mixtures of reactions of class A. In the following combination the 20 *[3+2] photocycloaddition of azirines to C=C* and the 20 *[4+2] and [4+4] photocycloaddition of olefins to carbon-only aromatic rings* yield 400 mixtures of reactions of class B, and so on until all possible combinations of reactions types were simulated. The final data set of mixtures of reactions consists in 12421 mixtures. From this data set, 8280 mixtures were randomly selected to the training set and the remaining 4141 were used as a test set, Partition 1. Figure 1 and Table 1 show the types of reactions and the number of reactions by type. Table 2 indicates the resulting number of mixtures.

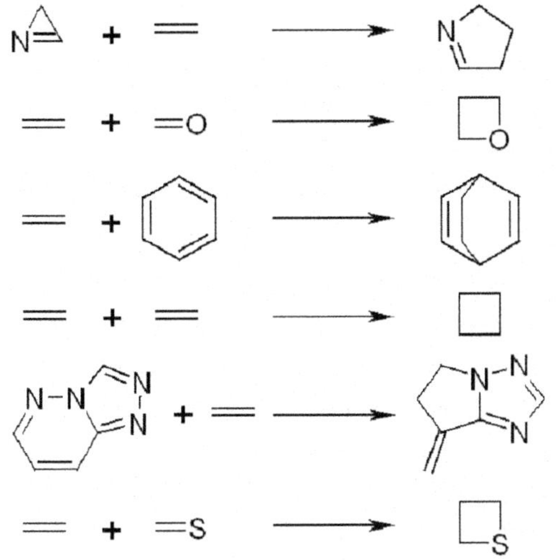

Figure 1. Types of photochemical reactions (from top): [3+2] photocycloaddition of azirines to C=C, [2+2] photocycloaddition of C=C to C=O, [4+2] and [4+4] photocycloaddition of olefins to carbon-only aromatic rings, [2+2] photocycloaddition of C=C to C=C, [3+2] photocycloaddition of s-triazolo[4,3-b]pyridazine to C=C, and [2+2] photocycloaddition of C=C to C=S.

Table 1: Number of reactions by reaction type and partition to be used to generate training and test sets of Partition 2 of mixtures of reactions

Types of Reactions	Number of Reactions	For Partition 2*
[3+2] photocycloaddition of azirines to C=C	20	16/4
[2+2] photocycloaddition of C=C to C=O	31	23/8
[4+2] and [4+4] photocycloaddition of olefins to carbon-only aromatic rings	20	16/4
[2+2] photocycloaddition of C=C to C=C	73	56/17
[3+2] photocycloaddition of s-triazolo[4,3-b]pyridazine to C=C	10	8/2
[2+2] photocycloaddition of C=C to C=S	27	21/6
Total	181	140/41

*Number of reactions in the training/test set to be used to generate Partition 2 of mixtures of reactions.
doi:10.1371/journal.pone.0088499.t001

Table 2: Number of reaction mixtures in each mixture class (mixture of two reactions of different types) for the two partitions of the data set

Class of mixture	Reaction 1	Reaction 2	Partition 1*	Partition 2*
A	[3+2] photocycloaddition of azirines to C=C	[2+2] photocycloaddition of C=C to C=O	413/207	368/32
B	[3+2] photocycloaddition of azirines to C=C	[4+2] and [4+4] photocycloaddition of olefins to carbon-only aromatic rings	267/133	256/16
C	[3+2] photocycloaddition of azirines to C=C	[2+2] photocycloaddition of C=C to C=C	975/487	896/68
D	[3+2] photocycloaddition of azirines to C=C	[3+2] photocycloaddition of s-triazolo[4,3-b]pyridazine to C=C	132/67	128/8
E	[3+2] photocycloaddition of azirines to C=C	[2+2] photocycloaddition of C=C to C=S	360/180	352/20
F	[2+2] photocycloaddition of C=C to C=O	[4+2] and [4+4] photocycloaddition of olefins to carbon-only aromatic rings	413/206	368/32
G	[2+2] photocycloaddition of C=C to C=O	[2+2] photocycloaddition of C=C to C=C	1510/754	1288/136
H	[2+2] photocycloaddition of C=C to C=O	[3+2] photocycloaddition of s-triazolo[4,3-b]pyridazine to C=C	206/104	184/16
I	[2+2] photocycloaddition of C=C to C=O	[2+2] photocycloaddition of C=C to C=S	558/279	506/40
J	[4+2] and [4+4] photocycloaddition of olefins to carbon-only aromatic rings	[2+2] photocycloaddition of C=C to C=C	974/486	896/68
K	[4+2] and [4+4] photocycloaddition of olefins to carbon-only aromatic rings	[3+2] photocycloaddition of s-triazolo[4,3-b]pyridazine to C=C	133/67	127/8
L	[4+2] and [4+4] photocycloaddition of olefins to carbon-only aromatic rings	[2+2] photocycloaddition of C=C to C=S	360/180	353/20
M	[2+2] photocycloaddition of C=C to C=C	[3+2] photocycloaddition of s-triazolo[4,3-b]pyridazine to C=C	498/250	448/34
N	[2+2] photocycloaddition of C=C to C=C	[2+2] photocycloaddition of C=C to C=S	1302/651	1232/85
O	[3+2] photocycloaddition of s-triazolo[4,3-b]pyridazine to C=C	[2+2] photocycloaddition of C=C to C=S	180/90	176/10

*Number of reactions in the training/test sets.
doi:10.1371/journal.pone.0088499.t002

Another more challenging partition of the data set was also used in which the data set of 181 reactions was randomly partitioned into subsets of 140 and 41 reactions, assuring that both sets cover the whole range of reactions, (see Table 1 for training and test set partition by reaction type) and the combinations were generated within each data set. For example the 16 *[3+2] photocycloaddition of azirines to C=C* and the 23 *[2+2] photocycloaddition of C=C to C=O* yield 368 mixtures of reactions of class A for the training set then, in the following combination, the 16 *[3+2] photocycloaddition of azirines to C=C* and the 16 *[4+2] and [4+4] photocycloaddition of olefins to carbon-only aromatic rings* yield 256 mixtures of reactions of class B for the training set of Partition 2, and so on until all possible combinations of reactions were simulated. The same was performed for the test set. The larger subset (7578 mixtures) was used as a training set and the smaller (593 mixtures) as a test set consisting in Partition 2 (see Figure 1 and Table 1 for types of reactions and number of reactions by type, and Table 2 for the resulting number of mixtures). Table 2 shows the constitution of each data set and the labels used in the experiments with Kohonen SOMs. Mixture classes (A to O) correspond to combinations of two reaction types. It is to emphasize that a 15-class classification problem like this (15 different mixtures of reactions) is a challenging modelling problem even using supervised learning techniques.

[1B1]H NMR Spectra of Reactants and Products

^{1}H NMR chemical shifts were predicted by the SPINUS program (v2) [22]–[24] from the molecular structures of the reactants and products. Only hydrogen atoms bonded to carbon atoms were predicted. The predicted chemical shifts were fuzzified with a triangular function and widths 0.1 ppm at each side of the chemical shift, which approximate the observed mean absolute error of SPINUS predictions (0.2–0.3 ppm). [22]

^{1}H NMR Spectra of Mixtures before and after the Reactions

All the signals, integrating proportionally to the number of protons, arising from all reactants of one reaction were taken together (spectrum of the reactants). The spectrum of the reactants was subtracted from the spectrum of the product. This is the difference between the spectra after and before the reaction, assuming full conversion.

The difference spectrum ("reaction spectrum") was binned in the range 0–12 ppm using 0.1 ppm wide intervals resulting in 120 variables (each variable integrating the intensities within an interval of 0.1 ppm). Experiments concerning the optimization of the binning procedure and integration of the intensities and their relation with the mean absolute error was performed in a previous publication [21].

For a mixture of two reactions, the reactions spectra are summed. The result corresponds to the difference of the spectra before (only reactants) and after (only products) the two reactions occur. Simultaneousness of the two reactions is assumed. Unless otherwise specified, the two reactions of each mixture were simulated in a 1:1 ratio and with full conversion.

In this way, we generate an NMR reaction descriptor for each reaction mixture of the data set, which is used as the input to the machine learning techniques.

Machine Learning Methods

Kohonen Self-Organizing Maps (Kohonen SOMs) [25], [26].

Kohonen SOMs learn by unsupervised training, distributing objects through a grid of so-called neurons, on the basis of the objects' features. This is an unsupervised method that projects multidimensional objects into a 2D surface (a map). SOM can reveal similarities between objects, mapped into the same or neighbor neurons. Each neuron of the map contains as many elements (weights) as the number of input variables (objects features). Before the training starts, the weights take random values. During the training, each individual object is mapped into the neuron with the most similar weights compared to its features (shortest Euclidean distance between weights and input). This winning neuron is excited (or activated), and its weights are corrected to make them even more similar to the object features. The neurons in its neighborhood also have their weights adjusted. The extent of adjustment depends on the topological distance to the winning neuron – the closer a neuron is to the winning neuron the more it is adjusted – and on the stage of training. The objects of the training set are iteratively fed to the map, and the weights corrected, until a pre-defined number of cycles is attained. A trained Kohonen SOM reveals similarities between objects of a data set in the sense that similar objects are mapped into the same or closely adjacent neurons.

In the investigations described here, the input variables are the 120 NMR reaction descriptors derived from the spectra of the reactants and products of two reactions. SOMs with toroidal topology and dimension 49×49 were trained and tested using the two different partitions of the data set. The maximum size was chosen such that the number of mixtures of reactions was at least twice the number of neurons. The toroidal topology means that neurons occupy the surface of a torus, so that all neurons have 8 neighbors – in the 2D representation of the map the neurons at the left edge are neighbors of those at the right edge, and the same happens for those at the top and bottom edges. After the training, each neuron is labeled (colored) according to the classes of reaction mixtures that activate it (see Table 1), which facilitates visualization, and enables the classification of new reaction mixtures.

Training was performed by using a linear decreasing triangular scaling function with an initial learning rate of 0.1. The weights were initialized with random numbers that are calculated using the mean and the standard deviation of each variable in the input data set. For the selection of the winning neuron, the minimum Euclidean distance between the input vector and neuron weights was used. The training was performed over 50–100 cycles, with the learning span and the learning rate linearly decreasing until zero. These parameters appeared as a reasonable balance between network stability and computation time. Kohonen SOM were implemented with in-house-developed software based on JATOON Java applets. [27] To overcome fluctuations induced by the random factors influencing the training, five or ten independent SOM were trained with the same objects, generating an ensemble of SOM. Ensemble predictions were obtained for new objects by majority vote of the individual SOMs.

Counter-Propagation Neural Networks (CPNNs) [26].

A Counter-Propagation Neural Networks incorporate a Kohonen SOM linked to a second layer of neurons (output layer) that acts as a look-up table and stores output data (the classification of the mixture of reactions). The

CPNN method is considered a semi-supervised technique. During the training, the winning neuron is determined exclusively on the basis of the Kohonen layer (input layer), but the weights of the corresponding output neuron are adjusted to become closer to the output values of the object – semi-supervised learning. After the training, the CPNN can produce an output for an object – the winning neuron is chosen and the corresponding weights in the output layer are taken as the prediction.

In this work, the types of the mixture reactions were encoded into a vector (output) with dimension six (the number of reaction types). The two components of the vector corresponding to the types of the reactions present in the mixture take the value one, the others take the value zero. After the training, when working in prediction mode, CPNN produce a six-values output for a reaction mixture, which is interpreted as a prediction of the two types obtaining the highest values.

Software and training settings were the same as in the experiments with Kohonen SOMs. Ensembles of five or ten independent CPNN were trained, and predictions were obtained by majority vote of the individual maps.

Random Forests (RF) [28], [29].

A random forest is an ensemble of unpruned classification trees created by using bootstrap samples of the training data and random subsets of variables to define the best split at each node. It is a high-dimensional nonparametric method that works well on large numbers of variables. The predictions are made by majority voting of the individual trees. The performance is internally assessed with the prediction error for the objects left out in the bootstrap procedure (out-of-bag estimation, OOB). Here, RFs were grown with the R program version 2.0.1, [30]using the randomForest library, [31] and were used to classify the reactions present in a mixture of reactions on the basis of NMR reaction descriptors. The models were built to classify objects (mixtures of reactions) according to the 15 classes of mixtures (classes A to O, Table 1). The number of trees in the forest was set to 1000, and the number of variables tested for each split was set to default (square root of the number of variables). The voting system of a RF allows the association of a probability to each prediction that reflects the percentage of votes obtained by the winning class. This probability was investigated as a measure of reliability.

RESULTS AND DISCUSSION

Previous to this work, the chemical shifts predicted by SPINUS had been validated for a subset of reactants and products in our data set of reactions, for which experimental chemical shifts were available [21]. A mean absolute error (MAE) of 0.24 ppm was obtained for the 349 chemical shifts of the subset, which was similar to earlier tests [23].

The following subsections present results concerning the ability of various machine learning methods (including unsupervised and supervised learning) to recognize patterns of changes in the ^{1}H NMR spectra corresponding to types of reactions, when two reactions occur simultaneously. Experiments were performed with two different partitions of the data set.

Mapping of Mixtures of Reactions on a Kohonen SOM

Kohonen SOMs of size 49×49 were trained using the NMR reaction descriptors for mixtures in the training set. In the learning procedure, the network made no use of the information related to mixture classes. After the training, each neuron of the surface was assigned to a mixture class (one of the 15 possible combinations of two reactions from six types). Figure 2 shows a Kohonen SOM of size 49×49 trained with 8280 mixtures corresponding to the training set of partition 1.

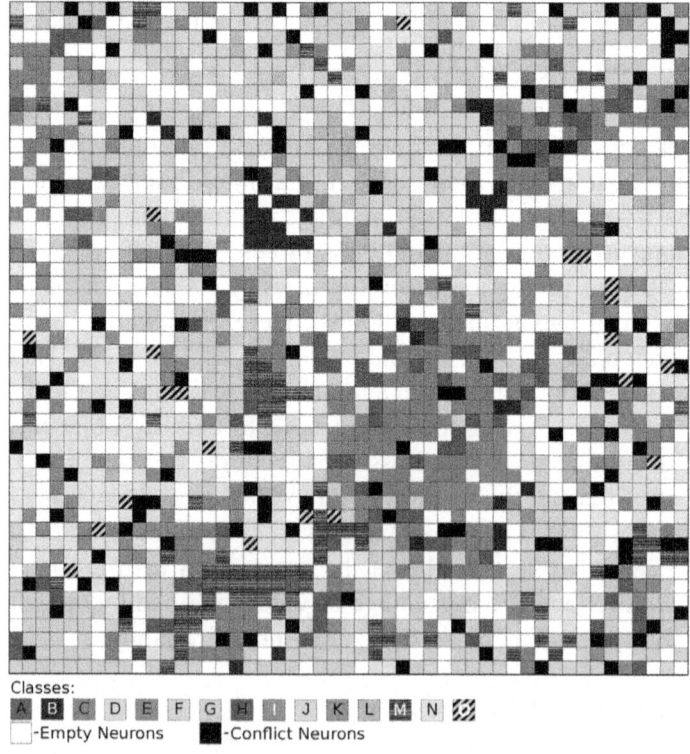

Classes:
A B C D E F G H I J K L M N
☐ -Empty Neurons   ■ -Conflict Neurons

Figure 2. Toroidal surface of a 49×49 Kohonen SOM trained with 8280 mixtures of two photochemical reactions encoded by the ^{1}H NMR descriptor.

After the training, each neuron was colored according to the reaction mixtures of the training set that are mapped onto it. The colors correspond to the classes in Table 1. Black neurons correspond to conflicts.

doi:10.1371/journal.pone.0088499.g002

The results show a trend for some classes of mixtures to cluster, namely class B, class C, class J, class L, class M and class N (see Table 1 for detailed information concerning the types of reactions in each class of mixtures). The 15 classes of mixtures correspond to combinations of two reactions from six different types. In fact, classes of mixtures sharing one type of reaction tend to be mapped on the same region of the map.

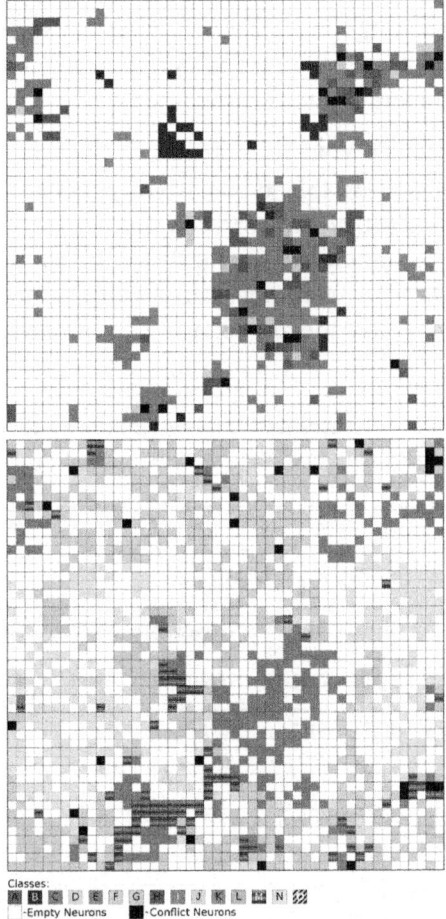

Classes:
A B C D E F G H I J K L M N ▨
□ -Empty Neurons   ■ -Conflict Neurons

Figure 3: The same map of Figure 2, with two different filters applied: top – only colored neurons belonging to mixtures of classes A, B, C, D, and E; bottom – only colored neurons belonging to mixtures of classes C, G, J, M, and N.

This is illustrated in Figure 3, which results from applying two different filters to Figure 2. In the first map, only neurons were colored that correspond to mixture classes A, B, C, D, or E (in all these mixtures is present a reaction of type [3+2] photocycloaddition of azirines to C=C). These mixtures concentrate on certain regions of the map and are not well separated from each other (the exception is class B). The second map only shows colored neurons of mixture classes C, G, J, M, or N (these are the classes sharing a reaction of type [2+2] photocycloaddition of C=C to C=C). These mixtures are much spread through the map (because of the large number of reactions) and are well distinguished from each other. The two images illustrate how the overlap of mixture classes on the map corresponds to the overlap of types of reactions in the mixtures.

The colors correspond to the classes in Table 1. Black neurons correspond to conflicts between these classes and white neurons correspond to empty neurons or neurons belonging to other classes.

doi:10.1371/journal.pone.0088499.g003

An individual SOM was able to consistently classify 80.6% of the reaction mixtures in the training set, and to correctly predict 71.1% of the test set (Table 3). Improvement in accuracy was achieved with ensembles of five and ten SOMs. Correct predictions were obtained for 86.7% and 89% of the training set, and 77.4% and 79.6% of the test set using ensembles of five and ten SOMs, respectively.

Table 3: Classification of mixtures of reactions (mixtures of two reactions) by Kohonen SOMs and Counter-Propagation Neural Networks of dimension 49×49

Data sets*		% Correct predictions					
		Best ind.		Ensemble of five		Ensemble of ten	
		SOM	CPNN	SOM	CPNN	SOM	CPNN
Partition	Training	80.6	61.3	86.7	73.0	89.0	75.6
1	Test	71.1	57.7	77.4	69.1	79.6	71.8
Partition	Training	82.9	68.4	89.4	77.4	91.4	78.6
2	Test	52.6	47.2	59.4	57.2	62.6	57.5

*Partition 1–8280 and 4141 mixtures of reactions in training and test set, respectively; Partition 2–7578 and 593 mixtures of reactions in training and test set, respectively.
doi:10.1371/journal.pone.0088499.t003

Then, experiments were performed using partition 2, with lower similarities between training and test sets. With partition 1, all mixtures are different, but the same reaction is often present in a mixture of the training and in a mixture of the test set (combined with another reaction). Differently, with partition 2 no reaction in mixtures of the test set was present in a mixture of the training set. Not surprisingly, the prediction accuracy decreased considerably – an ensemble of ten SOMs was able to correctly classify 62.3% of the mixtures in the test set.

Mapping of Mixtures of Reactions on a CPNN

CPNN process input data similarly to Kohonen SOM, but uses a different mechanism for producing classifications. A reaction type is identified in the mixture of the reactions if the activated neuron exhibits a high value for the output weight corresponding to that reaction type. Based on the six-values output, a mixture class is predicted if two and only two of the output values are higher than 0.5 (the mixture class corresponding to the combination of those two reaction types). Otherwise, the mixture is classified as undecided.

Figure 4 shows the six output layers (corresponding to the six possible types of reactions in the mixtures) of a 49×49 CPNN trained with 7578 mixtures (partition 2). High values of the weights at each output layer are represented by blue, and low values by red. It can be seen that mixtures including reactions from certain types cluster in typical regions while other types are more spread on the map and not so dominant in mixtures. The inspection of the six output layers reveals some correlation between the number of blue neurons in a layer, and the number of reactions of that type. For example, the most populated type of reaction in the data set is the [2+2] photocycloaddition of C=C to C=C with 73 reactions – it corresponds to output layer 4 with large regions of blue neurons. In fact, 4766 out of the 7578 mixtures in the training set (~63%) include this type of reaction. In the opposite side, output layer 5 is mostly red and the corresponding reaction type ([3+2] photocycloaddition of s-triazolo[4,3-b]pyridazine to C=C) is the least populated reaction type with only 10 reactions, present in 1063 mixtures (~14% of the training set).

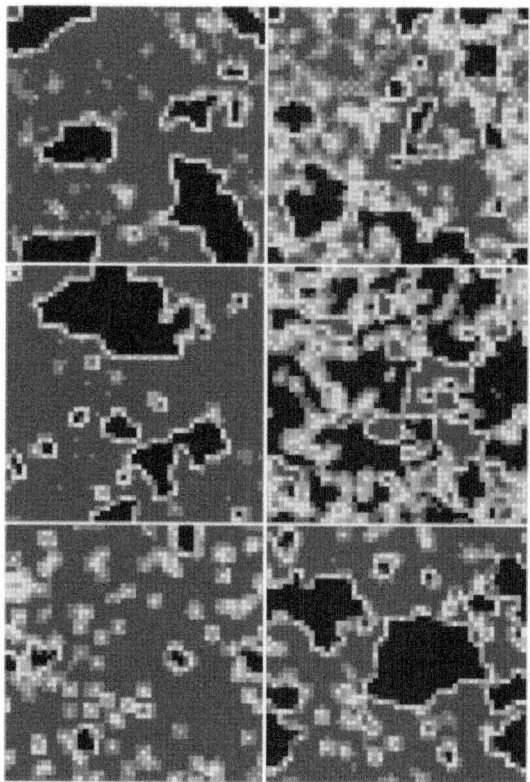

Figure 4. Representation of the six output layers of a 49×49 CPNN trained with 7578 mixtures of two reactions.

High values of the weights in each output layer are represented by blue, and low values by red. Output layers corresponding to the following reaction types, from left to right: First row – [3+2] photocycloaddition of azirines to C=C and [2+2] photocycloaddition of C=C to C=O reaction types. Second row – [4+2] and [4+4] photocycloaddition of olefins to carbon-only aromatic rings, and [2+2] photocycloaddition of C=C to C=C reaction types. Third row – [3+2] photocycloaddition of s-triazolo[4,3-b]pyridazine to C=C and [2+2] photocycloaddition of C=C to C=S reaction types.

doi:10.1371/journal.pone.0088499.g004

CPNN did not yield superior predictions to Kohonen SOMs (Table 3). For partition 1 an ensemble of ten CPNNs were only able to correctly classify 75.6% and 71.8% of the mixtures of the training and test set respectively. The prediction accuracy for the test set decreased in partition 2 to 57.5%. It is important to point out that in CPNNs, with six output layers, a class is only assigned to a mixture when two and only two of the output values are higher than 0.5 (the mixture class corresponding to the combination of those two reaction types). If such an assignment is not possible, the mixture is predicted as undecided. This strict condition gives rise to a large number of mixtures with no class assigned. For example, the best individual CPNN does not assign ~26% of the mixtures in the test set of partition 1. If only assigned mixtures are considered, the true classifications are 78%. Kohonen SOM is an unsupervised learning method, and CPNN is semi-supervised. Both present the advantage of an easy visualization of the objects in a map, and reveal relationships between similarities of descriptors and classes. However, they are based on global comparisons of the descriptor profile, and are not expected to learn associations between classes and reduced numbers of specific descriptors. Such associations may well occur in the studied data set – some regions of the spectrum are likely to be more relevant than others. Therefore, experiments were next performed with a supervised learning method.

Assignment of Reaction Types in Mixtures of Reactions by RFs

The results obtained with Random Forests are displayed in Table 4 for the two partitions. Predictions for training sets were from the internal cross validation obtained by out-of-bag (OOB) estimation. The accuracies of the predictions for partition 1 reached 99% both for OOB estimation of the training set and for the test set. A 10-fold cross-validation experiment was also performed with the training set of Partition 2. The obtained accuracy was similar to the OOB estimation and reached 99.5% of correct predictions. With a totally independent test set (partition 2) the accuracy of the predictions was 80%. RFs performed clearly better than the self-organizing maps. Table 5 shows the confusion matrix obtained for the test set of partition 2.

Table 4: Classification of mixtures of reactions (mixtures of two reactions) by Random Forests

Data sets*		% Correct predictions
Partition	Training	99.2
1	Test	99.1
Partition	Training	99.6
2	Test	80.3

*Partition 1–8280 and 4141 mixtures of reactions in training and test set, respectively; Partition 2–7578 and 593 mixtures of reactions in training and test set, respectively.
doi:10.1371/journal.pone.0088499.t004

Table 5: Confusion matrix for the classification of mixtures obtained by RF for the test set of partition 2

	A	B	C	D	E	F	G	H	I	J	K	L	M	N	O	%
A	25	–	5	–	–	–	2	–	–	–	–	–	–	–	–	78.1
B	–	15	–	–	–	–	–	–	–	1	–	–	–	–	–	93.8
C	5	–	60	–	–	–	3	–	–	–	–	–	–	–	–	88.2
D	–	–	–	8	–	–	–	–	–	–	–	–	–	–	–	100.0
E	1	–	7	–	10	–	–	–	–	–	–	–	–	2	–	50.0
F	–	–	–	–	–	28	1	–	–	3	–	–	–	–	–	87.5
G	–	–	1	–	–	–	135	–	–	–	–	–	–	–	–	99.3
H	–	–	–	–	–	–	–	16	–	–	–	–	–	–	–	100.0
I	–	–	–	–	–	–	16	–	21	–	–	–	–	3	–	52.5
J	–	–	–	–	–	8	1	–	–	59	–	–	–	–	–	86.8
K	–	–	–	–	–	–	–	–	–	–	8	–	–	–	–	100.0
L	–	–	–	–	–	2	–	–	–	7	–	11	–	–	–	55.0
M	–	–	–	–	–	–	–	–	–	–	–	–	34	–	–	100
N	–	–	3	–	–	–	34	–	7	–	–	–	–	41	–	48.2
O	–	–	–	–	–	–	–	–	–	–	–	–	5	–	5	50.0

doi:10.1371/journal.pone.0088499.t005

The confusion matrix shows a high prediction accuracy not only for the most populated classes like class C, G and J, but also for some less populated like classes D, H and K with 100% of correct classifications. The mixtures of classes E, I, L, N and O are the most difficult to classify. For them, true positives are only ca. 50% of the number of mixtures for these classes (last column of Table 5), although the counts of false positives are relatively low (inspection of Table 5 by columns). Classes E, I, L. N and O result from the combination of [2+2] photocycloadditions of C=C to C=S with the remaining five types of reactions, which indicates that the patterns of this reaction type, encoded in our "reaction spectrum", is from all types of reactions the most difficult to learn – a consequence of a lack of hydrogen atoms bonded to the atoms of the reaction center in our data set. The difficulty to learn this type of reaction was also found in our previous studies [21] for the classification of reactions outside of mixtures. This approach cannot properly encode mixtures of reactions where the reactants and products have no hydrogen atoms bonded to the atoms of the reaction center. The prediction ability increase from 80% to 93% if the mentioned classes are not considered.

RF associate a probability to each prediction reflecting the proportion of votes obtained by the winning class. Table 6 presents the relationship between the prediction accuracy and the probability of the predictions.

Table 6: Relationship between the prediction accuracy and the probability associated to each prediction by RFs for test set of partition 2

Classes[a]	Probability							
	No Selection		≥0.5		≥0.6		≥0.8	
	N.of Mixtures[b]	N. of Correct[c]	N.of Mixtures[b]	N. of Correct[c]	N.of Mixtures[b]	N. of Correct[c]	N.of Mixtures[b]	N. of Correct[c]
A (32)	31	25 (80.7)	12	11 (91.7)	4	4 (100.0)	2	2 (100.0)
B (16)	15	15 (100.0)	7	7 (100.0)	4	4 (100.0)	1	1 (100.0)
C (68)	76	60 (79.0)	52	50 (96.1)	41	41 (100.0)	14	14 (100.0)
D (8)	8	8 (100)	6	6 (100.0)	4	4 (100.0)	2	2 (100.0)
E (20)	10	10 (100)	3	3 (100.0)	3	3 (100.0)	–	–
F (32)	38	28 (73.7)	24	21 (87.5)	17	16 (94.1)	6	6 (100.0)
G (136)	192	135 (87.5)	130	115 (94.1)	91	88 (96.7)	37	37 (100.0)
H (16)	16	16 (100.0)	14	14 (100.0)	14	14 (100.0)	5	5 (100.0)
I (40)	28	21 (75.0)	17	15 (88.2)	10	8 (80.0)	–	–
J (68)	70	59 (84.3)	43	41 (95.4)	27	26 (96.3)	11	10 (90.9)
K (8)	8	8 (100.0)	8	8 (100.0)	7	7 (100.0)	6	6 (100.0)
L (20)	11	11 (100.0)	7	7 (100.0)	5	5 (100.0)	1	1 (100.0)
M (34)	39	34 (87.2)	30	30 (100.0)	28	28 (100.0)	14	14 (100.0)
N (85)	46	41 (89.1)	25	25 (100.0)	13	13 (100.0)	6	6 (100.0)
O (10)	5	5 (100.0)	5	5 (100.0)	3	3 (100.0)	2	2 (100.0)
Total	593	476 (80.3)	383	358 (93.5)	271	264 (97.4)	107	106 (99.1)

[a]Class labels and number of reactions in each class.
[b]Number of mixtures predicted to belong to each class.
[c]Number of true positives for each class and (in parenthesis) its percentage among the number of mixtures predicted to belong to that class.
doi:10.1371/journal.pone.0088499.t006

The results support the use of the probability of each prediction as a measure of reliability of the class assignment. For the test set of the second partition, 383 mixtures out of 593 (65%) were predicted with probability higher than 0.5 and, from these, 358 (94%) were correctly classified. If we consider only mixtures predicted with probability higher than 0.6, the number of predicted mixtures decreases to 271 (46%) but the percentage of correctly classified mixtures (among these) increases to 97%. With a probability higher than 0.8, almost all mixtures of reactions were correctly classified (106 out of 107).

Filtering predictions by the RF probability also improves the results for the more problematic mixture classes E, I, L N, and O (Table 7). The percentage of true positive predictions among the mixtures of each class predicted with probability above 0.5 increased to 60%, 60%, 88%, 78%, and 100% respectively – Tables 5 and 6.

Table 7: Confusion matrix for the classification of mixtures with probability higher than 0.5 obtained by RF for the test set of partition 2

	A	B	C	D	E	F	G	H	I	J	K	L	M	N	O	%
A	11	-	-	-	-	-	-	-	-	-	-	-	-	-	-	100.0
B	-	7	-	-	-	-	-	-	-	-	-	-	-	-	-	100.0
C	1	-	50	-	-	-	-	-	-	-	-	-	-	-	-	98.0
D	-	-	-	6	-	-	-	-	-	-	-	-	-	-	-	100.0
E	-	-	2	-	3	-	-	-	-	-	-	-	-	-	-	60.0
F	-	-	-	-	-	21	-	-	-	1	-	-	-	-	-	95.5
G	-	-	-	-	-	-	115	-	-	-	-	-	-	-	-	100.0
H	-	-	-	-	-	-	-	14	-	-	-	-	-	-	-	100.0
I	-	-	-	-	-	-	-	15	-	-	-	-	-	-	-	60.0
J	-	-	-	-	-	3	-	-	-	41	-	-	-	-	-	93.2
K	-	-	-	-	-	-	-	-	-	-	8	-	-	-	-	100.0
L	-	-	-	-	-	-	-	-	1	-	-	7	-	-	-	87.5
M	-	-	-	-	-	-	-	-	-	-	-	-	30	-	-	100.0
N	-	-	-	-	-	5	-	2	-	-	-	-	-	25	-	78.1
O	-	-	-	-	-	-	-	-	-	-	-	-	-	-	5	100.0

doi:10.1371/journal.pone.0088499.t007

The best RF model developed with partition 1 was further validated using the y-randomization technique. The model was retrained using a modified training set where the Y-column values – the column corresponding to the classification of the mixtures - was scrambled and the descriptor matrix was kept unchanged. Scrambling was performed 5 times. Each randomized model was used to make predictions for the test set. A considerable decrease in the % of correct predictions in comparison with the non-randomized model was observed for the five random models (% of correct predictions: 14.5–15.7%) which supports the reliability and robustness of the original model.

To check the impact of the random partition in training and test sets, five random alternative partitions to partition 1, with the same sizes of the training and test sets, were used to train RF models. The results, both for training sets (OOB estimation) and test sets were similar to those of Table 4 for partition 1 (99.1% for the original test set partition and a range of 99.2–99.5% of correct predictions for the new five randomly selected partitions).

In order to better simulate realistic situations and possible experimental conditions, the RF model trained with partition 1 was further validated using more challenging test sets, generated with partial conversion of the reactants into products and different ratios of the two reactions in the mixture. The accuracy of the predictions for the test set of partition 1 with 70%, 80% and 90% simulated reaction yields (for both reactions of the mixture) was 81.7%, 96.4% and 98.8% respectively. These compare with 99.1% of correct predictions for the test set with full conversion (Table 4).

The test set of partition 1 was also re-used to simulate different ratios of the two reactions in a mixture and different normalizations of the spectra integration.

A MIXTURE i was generated by the formula $MIXTURE_i = NORM*(RATIO * A_i + B_i)$, where A_i and B_i are the reactions of the mixture, RATIO took values 2 and 5, and NORM took random values between 0.2 and 1.0. Table 8 shows how the RF model developed with the training set of partition 1 (where NORM and RATIO were always 1) predicted the new test set (consisting of 8,282 mixtures).

Table 8. Impact of the ratio of the two reactions in the mixture and the integration normalization on the % of correct predictions.[a]

RATIO A_i/B_i	% Correct Predictions	
	NORM = 1	0.2≤NORM≤1
1 (Table 3)	99.1	–
2	96.2	75.3
5	82.7	70.6

[a]The same mixtures of the test set of partition 1 were used, but with different ratios between the two reactions, and different normalization factors in the spectra integration.
doi:10.1371/journal.pone.0088499.t008

Finally, the test set was simulated with simultaneous random variation of the three parameters – yields, NORM (range 0.2–1.0) and RATIO (range 1–4). The percentage of correct predictions for test sets with reaction yields ranges of 50–100%, 60–100% and 70–100% were 62, 65 and 68% respectively.

A relationship between the probability of the RF predictions and the prediction accuracy was observed again, for the most challenging test set – yields (range 50%–100%), NORM (range 0.2–1.0) and RATIO (range 1–4) – Table 9.

Table 9: Relationship between the prediction accuracy and the probability associated to each prediction by RFs for the test set of partition 1 simulated with simultaneous random variation of the three parameters – yields (range 50–100%), NORM (range 0.2–1.0) and RATIO (range 1–4)

In this test set, 6208 mixtures out of 8282 (75%) were predicted with probability higher than 0.5 and, from these, 4191 (68%) were correctly classified. If we consider only mixtures predicted with probability higher than 0.8, the number of predicted reactions decreases to 2452 (30%) but the percentage of correctly classified mixtures (among these) increases to 79%. It is to point out the results for the more difficult class G. From the 3327 mixtures classified as G, only 1476 (44%) were correctly classified, but the percentage increases to 72% among mixtures predicted with probability higher than 0.8.

| Classes[a] | Probability | | | | | | | |
| | No Selection | | ≥0.5 | | ≥0.6 | | ≥0.8 | |
	N.of Mixtures[b]	N. of Correct[c]	N.of Mixtures[b]	N. of Correct[c]	N.of Mixtures[b]	N. of Correct[c]	N.of Mixtures[b]	N. of Correct[c]
A (414)	179	178 (99.4)	141	141 (100)	95	95 (100)	17	17 (100)
B (266)	79	75 (94.9)	25	25 (100)	8	8 (100)	2	2 (100)
C (974)	1108	751 (67.8)	863	642 (74.4)	693	556 (80.2)	322	291 (90.4)
D (134)	62	61 (98.4)	44	44 (100)	36	36 (100)	9	9 (100)
E (360)	163	146 (89.6)	96	94 (97.9)	57	57 (100)	6	6 (100)
F (412)	116	106 (91.4)	68	66 (97.1)	48	48 (100)	11	11 (100)
G (1510)	3327	1476(44.4)	2600	1367(52.6)	2208	1269(57.5)	1250	895 (71.6)
H (206)	89	89 (100)	74	74 (100)	61	61 (100)	36	36 (100)
I (558)	224	221 (98.7)	153	153 (100)	96	96 (100)	21	21 (100)
J (972)	670	477 (71.2)	417	325 (77.9)	313	256 (81.8)	121	108 (89.3)
K (134)	36	26 (72.2)	29	22 (75.9)	23	21 (91.3)	9	9 (100)
L (360)	122	115 (94.3)	64	64 (100)	37	37 (100)	7	7 (100)
M (488)	449	288 (64.1)	352	243 (69)	283	200 (70.7)	196	155 (79.1)
N (1314)	1586	1046 (66)	1230	880 (71.5)	964	730 (75.7)	434	353 (81.3)
O (180)	72	67 (93.1)	52	51 (98.1)	33	33 (100)	11	11 (100)
Total	8282	5122(61.8)	6208	4191(67.5)	4955	3503(70.7)	2452	1931(78.8)

[a]Class labels and number of reactions in each class.
[b]Number of mixtures predicted to belong to each class.
[c]Number of true positives for each class and (in parenthesis) its percentage among the number of mixtures predicted to belong to that class.
doi:10.1371/journal.pone.0088499.t009

The results clearly show that the model learns the key patterns of NMR signals corresponding to classes of reactions in the mixtures and are reasonably capable of classifying new cases involving partial conversion of reactants, different ratios between reactions and different normalization of integrations, even without any re-parameterization of the initial model.

Clearly, if these more demanding situations are included in the training set, the ability to predict the test set are improved. For example, in an experiment where both training and test sets are simulated with simultaneous random variation of the three parameters – yields (range 25%–100%), NORM (range 0.2–1.0) and RATIO (range 1–7) – a RF correctly predicted 96% of the mixtures, which reinforces the conclusion that the model learns the classes of reactions by the presence of key patterns of NMR signals.

CONCLUSIONS

This study demonstrates the possibility of applying machine learning methods to automatically identify types of co-occurring chemical reactions from the differences between the ^{1}H NMR spectra of reactants and products. These results also illustrate the usefulness of SPINUS predictions of NMR data in that context, for the generation of training sets.

The fact that a supervised learning method yielded significantly better predictions suggests that changes in very specific ranges of the ^{1}H NMR spectra are markers of reaction types. The extremely high percentages of correct predictions for the test set of partition 1 with supervised learning, and for the random forest OOB estimation within training sets of both partitions, indicate that the same should happen for individual reactions.

In most practical situations, a reaction is accompanied by side reactions, and can proceed to different yields, which would require that the NMR interpretation system is able to identify reaction types even in the presence of a complex mixture of reactions with different conversions. Experiments simulating mixtures of reactions with a diversity of product yields, different proportions of reactions, and different normalization of integrations corroborated this possibility.

This study relies on ^{1}H NMR spectra, and is therefore limited by the availability of hydrogen atoms in the neighborhood of the reaction center and by the sensitivity of their chemical shifts to the changes resulting from the reactions. But, in principle, the method can be used with other types of spectra, e.g., ^{13}C NMR or IR. It must be emphasized that this approach does not require structural information on the reactions participants – it performs "reaction elucidation" without structure elucidation of the molecules in the mixtures.

ACKNOWLEDGMENTS

The authors thank InfoChem GmbH (Munich, Germany) for sharing the dataset of photochemical reactions from the SPRESI database. The assistance of Dr. Yuri Binev with the use of SPINUS software and simulation of NMR data is gratefully acknowledged.

AUTHOR CONTRIBUTIONS

Conceived and designed the experiments: DARSL JAS. Performed the experiments: DARSL. Analyzed the data: DARSL JAS. Wrote the paper: DARSL JAS.

REFERENCES

1. McKenzie JS, Donarski JA, Wilson JC, Charlton AJ (2011) Analysis of complex mixtures using high-resolution nuclear magnetic resonance spectroscopy and chemometrics. Prog Nucl Magn Reson Spectrosc 59: 336–359. doi: 10.1016/j.pnmrs.2011.04.003

2. Sykes BD (2011) Editorial J Biomol NMR. 49: 163–164. doi: 10.1007/s10858-011-9479-3

3. Brougham DF, Ivanova G, Gottschalk G, Collins DM, Eustace AJ, et al. (2011) Artificial Neural Networks for Classification in Metabolomic

Studies of Whole Cells Using 1H Nuclear Magnetic Resonance. J Biomed Biotechnol Volume 2011, Article ID 158094, 8 pages.

4.   Lloyd GV, Wongravee K, Silwood CJL, Grootveld M, Brereton RG (2009) Self Organising Maps for variable selection: Application to human saliva analysed by nuclear magnetic resonance spectroscopy to investigate the effect of an oral healthcare product. Chemom Intell Lab Syst 98: 49–161. doi: 10.1016/j.chemolab.2009.06.002

5.   Wongravee K, Lloyd GR, Silwood CJ, Grootveld M, Brereton RG (2010) Supervised Self Organizing Maps for Classification and Determination of Potentially Discriminatory Variables: Illustrated by Application to Nuclear Magnetic Resonance Metabolomic Profiling. Anal Chem 82: 628–638. doi: 10.1021/ac9020566

6.   Cho H-W, Kim SB, Jeong MK, Park Y, Ziegler TR, et al. (2008) Genetic algorithm-based feature selection in high-resolution NMR spectra. Expert Syst Appl 35: 967–975. doi: 10.1016/j.eswa.2007.08.050

7.   Beckonert O, Monnerjahn J, Bonk U, Leibfritz D (2003) Visualizing metabolic changes in breast cancer tissue using 1H NMR spectroscopy and self-Organizing maps. NMR Biomed 16: 1–11. doi: 10.1002/nbm.797

8.   Suna T, Salminen A, Soininen P, Laatikainen R, Ingman P, et al. (2007) 1H NMR metabonomics of plasma lipoprotein subclasses: elucidation of metabolic clustering by self-organising maps. NMR Biomed 20: 658–672. doi: 10.1002/nbm.1123

9.   Alonso-Salces RM, Héerger K, Holland MV, Moreno-Rojas JM, Mariani C, et al. (2010) Multivariate analysis of NMR fingerprint of the unsaponifiable fraction of virgin olive oils for authentication purposes, Food Chem. 118: 956–965. doi: 10.1016/j.foodchem.2008.09.061

10.   Aursand M, Standal IB, Axelson DE (2007) High-Resolution 13C Nuclear Magnetic Resonance Spectroscopy Pattern Recognition of Fish Oil Capsules, J Agric Food Chem. 55: 38–47. doi: 10.1021/jf0617541

11.   Vallikivi I, Järving I, Pehk T, Samel N, Tõugu V, et al. (2004) NMR monitoring of lipase-catalyzed reactions of prostaglandins: preliminary estimation of reaction velocities. J Mol Catal B: Enzym 32: 15–19. doi: 10.1016/j.molcatb.2004.09.002

12.   Ebbels TMD, Keun HC, Beckonert OP, Bollard ME, Lindon JC, et al. (2007) Prediction and Classification of Drug Toxicity Using Probabilistic Modeling of Temporal Metabolic Data: The Consortium on Metabonomic Toxicology Screening Approach. J Proteome Res 6: 4407–4422. doi: 10.1021/pr0703021

13.   .Ballard M, Bown M, James S, Yang Q (2011) NMR studies of mixed

amines. Energy Procedia 4: 291–298. doi: 10.1016/j.egypro.2011.01.054

14. .Shey J-Y, Sun C-M (2002) Liquid-phase combinatorial reaction monitoring by conventional 1H NMR spectroscopy. Tetrahedron Lett 43: 1725–1729. doi: 10.1016/s0040-4039(02)00061-8

15. Kalelkar S, Dow ER, Grimes J, Clapham M, Hu H (2002) Automated Analysis of Proton NMR Spectra from Combinatorial Rapid Parallel Synthesis Using Self-Organizing Maps. J Comb Chem 4: 622–629. doi: 10.1021/cc020031l

16. Bernstein MA, Stefinovic M, Sleigh CJ (2007) Optimising reaction performance in the pharmaceutical industry by monitoring with NMR. Magn Reson Chem 45: 564–571. doi: 10.1002/mrc.2007

17. Khajeh M, Bernstein MA, Morris GA (2010) A simple flowcell for reaction monitoring by NMR. Magn Reson Chem 48: 516–522. doi: 10.1002/mrc.2610

18. Gomez MV, Verputten HHJ, Díaz-Ortíz A, Moreno A, de la Hoz A, et al. (2010) On-line monitoring of a microwave-assisted chemical reaction by nanolitre NMR-spectroscopy. Chem Commun 46: 4514–4516. doi: 10.1039/b924936b

19. Mix A, Jutzi P, Rummel B, Hagedorn K (2010) A Simple Double-Chamber NMR Tube for the Monitoring of Chemical Reactions by NMR Spectroscopy. Organometallics 29: 442–447. doi: 10.1021/om900919f

20. Foley DA, Doecke CW, Buser JY, Merritt JM, Murphy L, et al. (2011) ReactNMR and ReactIR as Reaction Monitoring and Mechanistic Elucidation Tools: The NCS Mediated Cascade Reaction of α-Thioamides to α-Thio-β-chloroacrylamides. J Org Chem 76: 9630–9640. doi: 10.1021/jo201212p

21. Latino DARSL, Aires-de-Sousa J (2007) Linking databases of chemical reactions to NMR data: An exploration of 1H NMR-based reaction classification. Anal Chem 79: 854–862. doi: 10.1021/ac060979s

22. Binev Y, Aires-de-Sousa J (2004) Structure-based predictions of 1H NMR chemical shifts using feed-forward neural networks. J Chem Inf Comput Sci 44: 940–945. doi: 10.1021/ci034228s

23. Binev Y, Corvo M, Aires-de-Sousa J (2004) The impact of available experimental data on the prediction of 1H NMR chemical shifts by neural networks. J Chem Inf Comput Sci 44: 946–949. doi: 10.1021/ci034229k

24. SPINUS website. Available: http://joao.airesdesousa.com/spinus. Acessed 2013 November 27.

25. Kohonen T (1988) Self-Organization and Associative Memory. Berlin:Springer.

26. Zupan J, Gasteiger J (1999) Neural Networks in Chemistry and Drug Design. Weinheim:Wiley-VCH.

27. Aires-de-Sousa J (2002) JATOON: Java tools for neural networks. Chemom Intell Lab Syst 61: 167–173. doi: 10.1016/s0169-7439(01)00171-x

28. Breiman L (2001) Random forests. Machine Learn 45: 5–32. doi: 10.1023/a:1010933404324

29. Svetnik V, Liaw A, Tong C, Culberson JC, Sheridan RP, et al. (2003) Random forest: A classification and regression tool for compound classification and QSAR modeling. J Chem Inf Comput Sci 43: 1947–1958. doi: 10.1021/ci034160g

30. R. D. C. Team (2004) R: A Language and Environment for Statistical Computing. Vienna. URL http://www.R-project.Org.

31. Fortran original by Leo Breiman and Adele Cutler, R port by Andy Liaw and Mathew Wiener. Leo Breiman website. Available: http://www.stat.berkeley.edu/users/breiman/. Acessed 2013 November 27.

Chapter 3

A LATTICE GAS AUTOMATA MODEL FOR THE COUPLED HEAT TRANSFER AND CHEMICAL REACTION OF GAS FLOW AROUND AND THROUGH A POROUS CIRCULAR CYLINDER

Hongsheng Chen [1], Zhong Zheng [1,], Zhiwei Chen [2,] and Xiaotao T. Bi [2]

[1]School of Materials Science and Engineering, Chongqing University, Chongqing 400044, China

[2]Fluidization Research Center, Department of Chemical and Biological Engineering, University of British Columbia, Vancouver V6T 1Z3, Canada

ABSTRACT

Coupled heat transfer and chemical reaction of fluid flow in complex boundaries are explored by introducing two additional properties, *i.e.* particle type and energy state into the Lattice gas automata (LGA) Frisch–Hasslacher–Pomeau (FHP-II) model. A mix-redistribute of energy and type of particles is also applied on top of collision rules to ensure randomness while maintaining the conservation of mass, momentum and energy. Simulations of heat transfer and heterogeneous reaction of gas flow passing a circular porous cylinder in a channel are presented. The effects of porosity of cylinder, gas inlet velocity, and reaction probability on the reaction process are further analyzed with respect to the characteristics of solid morphology, product concentration, and temperature profile. Numerical results indicate that the reaction rate increases with increasing reaction probability as well as gas inlet velocity. Cylinders with a higher value of porosity and more homogeneous structure also react with gas particles faster. These results agree well with the basic theories of gas–solid reactions, indicating the present model provides a method for describing gas–solid reactions in complex boundaries at mesoscopic level.

INTRODUCTION

The simulation of heat transfer and chemical reaction of fluid flow in porous media is of considerable importance in many practical applications such as combustion chambers, heat exchangers, food processing, catalytic reactors, refrigeration, air cooling and thermal energy storage devices. Simulation results can be found for porous catalyst particles, packed catalyst beds, and arranged pipes. Among these studies, methods based on conventional partial differential equations such as volume-averaging theory [1] and Darcy models [2,3,4], or discrete methods, e.g., lattice Boltzmann methods (LBM) [5,6,7,8] have been the major approaches. However, to the best of our knowledge, the porous media reported in the literature are relatively simple and usually consist of regular arranged solid cylinders. Few studies have been carried out to investigate the characteristics of heat transfer coupling by chemical reaction in porous media. Additionally, the conventional methods based on nonlinear partial differential equations (PDEs) also suffer difficulties such as truncation error and high sensitivity to boundary conditions, making it difficult to describe the detailed structure of porous media or simulate complex processes in porous media.

Lattice gas automata (LGA) is a mesoscopic simulation method from the viewpoint that fluids consist of a large number of particles that "live" on regular lattices with interactions conserving mass and momentum [9]. It is a "bottom-up" and "equation-free" method capturing both macroscopic and mesoscopic characteristics of complex/multi-scale systems, quite distinctive from molecular dynamics (MD), kinetic theory of gases and other methods based on the discretization of partial differential equations. Although LGA suffers drawbacks like statistical noise, the lake of Galilean invariance and velocity-dependent pressure, LGA preserves the particle nature and numerical stability compared with lattice Boltzmann methods [10]. More detailed microscopic interaction among particles or between particles and walls can be obtained when using LGA. Thus, various investigations on flow past obstacles or reaction using lattice gas automata have been carried out, such as flow over cylinders or plat plate [11,12,13], reaction and diffusion systems [14,15,16], first order reactions [17], motivation phenomena of atom and molecule [18,19], kinetically and thermodynamically controlled reactions [20,21,22], and Lindemann theory [23,24]. However, investigations on flow, heat transfer and chemical reaction around a porous cylinder using LGA were rarely reported.

Therefore, in this paper, we intend to explore the application of a LGA method to the heat transfer and chemical reaction of fluid flow around and through a porous circular cylinder in a channel. An algorithm based on the Frisch–Hasslacher–Pomeau (FHP-II) LGA model was developed to deal with the coupled heat transfer and chemical reaction. Quartet structure generation set (QSGS) was used to construct the porous circular cylinder. The influences of porous structure, *i.e.* porosity, pore size and homogeneity, as well as that of reaction probability and flow velocity were further discussed.

SIMULATION METHOD

Lattice Gas Automata Model for Heat Transfer and Chemical Reaction

In FHP-II, particles interact with each other according to a number of pre-defined collision and propagation rules detailed by Frish *et al.* [25], as shown in Figure 1. Based on these simple rules, LGA is capable of displaying complex fluid flow behavior, and consequently, it can be used as a simulation tool for describing physical phenomena. The mass and momentum conservation can be written as below, respectively,

$$\sum_i n_i(t+1, \mathbf{r} + \mathbf{c}_i) = \sum_i n_i(t, \mathbf{r})$$

(1)

$$\sum_i \mathbf{c}_i n_i(t+1, \mathbf{r} + \mathbf{c}_i) = \sum_i \mathbf{c}_i n_i(t, \mathbf{r})$$

(2)

where $n_i(t,r)$ is the occupation state of the cell in i-th direction at time t and place r if the cell is empty, its value is 0, otherwise, its value is 1. $\mathbf{c}_i$ is the lattice velocity in i-th direction, and

$$\mathbf{c}_i = \begin{cases} \left(\cos\frac{\pi}{3}i, \sin\frac{\pi}{3}i\right), & i = 1, \ldots, 6 \\ 0 & , & i = 0 \end{cases}$$

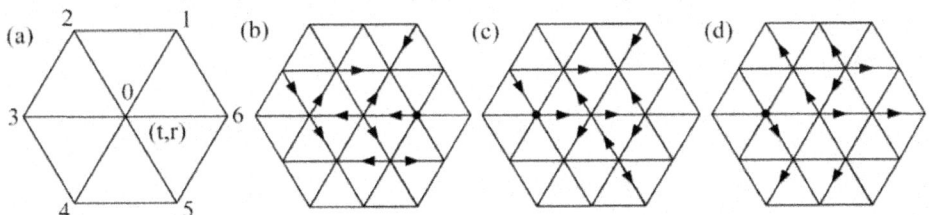

Figure 1. Evolution process of FHP-II, where arrows denote moving particles and points represent static particles: (a) node at (t, r); (b) initialization; (c) after collision; and (d) after propagation.

In the conventional LGA model, particles in the system are of same mass (assumed as 1) and velocity scale (assumed as 1), in other words, they are indistinguishable. In order to describe problems involving heat transfer and chemical reactions, the particles should be distinguishable on temperature and substance type. In current model, every particle is at either of the two energy states [26] ei, *i.e.*, 0 or 1, which represent low (minimum) and high (maximum) temperatures, respectively. Particularly, if the cell is empty, e_i is fixed as 0. Particles are also marked by finite kinds of substance types s_i, and in this paper, s_i is equal to 0 or 1, representing reactant and product, respectively. Besides mass and momentum, extra conservations are also taken into account with respect to energy state and substance type, given as

$$\sum_i n_i^{\alpha,\beta}(t+1, \mathbf{r}) = \sum_i n_i^{\alpha,\beta}(t, \mathbf{r})$$

(3)

where α and β are the substance type and energy state of the cell in i-th direction, respectively. To ensure the conservation of the number of particles with different energy states and substance types at each node during the collision step of FHP-II model, the substance type and energy state need to follow Equations (4) and (5), respectively, known as component conservations

$$\sum_i s_i^{\alpha}(t+1, \mathbf{r}) = \sum_i s_i^{\alpha}(t, \mathbf{r}) \ , \alpha = 1 \ or \ 0$$

(4)

$$\sum_i e_i^{\beta}(t+1, \mathbf{r}) = \sum_i e_i^{\beta}(t, \mathbf{r}) \ , \beta = 1 \ or \ 0$$

(5)

Afterwards, the energy states and substance types of each node are mixed and re-distributed. In fact, the energy states and substance types will be arbitrarily attached to the particles at the node after collision. The overall conservations of mass and momentum will still follow Equations (1) and (2). However, for the propagation process, the particle will move with energy state and substance type, which is described as Equation (6). The evolution process of this model is illustrated in Figure 2.

$$n_i^{s_i(t+1,\mathbf{r}+\mathbf{c}_i),e_i(t+1,\mathbf{r}+\mathbf{c}_i)}(t+1, \mathbf{r}+\mathbf{c}_i) = n_i^{s_i(t,\mathbf{r}),e_i(t,\mathbf{r})}(t, \mathbf{r})$$

(6)

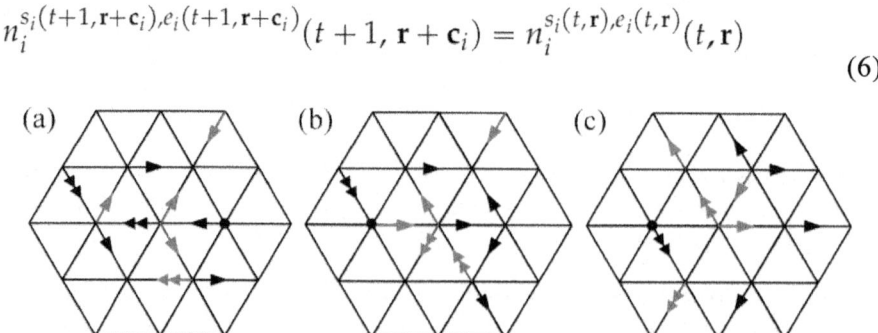

Figure 2. Evolution process of present model, where red represents particles with high energy state, black represents particles with low energy state, and double arrows mean product particles and single arrows denote reactant particles: (a) initialization; (b) after collision; and (c) after propagation.

In FHP-II, the density and momentum are defined as

$$\rho(t, \mathbf{r}) = \sum_{i=0}^{6} n_i(t, \mathbf{r})$$

(7)

$$\rho\mathbf{u} = \sum_{i=0}^{6} \mathbf{c}_i n_i(t, \mathbf{r})$$

(8)

The temperature and product concentration at a node are described as the proportion of particles with high energy state ($\beta = 1$) and the proportion of product particles ($\alpha = 1$), respectively, as the following dimensionless forms:

$$T = \frac{\sum\limits_{i} n_i^{\alpha,1}(t, \mathbf{r})}{\sum\limits_{i,\beta} n_i^{\alpha,\beta}(t, \mathbf{r})}$$

(9)

$$C = \frac{\sum\limits_{i} n_i^{1,\beta}(t, \mathbf{r})}{\sum\limits_{i,\alpha} n_i^{\alpha,\beta}(t, \mathbf{r})}$$

(10)

Chemical Reaction Scheme

The scheme of chemical reaction is based on the algorithm proposed by Bresolin and Oliveira [27] to simulate unimolecular and bimolecular reactions. For first-order reactions considering unimolecular collision, the rate constant, k, is described as

$$k = \int_0^\infty v(E)P(E)dE$$

(11)

where v(E) is the frequency of collisions with energy E above the minimum energy E*, P(E) is the energy distribution of the molecules, which is given by the Maxwell–Boltzmann distribution, and for a molecule with n classic energy states, the fraction of molecules with energy states $E_1, E_2, ..., E_n$ can be written as

$$P(E_1, E_2, ... E_n) = \frac{e^{\frac{-(E_1 + E_2 + ... E_n)}{RT}}}{RT} = \frac{e^{\frac{-E}{RT}}}{RT}$$

(12)

Integrating Equation (12) over all energy values yields:

$$P(E) = \left(\frac{E}{RT}\right)^{n-1} \frac{e^{\frac{-E}{RT}}}{(n-1)!RT} \tag{13}$$

According to Rice–Ramsperger–Kassel (RRK) model [28,29,30], the frequency of collisions v(E) is suggested to be the formula as follows:

$$\begin{cases} \overline{v(E)} = 0 & if\ E < E^* \\ \overline{v(E)} = C(1 - \frac{E^*}{E})^{n-1} & if\ E \geqslant E^* \end{cases} \tag{14}$$

where C is a constant. For the simulation of chemical reaction in LGA, the collisions between molecules can be interpreted as taking place among particles at a node.

In this paper, each reactant particle propagates with an associated probability r deciding it reacts and converts to a product particle or not, generally described as the probability of effective collision. This probability is determined by a probability distribution function and a threshold for reaction. Herein, for simplification, the standard Gaussian distribution was used as the probability distribution function instead of Maxwell–Boltzmann distribution, a threshold K^* was used instead of E^* to decide the critical energy state of a reactant particle. During the collision step of LGA, a random number K following the probability distribution function is generated and compared to K^*; if $K > K^*$, the reaction is consider to be able to take place, otherwise, no reaction occurs.

Moreover, the frequency of collisions in LGA is supposed to be 1.0 per iteration when $K > K^*$, otherwise its value is 0, similar to Equation (14). Thus, the specific reaction constant equals to the integration of energy distribution function above K^* according to Equation (11). Therefore, reaction probability r is equal to the proportion of the particles with energy above K^*, as well as the specific reaction constant. As shown in Figure 3, the curve represents the probability distribution function (PDF), and K^* is a threshold for reaction, the shaded area (beyond K^*) represents the frequency of collisions of hot populations, i.e., reaction probability r. For a given K^*, reaction probability is obtained, and vice versa. In this work, K^* was determined by a given reaction probability from the inverse of normal distribution [31]. The applications of the chemical reaction scheme will be further discussed.

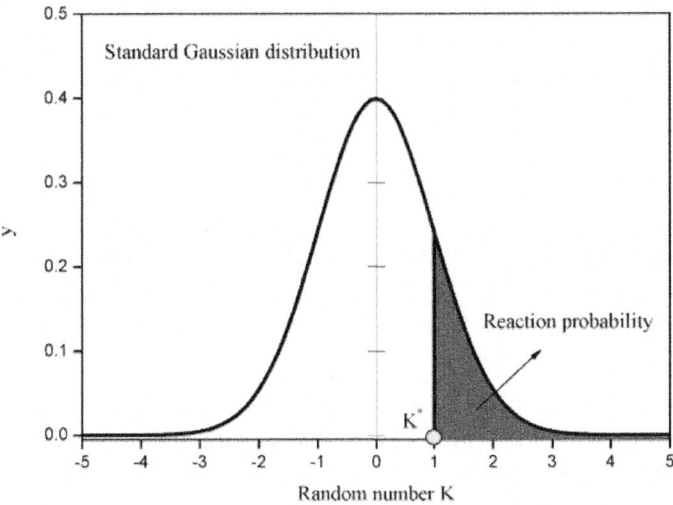

Figure 3. Standard Gaussian distribution function for determining the probability of chemical reaction.

RESULTS AND DISCUSSION

Validation of the Chemical Reaction Scheme

The chemical reaction scheme was applied to irreversible first-order reaction A→B and reversible first-order reaction A⇌B, such as nuclear decay and transformation of isomers. The reactions take place in a square enclosure with area 200 × 200 (lattice units). Every site was initially filled with six A-type particles and bounce-back type boundary condition was employed to all walls, where the momentum of particle is directly reversed while the substance state keeps unchanged during the collision process. Additionally, no heat transfer was taken into consideration in these two cases.

Molecules collide and react at random. Nevertheless, the time evolution of macroscopic amounts or concentrations of molecules is usually quite reproducible due to the reproducibility of experimental conditions. Such laws are called deterministic. More detailed introduction can be found in reference [32]. For reaction A→B, the deterministic half-life period can be obtained by $t1/2 = \ln2/R$, where r represents the probability of one A-type particle changes to one B-type particle at each time, while the half-life period means the iterations needed for reducing the number of A-type particle by half during the simulation process. The reaction probability, r, from A to B was set to be 0.03, 0.04 and 0.05, and K^* was determined as 1.88, 1.75 and 1.65, respectively. During the collision process, a normal distributed random number was generated to compare with K^* to decide to react or not. As shown in Figure 4, it can be observed that the concentration of particle A, obtained by Equation (10), decreases with increasing reaction time, and the reaction rate, i.e., the slope of curve inFigure 4a, increases as reaction probability increases. Figure 4b shows reasonable agreement has been achieved between deterministic method and present simulation. Additionally, for a system with 100×100, and reduction probability $r = 0.001$, the deterministic half-life period is 693.1 iterations, Seybold et al. [17] reported 688.3 ± 9.7 iterations, and 691 iterations using present model, which also indicates the feasibility of present model in reaction systems.

While $A \rightleftharpoons_{k2}^{k1} B$ is an equilibrium system, from the law of mass action, the deterministic equilibrium coefficient is defined as $Keq = k1/k2 = R(A,B)/R(B,A)$, and the equilibrium coefficient can be obtained by the ratio of the final concentration of B and A as $Keq = [B]/[A]$ in a stochastic system, like simulations using lattice gas automata. The reaction probability from A to B was set as 0.05, 0.06, and 0.07, corresponding to the reaction probability from B to A 0.04, 0.03, and 0.02, and for reaction probability of 0.02 and 0.07, K^* is 2.05 and 1.48, respectively. Similar results are obtained compared to $A \rightarrow B$, as shown inFigure 5, however, Figure 5a presents a platform as time advances, meaning an equilibrium state has been reached. Figure 5b indicates that good agreement has been achieved between the stochastic method and deterministic method. The chemical reaction scheme will be further used for the simulations of gas–solid reaction described latter.

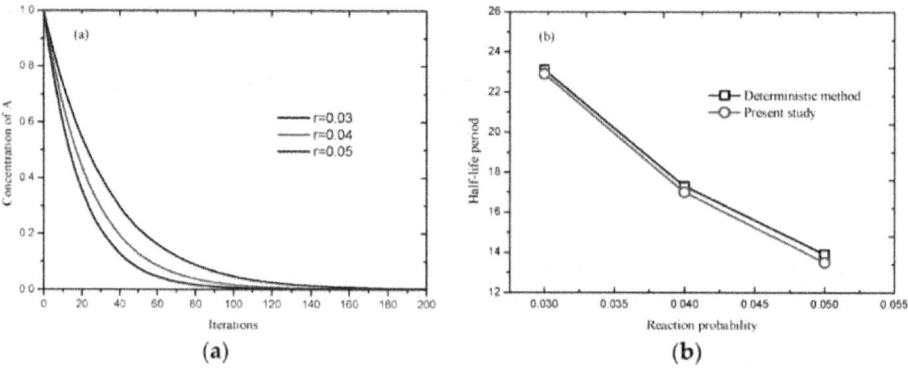

(a)   (b)

Figure 4. Simulation of $A \rightarrow B$, **(a)** processes at different reaction probabilities; and **(b)** comparison of results obtained by deterministic method and present model.

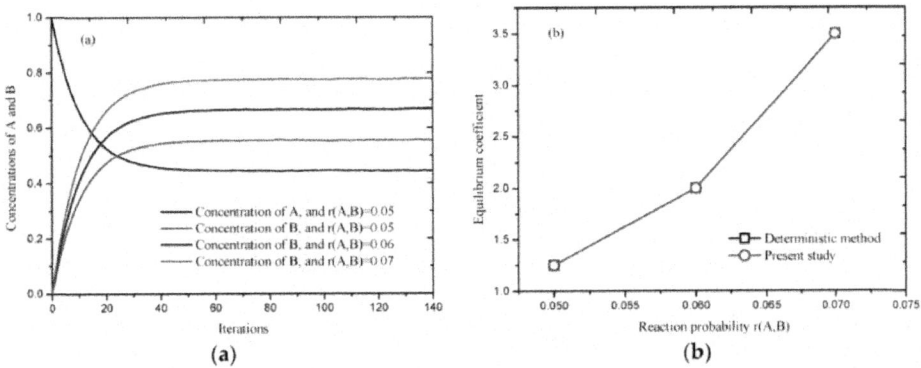

Figure 5. Simulation of A⇌B, (**a**) processes at different reaction probabilities; and (**b**) comparison of results obtained by deterministic method and present model.

Heat Transfer and Reaction across a Porous Circular Cylinder

The characteristics of the flow and heat transfer over a solid circular cylinder in a square enclosure and a rectangular channel have been investigated by us using this model previously, showing reasonable feasibility and reliability of the application of this model to the problems of flow and heat transfer, details can be found in reference [33]. Herein, simulations of flow, heat transfer and reaction around and through a porous circular cylinder in a channel were carried out. The schematic diagram of the simulated system is shown as Figure 6, where a porous circular cylinder with diameter $D = 1/4W$ is placed at the coordinates ($x = 1/3L$, $y = 1/2W$) of a channel with width $W = 400$ and length $L = 1200$ (lattice unit). Reactant gas entering from the inlet at a velocity u, flows around and through and react with the porous cylinder. The effects of porous structure, reaction probability and gas velocity at inlet on the characteristics of the system will be further discussed in detail.

The porous media investigated were generated by a comprehensive approach termed as quartet structure generation set (QSGS) [34,35], which has been demonstrated capable of generating morphological features close to many real porous media [34]. Following the steps illustrated by Wang *et al.* [23], a porous two-dimensional cylinder can be generated with a set of three construction parameters, including (i) growing phase (fluid) distribution probability, C_d, which decides initial number of fluid seeds in the system; (ii) directional growth probability of fluid, D_i, which is considered the same for all directions in this work; and (iii) fluid volume fraction P.

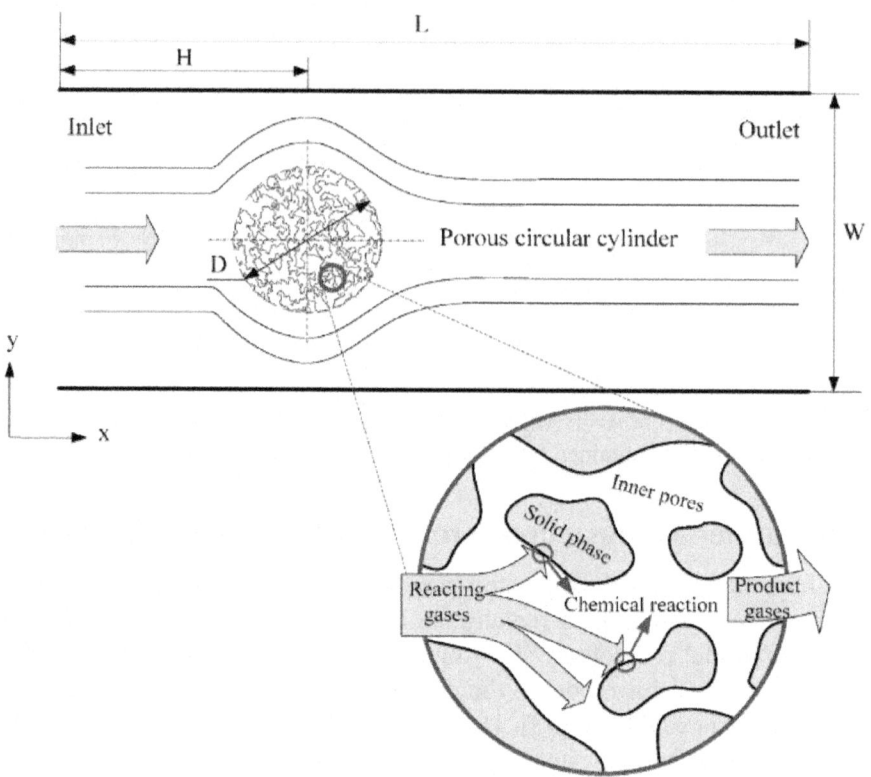

Figure 6. Schematic diagram of the processes happened in the circular cylinder with porous media structure.

Figure 7 presents the inner morphology of the porous circular cylinders of diameter D = 100 (lattice unit) resulted from different combinations of construction parameters, where shaded area is solid phase and the rest to be pores. As can be seen in Figure 7a–d, the solid phase disappears homogenously with increasing porosity, leading to larger pore sizes; and forFigure 7e–h, more agglomeration of solid phase, *i.e.*, less surface area and bigger pore size, appeared for larger directional growth probability; on the contrary, for Figure 7i–l, more homogenous structure as well as smaller pore size of porous media can be obtained as the distribution probability increases, as a result, more reactable surface area is generated. Porous cylinders of different porosity, pore size, and surface area could be obtained by adjusting the three parameters (C_d, D_i, P), for the purpose of comparison.

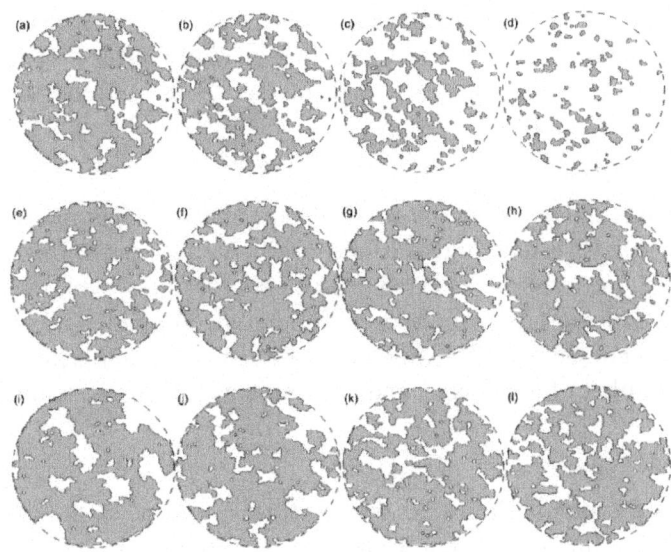

Figure 7. Porous cylinder generated with different construction parameter. (A–d) Effect of porosity on the morphology of cylinder, where $C_d = 0.02$, $D_i = 0.2$ and (a) P = 0.2, (b) P = 0.4, (c) P = 0.6, and (d) P = 0.8. (E–h) Effect of directional growth probability on the morphology of cylinder, where $C_d = 0.02$, P = 0.2 and (e) $D_i = 0.05$, (f) $D_i = 0.15$, (g) $D_i = 0.25$, and (h) $D_i = 0.35$. (I–l) Effect of distribution probability on the morphology of cylinder, where $D_i = 0.2$, P = 0.2 and (i) $C_d = 0.005$, (j) $C_d = 0.015$, (k) $C_d = 0.025$, and (l) $C_d = 0.035$.

Table 1: lists the simulation cases with different parameter sets, with Cases 1 to 4 selected from Figure 7 to investigate the influence of the parameters of QSGS algorithm on the reaction process, Cases 1, 5 and 6 to study the effect of reaction probability, and Cases 1, 7 and 8 are used to investigate the inlet velocity.

Table 1. Parameter sets for simulation cases

Case No.	Parameters of QSGS			r	u
	C_d	D_i	P		
1	0.02	0.2	0.2	0.1587	1.0
2	0.02	0.2	0.4	0.1587	1.0
3	0.035	0.2	0.2	0.1587	1.0
4	0.02	0.35	0.2	0.1587	1.0
5	0.02	0.2	0.2	0.0668	1.0
6	0.02	0.2	0.2	0.3085	1.0
7	0.02	0.2	0.2	0.1587	0.5
8	0.02	0.2	0.2	0.1587	0.8

A conceptual first order heterogeneous reaction between the reaction gas and the porous cylinder illustrated in Equation (10) is considered in this research, with A being the inlet reactant gas, B the porous solid material, and C as the product gas.

$$A(g)+B(s)\rightarrow C(g) \qquad (15)$$

This reaction happens at the solid surface and is simply interpreted as one reactant gas particle reacting with one solid site and generating one product gas particle at mesoscopic level, where the (microscopic level) inner structures or properties of the gas particle or solid site are ignored. The conversion of solid phase B can be defined by the ratio of the number of reacted solid sites to the initial number of solid sites, and a formula as follows is defined

$$X = 1 - \frac{N_t}{N_0} \qquad (16)$$

where N_t is the number of solid sites at time t, and N_0 is the initial number of solid sites. Thus, the reaction rate can be further obtained from the time derivative of Equation (11) as dX/dt.

Instead of heat generated during the reaction process, a simplified heat transfer case is considered, where solid sites are set as hot heat source with constant temperature (*i.e.* 1.0). Gaseous particles will be enhanced to high energy state at impact with solid surface. The gas particles initially entering the channel are of constant low temperature state as 0.

Bounce-back and adiabatic boundary conditions are applied to the channel walls except the outlet which is a free boundary where all particles are released. Solid sites are set as bounce-back boundaries, where gaseous particles will also decide to reaction with according to the chemical reaction scheme.

The simulation carried out in such an order that reaction is not considered in the first 2000 iterations to test the coupling of flow and heat transfer only, also to ensure that the reaction takes place at a stable flow field. Reaction is introduced thereafter to investigate its effect to the flow and heat transfer.

The statistical result of temperature, component concentration, and solid phase conversion are obtained from the simulation by space average. It is noted that the statistics level has impacts on the detail of macroscopic picture. This effect is, however, not discussed in this work. Instead, to ensure consistency, the space averaged results in every 2×2 grids are presented for all cases.

Effect of Inner Porous Structure

As discussed in the previous part, the inner porous structure will be influenced by the parameters of QSGS. In this section, the influence of inner structure on the behavior of flow, heat transfer and chemical reaction around and through the cylinder will be investigated. The inlet velocity is constant as 1.0 with a site density $\rho = 1.0$, and based on FHP-II model, $Re = 158.2$ using D as the characteristics parameter.

Figure 8 shows the temperature contours around and inside the porous cylinder before chemical reaction takes place at $t = 2000$ iterations. A large number of fluid particles are activated to high energy state when they flow over and in the porous circular cylinder, forming a high temperature zone at surrounding and inside the porous structure. As time elapses, the high temperature zone extends to the neighboring field gradually, and no clear eddies have developed behind the cylinder. The field of gas velocity also appeared to have impacts on the profile of temperature.

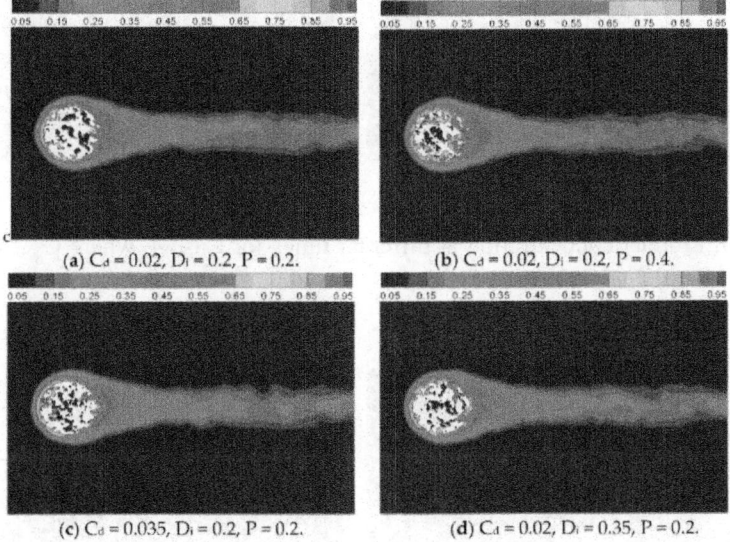

(a) $C_d = 0.02$, $D_i = 0.2$, P = 0.2.

(b) $C_d = 0.02$, $D_i = 0.2$, P = 0.4.

(c) $C_d = 0.035$, $D_i = 0.2$, P = 0.2.

(d) $C_d = 0.02$, $D_i = 0.35$, P = 0.2.

Figure 8. Temperature contour before reaction with different inner structures, $t = 2000$ iterations.

The heterogeneous reaction is started after 2000 iterations. A product zone is then formed as fluid particles flow over and through the porous cylinder and react with the solid sites according to the reaction scheme. Figure 9 shows the product concentration around the solid cylinder at different times for Case 4. The solid sites disappear gradually as time goes on. It can be observed that the product emerges mainly at some hot points, and distributes homogeneously around the cylinder at the beginning, and then diffuses off from the reaction interface to the surrounding. Figure 10 shows the corresponding temperature contour around the cylinder for Figure 9. The structure of cylinder changes with time, thus the heat transfer characteristics changes as a result.

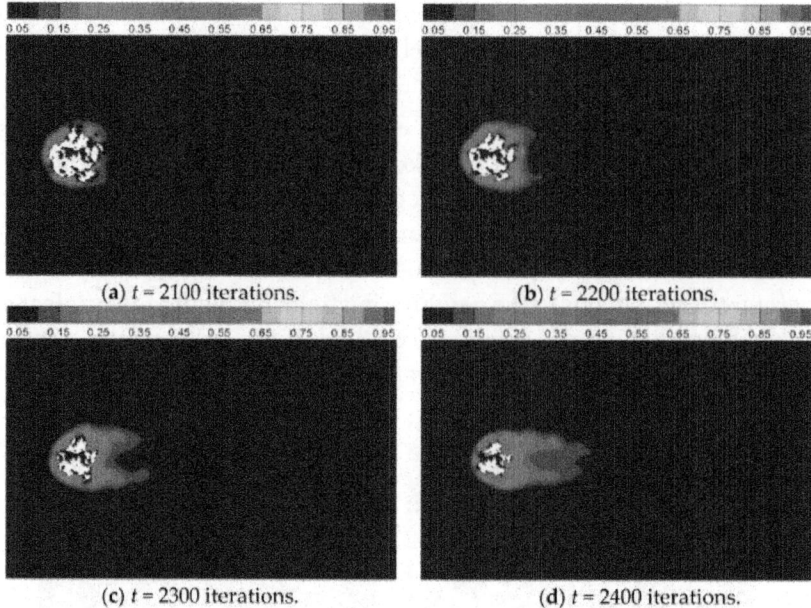

Figure 9. Product concentration at different times for Case 4, where C_d = 0.02, D_i = 0.35, and P = 0.2.

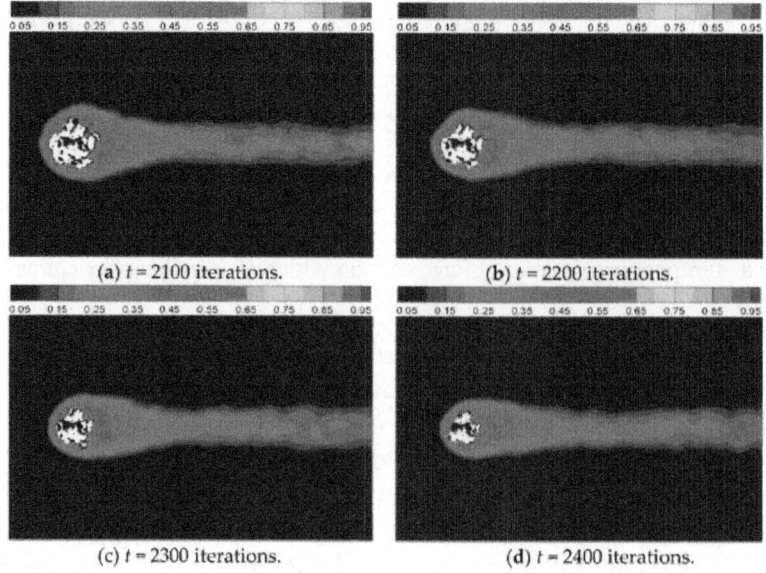

Figure 10. Temperature contour at different times for Case 4, where C_d = 0.02, D_i = 0.35, and P = 0.2.

In order to compare the reaction characteristics of computational cases with porous inner structure, contours of product concentration are shown as Figure 11, for $t = 2500$. It can be observed that the solid cylinder has disappeared completely at $t = 2500$ for Case 2 which has higher porosity than the other three cases. Figure 12 shows the conversion of porous cylinders with different inner structure, as a function of time. It is also notable that solid conversion of higher porosity (Case 2) is faster. This is considered due to the less amount of solid phase, as well as the larger pore size, which facilitates the diffusion of reactant and product. The other three cases with same porosity of 0.2 proceed similarly, but it can still be noted that Case 3 with larger distribution probability progress faster than the other two cases for the former part of time, about 2500 iterations. This can be attributed to the higher homogeneity resulting from larger distribution probability, leading to a larger surface area (reacting sites) to mass ratio. However, this improvement on reaction speed is limited by the relatively smaller pore size slowing down the gas diffusion. Knowing this, it is understandable that Case 4, with larger solid agglomeration and bigger pore size due to higher value of directional growth probability, has the exact opposite performance compare to Case 2.

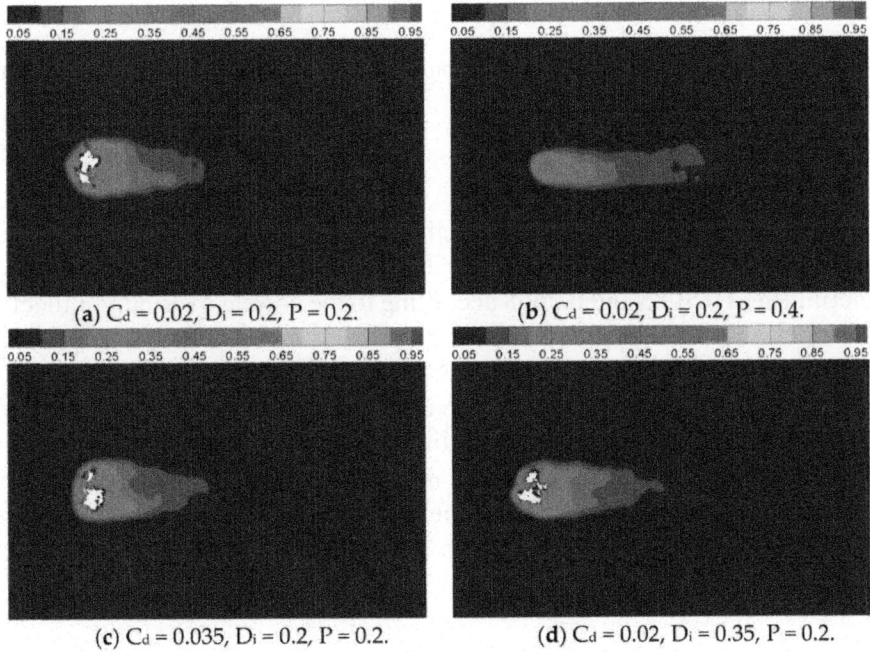

(a) $C_d = 0.02$, $D_i = 0.2$, $P = 0.2$.   (b) $C_d = 0.02$, $D_i = 0.2$, $P = 0.4$.

(c) $C_d = 0.035$, $D_i = 0.2$, $P = 0.2$.   (d) $C_d = 0.02$, $D_i = 0.35$, $P = 0.2$.

Figure 11. Product concentration with different inner structures, $t = 2500$.

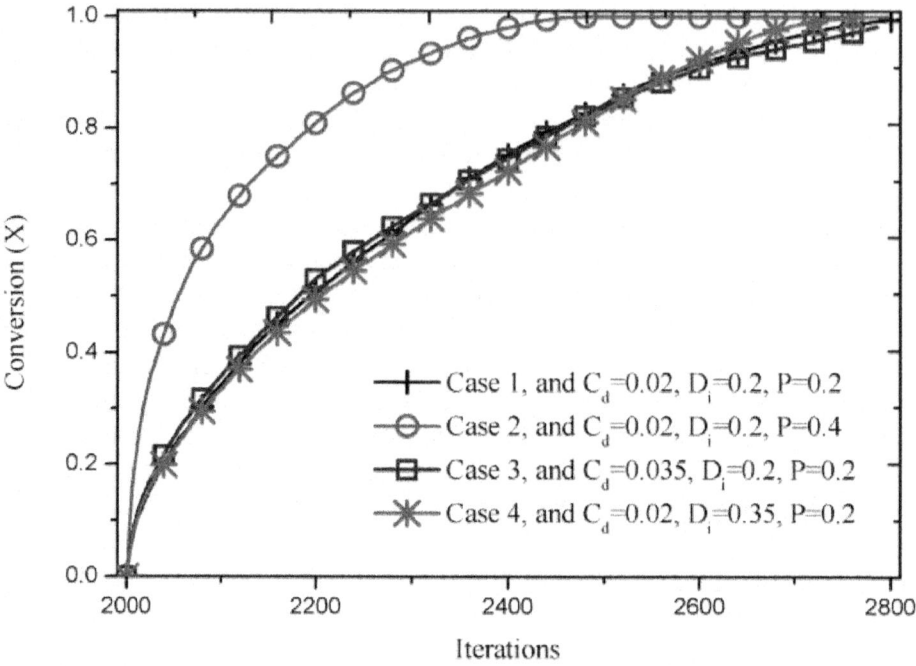

Figure 12. Dependence of conversion on reacting time for cylinders with different inner structure.

Effect of Reaction Probability

In order to obtain the influence of reaction probability on the process evolution, the value of K^* was set to be 0.5 and 1.5, and the reaction probability was determined as 0.3085 and 0.0668 according to the normal distribution function. The reaction probability decides the reaction velocity, as Equation (9). The product concentration and temperature contour around the cylinder at $t = 2500$ are shown as Figure 13. It can be obviously noted that the reaction processes faster with a higher value of probability, which also can be seen from Figure 14. The conversion increases as reacting time advances and reaction probability increases. For instance, the complete reaction time increases from 696 iterations to 1140 iterations when the reaction probability decreases from 0.3085 to 0.0668.

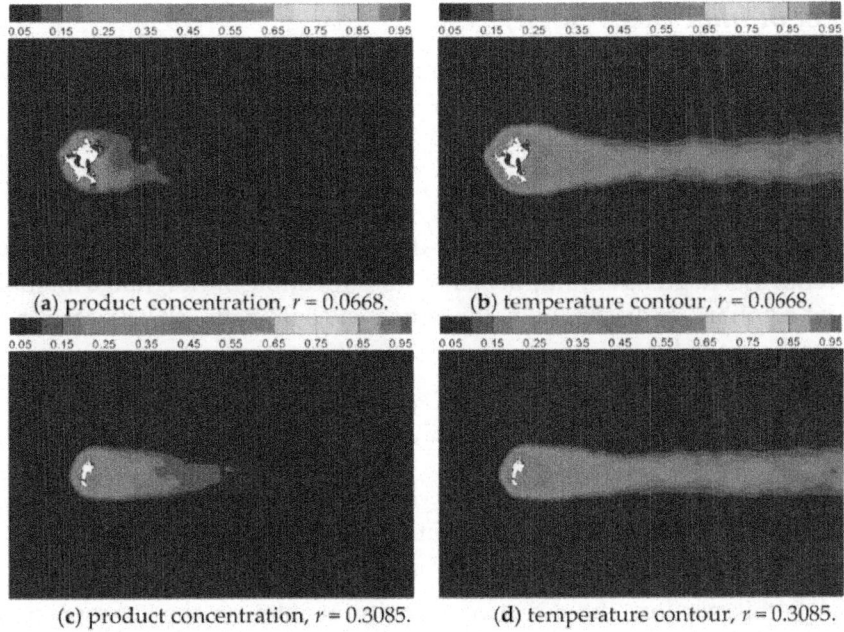

(a) product concentration, $r = 0.0668$. **(b)** temperature contour, $r = 0.0668$.

(c) product concentration, $r = 0.3085$. **(d)** temperature contour, $r = 0.3085$.

Figure 13. Product concentration and temperature contour of Case 5 and Case 6, where reaction probability is 0.0668 and 0.3085, respectively, and $t = 2500$.

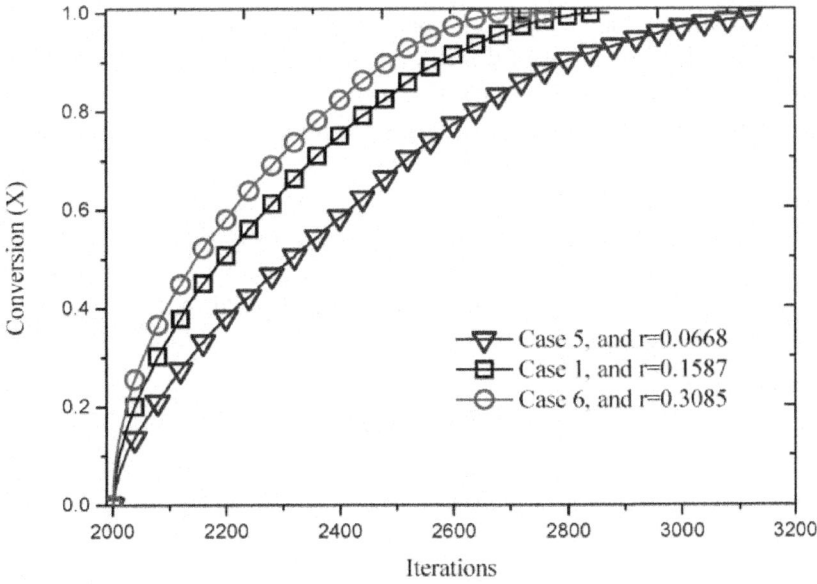

Figure 14. Reaction processes with different reaction probabilities.

Effect of Inlet Gas Velocity

The equilibrium mean occupation numbers are calculated by Fermi–Dirac distribution, as follows

$$n_i^{eq} = \frac{1}{1 + \exp(h + \mathbf{q} \cdot \mathbf{c}_i)} \qquad (17)$$

where h is a real number and q is a D-dimensional vector. The two parameters are termed as Lagrange multipliers. For simplification, the Lagrange multipliers of the equilibrium distributions for lattice gas automata can be obtain by the algebra formula [9]

$$n_i(\mathbf{u}) = d + 2d\mathbf{c}_i \cdot \mathbf{u} + 2d\frac{1 - 2d}{1 - d}c_{i\alpha}^2 u_\alpha^2 - d\frac{1 - 2d}{1 - d}\mathbf{u}^2 \qquad (18)$$

where d is equal to $\rho/7$ for FHP-II, and u is the node velocity. Thus, the variable ni at the inlet nodes can be initialized by Equation (18) with a given particle density d and a node velocity u.

For this section, the effect of velocity at inlet on the reaction process is discussed, and the parameters are listed in Table 1, as Cases 1, 7 and 8. Particle density ρ is fixed as 1.0 for all numerical computation cases, thus d is equal to 1/7. The product and temperature contour of cases with different inlet velocities are shown as Figure 15, it can be seen that Case 8 reacts faster than Case 7 and slower than Case 1 compared with Figure 11a, indicating that the reaction velocity increases as the velocity at inlet increases. This information can also be obtained from the dependence of conversion on the reacting time, as shown in Figure 16, which shows that the extent of conversion increases with increasing reacting time, as well as inlet velocity. This agrees well with the theories of surface reaction in gas–solid systems, such as unreacted shrinking core model [36] and pore model [37], which indicate that the increase of gas velocity promotes the collision between gas particles as well as between gas particles and solid sites, facilitating the diffusion and external dispersion of reactant and product.

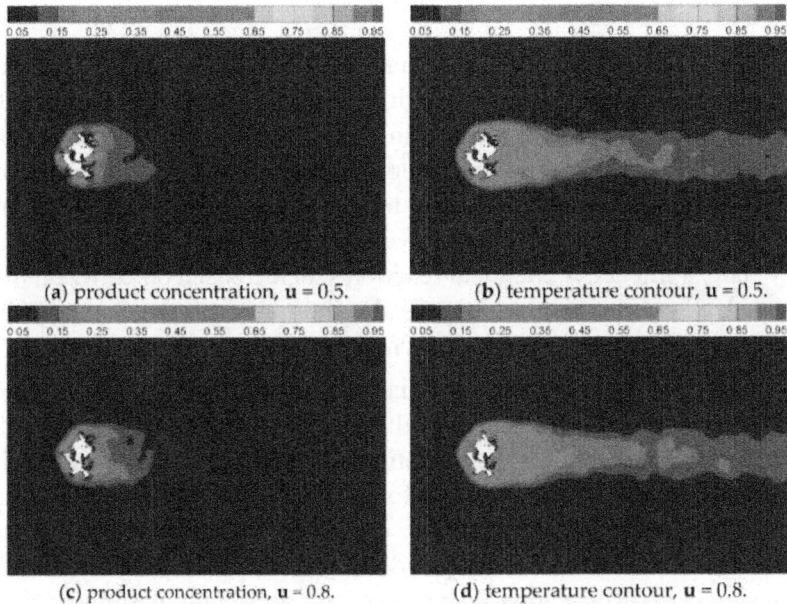

(a) product concentration, $u = 0.5$.

(b) temperature contour, $u = 0.5$.

(c) product concentration, $u = 0.8$.

(d) temperature contour, $u = 0.8$.

Figure 15. Product concentration and temperature contour of Case 7 and Case 8, where inlet gas velocity is 0.5 and 0.8, respectively, and $t = 2500$ iterations.

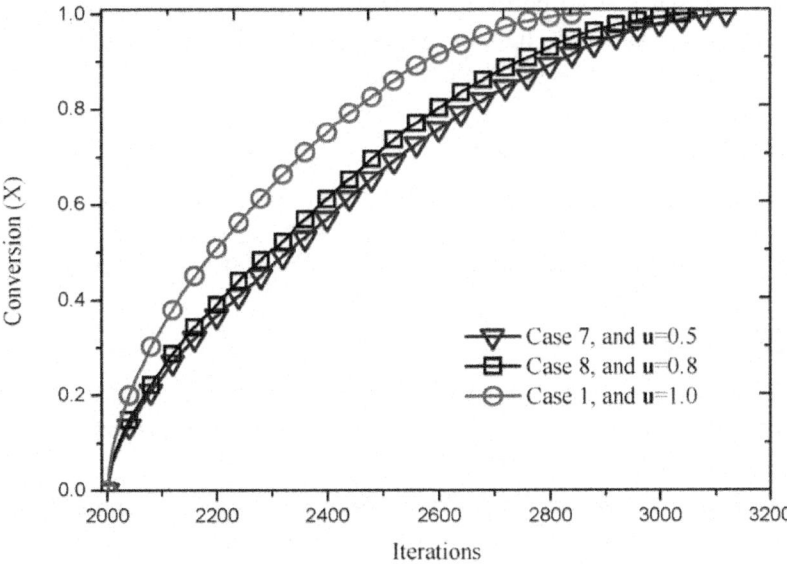

Figure 16. Reaction processes at different inlet velocities.

CONCLUSIONS

A two-dimensional lattice gas automata model is developed to simulate the flow, heat transfer and chemical reaction around and through porous cylinders constructed by QSGS algorithm. In this model, two additional particle properties, *i.e.*, particle type and energy state, are introduced to account for involved chemical species and fluid temperature, respectively. Heat transfer on the interface of gas–solid is then simulated by the change of gas particle energy state at impact. A chemical reaction scheme based on collision theory is developed, where chemical reaction is also interpreted as the change of particle type according to the probability of reaction.

By controlling the construction parameters of the QSGS method, porous cylinders of different pore sizes, solid agglomeration, porosity and surface to mass ratio are generated for the investigation. Their effects, together with that of reaction probability and inlet fluid velocity on the profiles of temperature, solid conversion rate, and reaction product concentration, are discussed. Numerical results indicate that cylinders with a higher porosity, larger pore size, and more surface area to mass ratio react with gas particles faster. Moreover, the reaction velocity increases with increasing reaction probability as well as gas velocity at inlet. These results agree well with the basic theories of the gas dispersion, pore diffusion, and solid surface reaction. The proposed LGA model is therefore believed to provide a prospective modeling strategy for describing gas–solid chemical reaction occurring at complex boundaries from the viewpoint of mesoscopic level.

ACKNOWLEDGMENTS

This work is made possible by the financial support from the National Natural Science Foundation of China (Grant No. 51074201 and No. 51274264), and the National Instrumentation Grant Program (No. 2011YQ120039).

REFERENCES

1.   Yang, C.; Thovert, J.-F.; Debenest, G. Upscaling of mass and thermal transports in porous media with heterogeneous combustion reactions. *Int. J. Heat Mass Transf.* 2015, *84*, 862–875.

2.   Mahmud, S.; Fraser, R.A. Free convection and irreversibility analysis inside a circular porous enclosure. *Entropy* 2003, *5*, 358–365.

3.   Liu, H.; Patil, P.R.; Narusawa, U. On Darcy-Brinkman equation: Viscous flow between two parallel plates packed with regular square arrays of cylinders. *Entropy* 2007, *9*, 118–131.

4. Makinde, O.D.; Eegunjobi, A.S. Entropy generation in a couple stress fluid flow through a vertical channel filled with saturated porous media. *Entropy* 2013, *15*, 4589–4606.

5. Li, M.; Wu, Y.; Zhao, Z. Effect of endothermic reaction mechanisms on the coupled heat and mass transfers in a porous packed bed with Soret and Dufour effects. *Int. J. Heat Mass Transf.* 2013, *67*, 164–172.

6. Li, X.; Cai, J.; Xin, F.; Huai, X.; Guo, J. Lattice Boltzmann simulation of endothermal catalytic reaction in catalyst porous media. *Appl. Therm. Eng.* 2013, *50*, 1194–1200.

7. Machado, R. Numerical simulations of surface reaction in porous media with lattice Boltzmann. *Chem. Eng. Sci.* 2012, *69*, 628–643.

8. Xin, F.; Li, X.-F.; Xu, M.; Huai, X.-L.; Cai, J.; Guo, Z.-X. Simulation of gas exothermic chemical reaction in porous media reactor with lattice Boltzmann method. *J. Therm. Sci.* 2013, *22*, 42–47.

9. Wolf-Gladrow, D.A. *Lattice-Gas Cellular Automata and Lattice Boltzmann Models*; Springer: Berlin/Heidelberg, Germany, 2000.

10. McNamara, G.R.; Garcia, A.L.; Alder, B.J. Stabilization of thermal lattice Boltzmann models. *J. Statist. Phys.* 1995, *81*, 395–408.

11. McCarthy, J.F. Flow through arrays of cylinders: Lattice gas cellular automata simulations. *Phys. Fluids* 1994, *6*, 435–437.

12. Vogeler, A.; Wolf-Gladrow, D.A. Pair interaction lattice gas simulations: Flow past obstacles in two and three dimensions. *J. Statist. Phys.* 1993, *71*, 163–190.

13. Eissler, W.; Drtina, P.; Frohn, A. Cellular automata simulation of flow around chains of cylinders. *Int. J. Numer. Methods Eng.* 1992, *34*, 773–791.

14. Dab, D.; Lawniczak, A.; Boon, J.-P.; Kapral, R. Cellular-automaton model for reactive systems. *Phys. Rev. Lett.* 1990, *64*, 2462–2465. [PubMed]

15. Boon, J.P.; Dab, D.; Kapral, R.; Lawniczak, A. Lattice gas automata for reactive systems. *Phys. Rep.* 1996, *273*, 55–147.

16. Weimar, J.R. Cellular automata for reaction-diffusion systems. *Parallel Comput.* 1997, *23*, 1699–1715.

17. Seybold, P.G.; Kier, L.B.; Cheng, C.-K. Simulation of first-order chemical kinetics using cellular automata. *J. Chem. Inf. Comput. Sci.* 1997, *37*, 386–391.

18. Seybold, P.G.; Kier, L.B.; Cheng, C.-K. Stochastic cellular automata models of molecular excited-state dynamics. *J. Phys. Chem. A* 1998, *102*, 886–891.

19. Seybold, P.G.; Kier, L.B.; Cheng, C.-K. Aurora borealis: Stochastic cellular automata simulations of the excited-state dynamics of oxygen atoms. *Int. J. Quantum Chem.* 1999, *75*, 751–756.

20. Neuforth, A.; Seybold, P.G.; Kier, L.B.; Cheng, C.-K. Cellular automata models of kinetically and thermodynamically controlled reactions. *Int. J. Chem. Kinet.* 2000, *32*, 529–534.

21. Roberts, J.D.; Caserio, M.C. *Basic Principles of Organic Chemistry*, 2nd ed.; WA Benjamin: Menlo Park, CA, USA, 1977.

22. Lin, K.-C. Understanding product optimization: Kinetic versus thermodynamic control. *J. Chem. Educ.* 1988, *65*, 857–860.

23. Hollingsworth, C.A.; Seybold, P.G.; Kier, L.B.; Cheng, C.-K. First-order stochastic cellular automata simulations of the Lindemann mechanism. *Int. J. Chem. Kinet.* 2004, *36*, 230–237.

24. Lindemann, F.A.; Arrhenius, S.; Langmuir, I.; Dhar, N.R.; Perrin, J.; McC. Lewis, W.C. Discussion on "the radiation theory of chemical action". *Trans. Faraday Soc.* 1922, *17*, 598–606.

25. Frisch, U.; d'Humières, D.; Hasslacher, B.; Lallemand, P.; Pomeau, Y.; Rivet, J.-P. Lattice gas hydrodynamics in two and three dimensions. *Complex Syst.* 1987, *1*, 649–707.

26. Zheng, Z.; Gao, X. Lattice gas automata method for modeling fluid flow and heat transfer in metallurgical porous media. *Acta Met. Sin.* 2000, *36*, 433–437.

27. Bresolin, C.S.; Oliveira, A.A.M. An algorithm based on collision theory for the lattice Boltzmann simulation of isothermal mass diffusion with chemical reaction. *Comput. Phys. Commun.* 2012, *183*, 2542–2549.

28. Baercor, T.; Mayer, P.M. Statistical Rice–Ramsperger–Kassel–Marcus quasiequilibrium theory calculations in mass spectrometry. *J. Am. Soc. Mass Spectrom.* 1997, *8*, 103–115.

29. Moon, J.H.; Oh, J.Y.; Kim, M.S. A systematic and efficient method to estimate the vibrational frequencies of linear peptide and protein ions with any amino acid sequence for the calculation of Rice–Ramsperger–Kassel–Marcus rate constant. *J. Am. Soc. Mass Spectrom.* 2006, *17*, 1749–1757. [PubMed]

30. Moon, J.H.; Sun, M.; Kim, M.S. Efficient and reliable calculation of Rice–Ramsperger–Kassel–Marcus unimolecular reaction rate constants for biopolymers: Modification of Beyer–Swinehart algorithm for degenerate vibrations. *J. Am. Soc. Mass Spectrom.* 2007, *18*, 1063–1069. [PubMed]

31. Wichura, M.J. Algorithm as 241: The percentage points of the normal

distribution. *J. R. Stat. Soc.* 1988, *37*, 477–484.

32. Lecca, P.; Laurenzi, I.; Jord, F. *Deterministic Versus Stochastic Modelling in Biochemistry and Systems Biology*; Woodhead Publishing: Cambridge, UK, 2013.

33. Chen, H.; Zheng, Z.; Chen, Z.; Bi, X.T. Simulation of flow and heat transfer around a heated stationary circular cylinder by lattice gas automata. *Powder Technol.* 2015.

34. Wang, M.; Wang, J.; Pan, N.; Chen, S. Mesoscopic predictions of the effective thermal conductivity for microscale random porous media. *Phys. Rev. E* 2007, *75*, 036702. [PubMed]

35. Chen, L.; Wu, G.; Holby, E.F.; Zelenay, P.; Tao, W.-Q.; Kang, Q. Lattice Boltzmann pore-scale investigation of coupled physical-electrochemical processes in C/Pt and non-precious metal cathode catalyst layers in proton exchange membrane fuel cells. *Electrochim. Acta* 2015, *158*, 175–186.

36. Homma, S.; Ogata, S.; Koga, J.; Matsumoto, S. Gas–solid reaction model for a shrinking spherical particle with unreacted shrinking core. *Chem. Eng. Sci.* 2005, *60*, 4971–4980.

37. Petersen, E.E. Reaction of porous solids. *AIChE J.* 1957, *3*, 443–448.

Chapter 4

CHEMICAL REACTIONS DIRECTED PEPTIDE SELF-ASSEMBLY

Dnyaneshwar B. Rasale and Apurba K. Das

Department of Chemistry, Indian Institute of Technology Indore, Khandwa Road, Indore 452017, India

ABSTRACT

Fabrication of self-assembled nanostructures is one of the important aspects in nanoscience and nanotechnology. The study of self-assembled soft materials remains an area of interest due to their potential applications in biomedicine. The versatile properties of soft materials can be tuned using a bottom up approach of small molecules. Peptide based self-assembly has significant impact in biology because of its unique features such as biocompatibility, straight peptide chain and the presence of different side chain functionality. These unique features explore peptides in various self-assembly process. In this review, we briefly introduce chemical reaction-mediated peptide self-assembly. Herein, we have emphasised enzymes, native chemical ligation and photochemical reactions in the exploration of peptide self-assembly.

INTRODUCTION

The spontaneous formation of ordered structures at the nanoscale is usually referred to as self-assembly [1]. When the constitutive components are molecules, the process is generally termed as molecular self-assembly. The molecular self-assembly process is again divided into intramolecular and intermolecular self-assembly. The term molecular self-assembly refers to intermolecular self-assembly and the intramolecular analogue is more commonly called folding. Several studies on the origin of life noted that there must have been processes by which prebiotic organic compounds were sufficiently concentrated [2] to undergo physical and chemical interactions. The physical properties of certain kind of molecules lead to the formation of complex structures with emergent properties. Such emergent phenomena are referred to as self-assembly processes or self-organization. Cellular life began when self-assembled membrane-bound polymers had ability to not only polymerize, but also replicate their linear sequence of monomers [3]. Thus, self-assembly is a basic process by which contemporary cellular life produces membranes, duplex DNA and folding proteins. It has been demonstrated that the first cell must have been formed by the same intermolecular interactions and self-assembled structures. Self-assembly is a prevalent process in nature which plays an important role in maintaining integrity of cells [4,5] to perform various functions of cells [6]. The cellular components such as actin filaments, microtubules, DNA, vesicles and micelles are the classic representation of molecular self-assembly in biological pools [7].

DEVELOPMENT OF MOLECULAR SELF-ASSEMBLY

Molecular self-assembly [8] is the spontaneous association of molecules under equilibrium conditions into stable, structurally well-defined aggregates joined by non-covalent interactions. Molecular self-assembly is a prevalent process in biological systems and underlies the formation of a wide variety of complex biological structures [9,10]. The understanding that the self-assembly process utilizes the association of non-covalent interactions [11,12] of molecular backbones in biological aggregates is a central concern in chemical biology. Besides the biomacromolecular nanostructures, certain small organic molecules are capable to self-assemble in a particular solvent, resulting in self-supporting gel [13,14,15,16,17]. If the self-assembly occurs in an aqueous medium, the resulting gel is referred to a supramolecular hydrogel [18,19]. The design of biomolecules that can self-assemble into higher order structures, have received increasing attention over the past few years, because of their applications in supramolecular electronics [20], drug delivery [21,22,23], wound healing, biosensing [24,25] and tissue engineering [26,27,28]. There are many weak interactions such as hydrogen bonding, hydrophobic interactions and π–π stacking interactions that govern the assembly of everything from DNA in its double helix to the triple helical structure in collagen fibers. Self-assembly is also the only practical approach to build a wide variety of nanostructures [29,30].

The development of nanoscale structures and devices can be accomplished through "bottom-up approach" or "top-down" methods. In the bottom-up approach, small building blocks assemble into larger structures [31,32] (Figure 1). Examples of this approach include chemical synthesis [33], molecular self-assembly [34], and colloidal aggregation [35,36,37]. Most of the self-assemblies are directed by small molecular weight organic molecules and bioactive molecules. With the increasing applications of supramolecular hydrogels in biomedicine, there is wide interest in the development of supramolecular soft materials [38]. Several, physical stimuli such as pH, temperature, light, enzymes, and sonication are used to control peptide self-assembly.

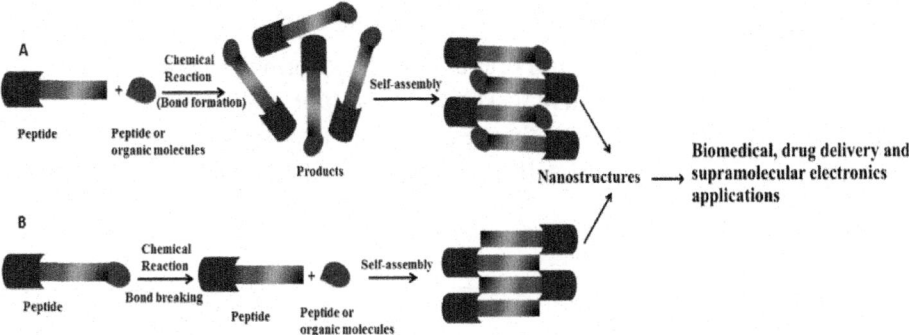

Figure 1. (A) Schematic representation shows simple peptide building blocks undergo chemical reactions with peptide or organic molecules to form self-assembled nanostructures via non-covalent interactions; **(B)** Peptides undergo bond breaking via chemical reactions to form peptide self-assembly.

In this review, we aim to describe various mild chemical reactions in the development of peptide self-assembly (Table 1). A chemical reaction is a process that leads to the transformation of one set of chemical substances to another, which are usually characterized by chemical changes. Chemical reactions are used for the synthesis of new compounds. It has a crucial role not only in everyday life but also in biology where it is referred as metabolism. In living organisms, biochemical reactions are mainly controlled by enzymes. One of the most important biochemical reactions is anabolism, in which different DNA and enzyme-controlled processes result in the production of large molecules such as proteins and carbohydrates from smaller units. There has been wide interest to achieve such processes in the laboratory to develop some complex architecture exhibiting structural complexity ranging from nano- to mesoscale which is of fundamental importance for various protein-related diseases but also holds great promise for various nano- and biotechnological applications [39]. Several physical perturbations are known to develop such complex self-assembled architectures. However, chemical reactions find wide scope in the development of self-assembled biomaterials due to its one pot propensity. Generally, the efforts are made to develop bioorthogonal chemical reactions to further explore in the biomedical applications [40,41]. The self-assembly is mainly governed by non-covalent interactions which could be achieved by adding or removing constraining moieties from the molecules [42]. Here we will emphasize some important chemical reactions in peptide self-assembly.

The self-assembly of short peptides can be controlled by imposing a conformational constraint that leads to the prevention of the β-sheet structure (Figure 2) [43]. Nilsson *et al.* flanked a short self-assembling peptide sequence with cysteine (Cys) residues that enabled the macrocyclization of these peptides [43]. Macrocyclization prevents β-sheet formation and self-assembly in the cyclic form. Thus, using TCEP, constraint was removed by simple reduction of the disulfide bond that resulted in relaxation to the stable β-strand and subsequent formation of self-assembly.

Table 1: Self-assembly driven by chemical reactions

Entry	Chemical Reactions	Catalyst/ Reaction Condition	Reaction Medium	References
1	Disulfide formation	Air	Aqueous	[43]
2	Photochemical reaction	Light	Aqueous	[44–46]
3	Enzymatic reactions	Enzyme	Aqueous	[47–50]
4	Enzymatic reactions	Enzyme	Organic	[51,52]
5	Thioester mediated native chemical ligation	4-Mercaptophenyl acetic acid	Aqueous	[53–55]
6	Oxo-ester mediated native chemical ligation	Heating 80 °C	Aqueous/MeOH	[56–58]
7	Seleno ester mediated native chemical ligation	No Catalyst	Aqueous/EtOH	[59]

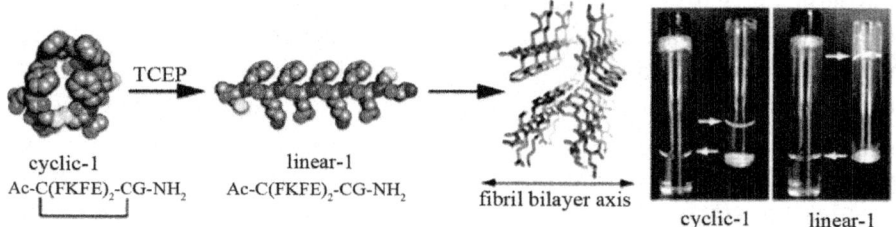

cyclic-1   linear-1
Ac-C(FKFE)$_2$-CG-NH$_2$ Ac-C(FKFE)$_2$-CG-NH$_2$ fibril bilayer axis   cyclic-1   linear-1

Figure 2. Cyclic to linear peptide conformational switch using a chemical reductive trigger, (adapted from reference [43] with permission from American Chemical Society).

Van Esch *et al.* reported a dissipative self-assembly system in which a synthetic DSA fibrous network uses chemical fuel as an energy source [60] (Figure 3). A gelator precursor dibenzoyl-(l)-cystine (DBC) is converted into self-supporting gel by reaction with a chemical fuel methyl iodide at pH = 7 leading to the formation of diester. Hydrolysis of the methyl esters of the gelator, which is labile under ambient conditions, leads to energy dissipation and disassembly of the formed structures. Yang *et al.* reported redox controllable self-assembly properties of selenium containing peptides [61].

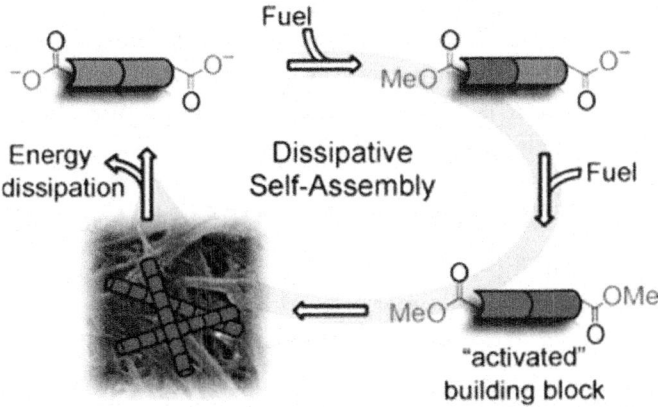

Figure 3. A monomeric building block (blue) is activated by fuel consumption and is able to assemble (forming red fibers). In the assembled state, it can dissipate its energy and revert to its monomeric state (blue), (adapted from reference [60] with permission from John Wiley and Sons).

An N-capped 4-phenyl-selenyl butanoic acid peptide was converted to selenoxide upon oxidation with H_2O_2, which is easily soluble in phosphate buffer saline [61]. However, chemical treatment with vitamin C converts selenoxide into less soluble selenide in aqueous medium leading to the formation of self-assembling nanostructures. Lehn and coworkers described that guanosine hydrazide yields a stable supramolecular hydrogel based on the formation of a guanine quartet (G-quartet) in the presence of metal cations. Guanosine hydrazide and its assemblies can be reversibly decorated by acylhydrazone formation upon reaction with various aldehydes, resulting in the formation of highly viscous dynamic hydrogels [62]. The dynamic system selects an aldehyde from the mixture of aldehydes, which leads to the formation of the most stable gel. Rao *et al.* demonstrated a biocompatible condensation reaction for controlled assembly of nanostructures in living cells using 1, 2 aminothiol and 2-cyanobenzothiazole [63]. Ajayan reported uniform and crystalline nanofibers of perylene-3,4,9,10-tetracarboxylic dianhydride (PTCDA), an insoluble organic semiconducting molecule which have been achieved by self-assembling molecules using chemical reaction mediated conversion of an appropriately designed soluble precursor perylene tetracarboxylic acid (PTCA) using carbodiimide chemistry [64]. Das *et al.* exploited a reversible esterification reaction that leads to the formation of a single predominant product among the library members using dimethyl sulfate (DMS) as chemical fuel [65]. The library members formed a self-supporting hydrogel and showed the formation of a single predominant product. Otto and coworkers developed two self-replicating peptide-derived macrocycles that emerge from a small dynamic combinatorial library through oxidative disulfide formation from their pendant thiol groups in presence of oxygen and compete for a common feedstock. Replication is driven by nanostructure formation resulting from self-assembly of peptide [66].

PHOTO-SWITCHED MOLECULAR SELF-ASSEMBLY

The self-assembly of bio-organic molecules into nanostructures is an attractive route to fabricate functional materials.

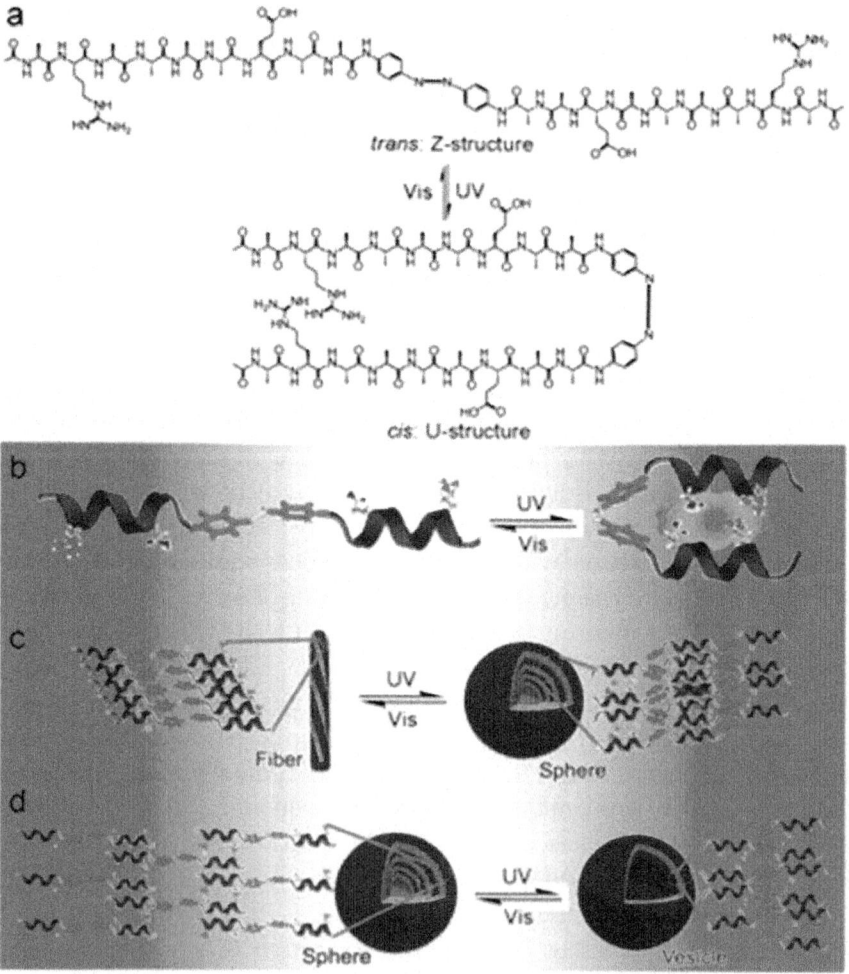

Figure 4. Schematic illustration of the light-switched self-assembly of the gemini α-helical peptide. (**a**) Molecular structure of the gemini α-helical peptide and light-

triggered reversible change between Z- and U-structures; (b) Model of light-triggered reversible structural change in the gemini α-helical peptide; (c) Light-switched self-assembly behaviors of the gemini α-helical peptide in acidic (pH 3.0); and (d) in basic (8.0) medium, (adapted from reference [44] with permission from Royal Society of Chemistry).

For example, diphenylalanine (Phe-Phe, FF), an aromatic dipeptide consisting of two covalently linked phenylalanine units, can form various nanostructures such as nanotubes [67,68], nanowires and nanosphere [69] under different processing conditions [70]. FF can readily self-assemble into different nanostructures in a simple way and possess the functional flexibility and molecular recognition capability suitable for a wide range of applications, such as biosensors [71], imaging, guest encapsulation and nanofabrication [72,73,74]. Zhang et al. reported an azobenzene-linked symmetrical gemini α-helical peptide which reversibly transforms between the trans- (Z-) and cis-structure (U-structure) under UV (λ = 365 nm) and subsequently visible light irradiation [44] (Figure 4). This also affects self-assembly behavior of the gemini α-helical peptide. Park et al. developed light-harvesting peptide nanotubes that integrate photosynthetic units for mimicking natural photosynthesis [45]. Zinic et al. demonstrated that a bis(phenylalanine) maleic acid shows irreversible photoinduced gelation in water that works on photochemical isomerization of nongelling maleic acid amide to gelling fumaric acid amide [46].

ENZYME CATALYZED PEPTIDE SELF-ASSEMBLY

Enzymes are a class of highly efficient and specific catalysts in nature. Enzyme-regulated molecular self-assembly plays a critical role in many cell processes [47]. The formation of microtubules, which governs mitosis, is one example [75]. The polymerization of actins which governs the focal adhesion of cells, is essentially an enzyme regulated self-assembly process of seemingly miraculous sophistication. These natural self-assemblies inspire the development of enzymatic hydrogelation of small molecules. Compared with physical or conventional chemical perturbations, enzymatic regulation promises a unique opportunity to integrate molecular self-assembly in water with natural biological processes. Moreover, as a new method to make biomaterials, the enzyme-catalyzed formation of hydrogels of small molecules has already shown promise in biomedical applications [48,49]. Xu and coworkers reported a method to image enzyme-triggered self-assembly of small molecules inside live cells [76]. George et al. have focused on the enzyme catalyzed self-assembly strategy to develop molecularly defined and functional materials [77]. Ulijn et al. exploited several enzymes including phosphatases, esterases and proteases, to trigger the self-assembly of aromatic peptide amphiphiles by converting non-assembling precursors into self-assembling components [78,79,80,81] (Figure 5). Saiani *et al.* reported the effect of enzyme concentration on the morphology and properties of enzymatically triggered peptide hydrogels [82].

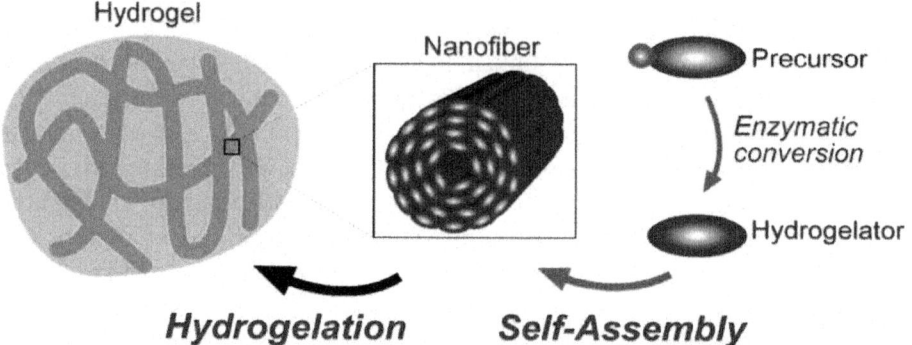

Figure 5. The essential steps in the enzymatic hydrogelation of small molecules, (adapted from reference [78] with permission from American Chemical Society).

SELF-ASSEMBLY DRIVEN BY PEPTIDE HYDROLYSIS

Generally, hydrolysis is a chemical process in which a molecule of water is added to a substance. This addition causes both substance and water molecule to split into two parts. Acid-base-catalysed hydrolyses are very common. One such example is the hydrolysis of amides or esters [83]. The amide bond in a peptide is more rigid due to significant delocalisation of the lone pair of electrons on the nitrogen atom giving the bond a partial double bond character. Therefore, the hydrolysis reactions are very slow in water. Usually in the laboratory, hydrolysis of proteins or peptides is being carried out by using 6 N HCl at 110 °C [84]. However, use of this method leads to partial destruction of many amino acids including serine, threonine and cysteine. Tryptophan is totally destroyed by this procedure. In a living system, most biochemical reactions including ATP hydrolysis take place by the catalysis of enzymes. The catalytic action of enzymes allows the hydrolysis of proteins, fats, oils, and carbohydrates. Protease enzymes aid digestion by causing hydrolysis of peptide bond in proteins [85]. In recent years, there has been a growing interest in the development of peptide nanostructures using biocatalytic methods via peptide bond hydrolysis. Researchers are interested to explore enzymatic development of peptide self-assembly for further biomedical applications without destruction of any amino acids. Shao *et al.* reported self-assembly of a peptide amphiphile based on hydrolysed Bombyx mori silk fibroin usin α-chymotrypsin [50]. Das *et al.* described a general strategy to control the state of molecular self-assembly under thermodynamic control using the protease thermolysin. Self-assembly drives the formation of a single π-stacked predominating product in a dynamic library [86].

Moore and coworkers described a chymotrypsin responsive hydrogel in which the tetrapeptide CYKC was used as a cross-linker to create a poly(acrylamide) hydrogel [87] (Figure 6); the sequence of CYKC hydrolysis by chymotrypsin leads to degradation of the hydrogel.

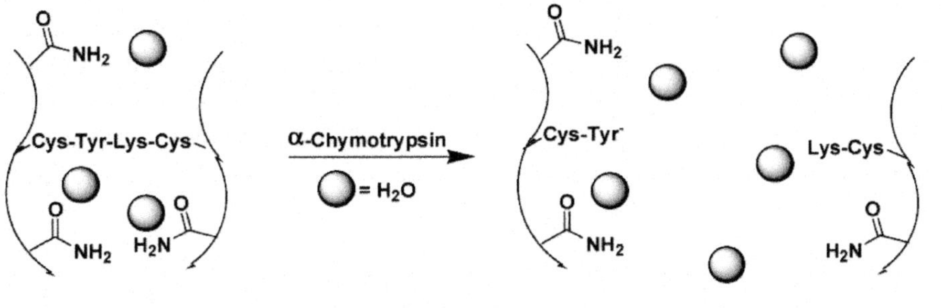

Figure 6. A CYKC-cross-linked hydrogel degraded with α-chymotrypsin, (adapted from reference [87] with permission from American Chemical Society).

PEPTIDE SELF-ASSEMBLY DRIVEN BY AMIDE BOND FORMATION

Amide bond formation of peptides is exactly opposite to the hydrolysis of amide bonds in peptides. Usually some proteases show the ability to reversibly synthesize the amide bond as well as hydrolyze the amide bond in dilute aqueous conditions. It has been reported that self-assembly drives the reverse hydrolysis of peptides and prefers the most stable product formation in gel phase medium. Thus, the enzyme thermolysin is preferably used to exploit peptide self-assembly. Self-assembly of macroscopic materials from small molecular building blocks provides a route to design molecular biomaterials. The controlled development of biomaterials is in high demand in the context of biomedical applications. The stimuli that trigger self-assembly include various physical chemical perturbations. Enzyme triggered self-assembly is particularly interesting due to mild and physiological reaction conditions. Enzymes allow biocatalytic reactions in a biological environment which could lead to hydrogelation of peptides. Herein, some examples demonstrate the hydrogelation of peptides via reverse hydrolysis of amide bonds in peptides.

Ulijn *et al.* have used a protease enzyme that normally hydrolyzes peptide bonds in aqueous medium. They described a conceptually novel approach by using thermolysin to perform the reverse reaction (*i.e.*, peptide synthesis or amide formation), which can produce amphiphilic peptide hydrogelators and self-assembles to form nanofibrous structures [88] (Figure 7). The Ulijn group also reported the use of reversible enzyme-catalysed reactions to drive self-assembly. They demonstrated that this system combines three features such as (i) self-correction-fully reversible self-assembly under thermodynamic control; (ii) component-selection ability to amplify the most stable molecular self-assembly structures in dynamic combinatorial libraries and (iii) spatiotemporal confinement of nucleation and structure growth [89]. Enzyme-assisted self-assembly therefore provides control in bottom-up fabrication of nanomaterials that could ultimately lead to functional nanostructures with enhanced complexities and fewer defects. Das *et al.* reported bio-catalytic evolution of dynamic combinatorial libraries via amide bond formation. Self-assembly evolves thermodynamically downhill library members via enzyme-catalysed amide bond formation [90].

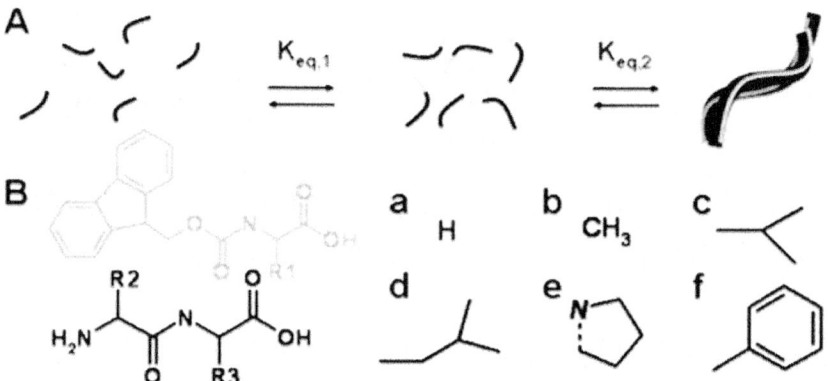

Figure 7. (A) Proposed mechanism: Fmoc amino acids (gray) are enzymatically coupled to dipeptides (black) by a protease to form Fmoc-tripeptides that self-assemble to higher-order aggregates driven by π-π interactions between fluorenyl groups. $K_{eq,1}$ represents the equilibrium constant for peptide synthesis/hydrolysis, $K_{eq,2}$ for self-assembly; (B) Chemical structures of Fmoc-amino acids, dipeptide precursors and amino acid side chains: a Gly, b Ala, c Val, d Leu, e Pro, f Phe, (adapted from reference [88] with permission from American Chemical Society).

LIPASE CATALYZED PEPTIDE SELF-ASSEMBLY

Lipase is an important class of enzyme that catalyses the hydrolysis of fats (lipid). Lipases are the most broadly deployed biocatalysts because of their ability to produce chiral products [91] with high enantiomeric purity. Lipases are used as catalysts for hydrolysis, alcoholysis, esterification and transestrification of carboxylic acids or esters reactions [92,93]. They also work in organic as well as in aqueous medium, which makes it a versatile candidate in the broad research area. One of the important aspects in chemical synthesis of drugs is to retain a single enantiomer, which is often a difficult task. Thus, one of the commercially attractive and environmentally compatible ways of making enantiomerically pure drugs is biotransformation. In addition, peptides acylated with fatty acids become capable of being anchored to liposomes, translocating across lipid membranes, penetrating intact cells, and penetrating through the blood-brain barrier. However, selective acylation is a formidable task to a chemist due to the presence of numerous reactive groups in peptides. Klibanov *et al.* reported a lipase catalyze selective acylation of a dipeptide l-Phe-α-l-Lys-O^{t}Bu [94]. It has two primary amino groups. The α-NH$_2$ group of Phe and the ω-NH$_2$ group of Lys offer a challenge for selective acylation. Thus, lipase selectively acylates the ω-NH$_2$ group of Lys in dipeptides in the presence of excess trifluoroethyl acetate. Besides its useful applications in organic synthesis, lipases are being used in the development of molecular self-assembly. Recently, the use of enzymes for the fabrication of biomaterials, starting from small molecular building blocks, to promote the synthesis of self-assembling materials has become an emerging area of research activity. Additionally, the structure of these (bio) materials can be easily controlled and tuned by using biological catalysts taking advantage of chemo-, regio- and enantioselective synthesis and of mild reaction conditions. In general, there are two routes to drive enzymatic molecular self-assembly through either breaking or making of covalent bonds. Interestingly, both routes lead to self-assembly by maintaining the hydrophobic and hydrophilic balance between the self-assembling molecules. Important efforts have been made to gain more insight into molecular self-assembly.

These studies have proved that non-covalent interactions such as π-π stacking, hydrogen bonding and hydrophobic interactions play a key role in the development of such systems. Recently, thermolysin was used to generate dynamic combinatorial libraries for the discovery of stable self-assembling nanostructures. Although lipases are primarily used in esterification and transesterification of carboxylic acids, there are few examples where lipases have been used to develop peptide self-assembly through amide bond synthesis. Palocci and colleagues reported peptide self-assembly via coupling of Fmoc-phenylalanine and diphenylalanine using lipase as the catalyst at physiological conditions [95] (Figure 8). This is an excellent example of synthesis of a peptide bond instead of using expensive protease. Lipase is known as an industrial biocatalyst and can be exploited to study such self-assembling systems. In another work, the Palocci group described the self-assembly of homochiral and heterochiral lipase catalyzed Fmoc-based peptides with control drug release from hydrogel metrix [51].

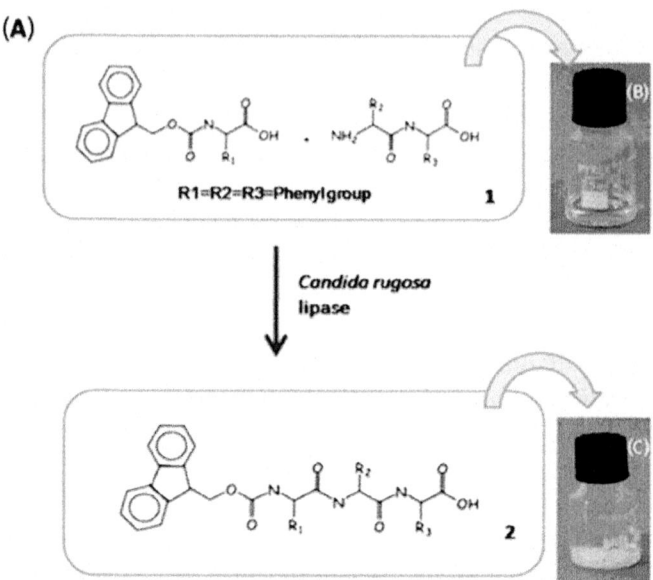

Figure 8. (A) Chemical structures of the precursor (1) and its corresponding hydro-gelator (2) and the schematic gelation process; (B) optical images of a solution of 1 in phosphate buffer (pH 7.4) and (**C**) the hydrogel of 2 formed by adding lipase to a solution of 1, (adapted from reference [95] with permission from Royal Society of Chemistry).

Dordick *et al.* demonstrated lipase catalyzed sugar-containing self-assembled organogels with nanostructured morphologies [52] (Figure 9). The lipase activity and thermostability can be increased upon immobilization with many supports. The self-assembled peptide architectures has also been used for the improved activity of lipase. Matsui *et al.* reported lipase incorporated peptide nanotubes [96]. However, there are some nonspecific proteases that have been used for ester hydrolysis followed by self-assembly. Although, lipases are widely used in ester hydrolysis and transesterification, it is less explored in molecular self-assembly. Das *et al.* described lipase-catalysed incorporation of gastrodigenin (*p*-hydroxybenzyl alcohol) to Nmoc-protected peptides [97]. The lipase catalysed esterification reaction results in the formation of blue light emitting peptide nanofibers in aqueous medium. Self-assembly of peptides evolves blue light emission upon illumination under UV light. They also reported that *p*-hydroxybenzyl alcohol can efficiently be incorporated into peptide bolaamphiphiles [98]. The activated esterified products self-assemble to form thixotropic hydrogel. A self-assembled hydrogel matrix was used for 3D cell culture. Significant cell support and proliferation of human umbilical cord mesenchymal stem cells were observed.

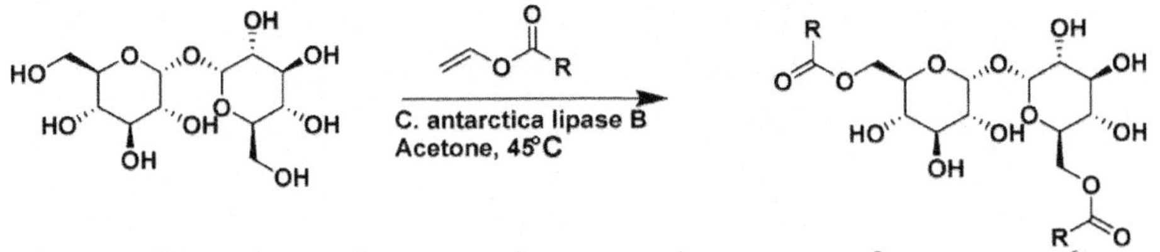

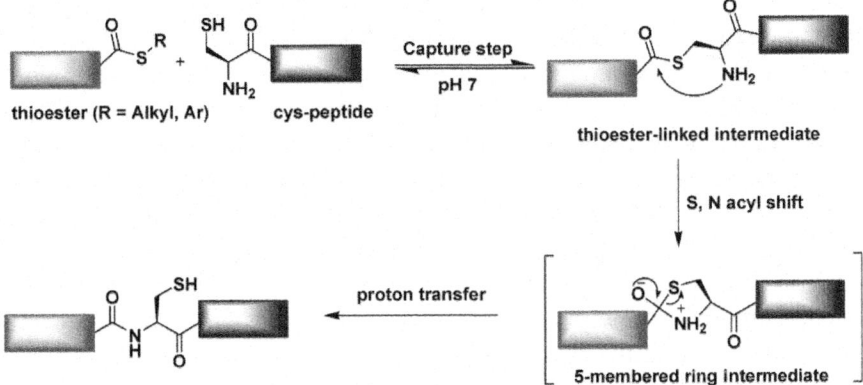

Compounds   1,   2,   3,   4,   5,   6,

R=   -CH$_3$, -(CH$_2$)$_2$CH, -(CH$_2$)$_8$CH$_3$, -(CH$_2$)$_{12}$CH$_3$, -(CH$_2$)$_{16}$CH$_3$, -CH=CH

Figure 9. Lipase catalyzed acylation of the disaccharide trehalose generated a family of low-molecular-weight gelators via transesterification, (adapted from reference [52] with permission from John Wiley and Sons).

NATIVE CHEMICAL LIGATION

Native chemical ligation (NCL) is an important extension of the chemical ligation discovered by Kent and co-workers in 1994 [99]. It is a widely used method for the total or semi-synthesis of proteins. Native chemical ligation involves the chemoselective reaction of two unprotected peptides in aqueous solution to give a single covalently linked ligated product (Figure 10). In fact, in 1953, Wieland and co-workers discovered the chemical foundation of this reaction [100]. The reaction of valine-thioester and cysteine amino acid in aqueous buffer was shown to yield the dipeptide valine-cysteine. The reaction proceeds through the intermediacy of a thioester containing the sulfur of the cysteine residue. Wieland's work led to the "active ester" method for making protected peptide segments in conventional solution synthesis in organic solvents.

Figure 10. Native chemical ligation (NCL) mechanism.

PEPTIDE SELF-ASSEMBLY DRIVEN BY THIOESTER MEDI-ATED NATIVE CHEMICAL LIGATION

As posited in a "Thioester World", thioesters are possible precursors to life [101]. It is revealed that thioesters are obligatory intermediates in several key processes in which ATP is either used or regenerated. Thioesters are involved in the synthesis of all esters, including those found in complex lipids. The biosynthesis of lignin, which comprises a large fraction of biomass, proceeds via thioester derivative of caffeic acid. Acetyl CoA is an important molecule in metabolism. The chemical structure of Acetyl CoA includes a thioester between coenzyme A (a thiol) and acetic acid that is produced during the second step of aerobic cellular respiration in the Krebs cycle. The *C*-terminal peptides or protein thioesters are essential in NCL. The synthetic thioester preparation is generally carried out via conventional carboxyl activation chemistry rather than the N→S route so elegantly directed by intein. Thioesters can also be prepared via *tert*-butyloxycarbonyl (Boc)-based solid-phase peptide synthesis (SPPS) on a thioester resin or 9-fluorenylmethoxycarbonyl (Fmoc)-based SPPS, which involves mild bases during synthesis [102]. Melnyk and co-workers reported a solid-phase N→S acyl transfer for thioester synthesis after peptide chain assembly using Fmoc/t-Bu chemistry in combination with the sulfonamide safety-catch linker [103]. Once, the thioester is synthesized, it can easily be subjected to NCL reaction at physiological conditions. Since the beginning of thioester-mediated NCL reactions, it has been exploited for total synthesis of proteins or peptides and dendrimers [53] and is combined with peptide self-assembly. Recently, NCL reactions have attracted broad attention from research groups for the development of functional biomaterials.

Woolfson *et al.* described the combination of chemical ligation with peptide self-assembly to deliver extremely long polypeptide chains with stipulated, repeated sequences. The self-assembling fibers were used to align peptide from their *N*-to *C*-terminals, which facilitates the ligation reactions without the usual requirement of an *N*-terminal catalytic cysteine residue [54]. Collier and co-workers investigated a novel method for rapidly increasing the stiffness of self-assembled β-sheet fibrillar peptide hydrogels using native chemical ligation (NCL) [55]. Messersmith *et al.* illustrated the use of NCL as a strategy to form covalently cross-linked polymer hydrogels under mild conditions and in the absence of catalysts [104] (Figure 11). The thioester based polymer and *N*-terminal cysteine polymer bioconjugate were synthesized to bring up hydrolgelation by native chemical ligation reaction and the viscoelastic nature of hydrogel was studied by oscillatory rheology.

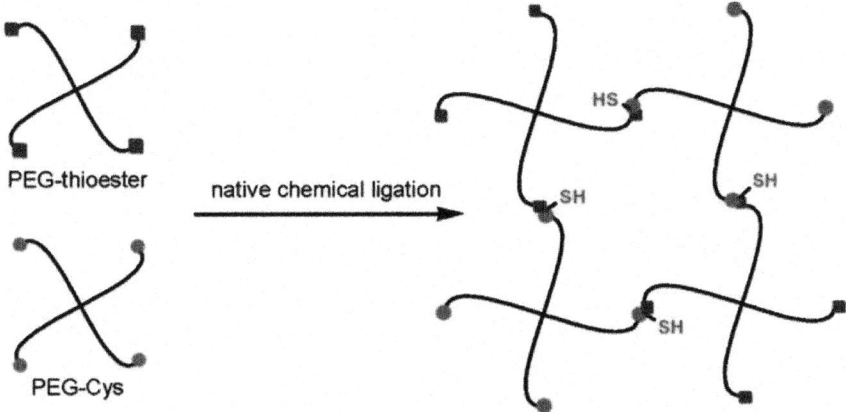

Figure 11. Native chemical ligation is used in cross-linking of hydrogel, (adapted from reference [104] with permission from American Chemical Society).

PEPTIDE SELF-ASSEMBLY DRIVEN BY OXO-ESTER MEDI-ATED NATIVE CHEMICAL LIGATION

Amino acid and peptide oxo-esters have played an important role in peptide chemistry for many years. Various amino acid esters are commonly used as protecting groups for peptide synthesis in conventional solution phase methodology. Moreover, activated phenyl ester derivatives have been used for chemical ligation of peptides. Several modifications of thioester mediated NCL reactions have been made in an effort to expand the utility of the method. Danishefsky *et al.* first reported the use of oxo-ester in NCL reactions through indirect approach involving o-thiophenolic ester [105], which was followed later by a direct approach utilizing *p*-nitrophenyl (pNP) activated *C*-terminal ester (Figure 12). The oxo-ester mediated native chemical ligation reactions were successfully carried out with sterically hindered *C*-terminal amino acids which is rather problematic with thioester mediated NCL. Hackeng *et al.* observed that β-branched amino acids such as Thr, Val and Ile in this position react extremely slowly under standard NCL reaction conditions (>48 h) [106]. Long NCL reaction times are generally discouraged, due to potential side-reactions (thioester hydrolysis, desulfurization of cysteine and methionine oxidation) [107] under the NCL conditions employed. Moreover, proline in this position was found to react even more sluggishly with, at best, conversions of around 20% after 48 h [108,109]. However, the *p*-nitrophenol oxo-esters are more labile acyl donors than thioester in NCL reactions.

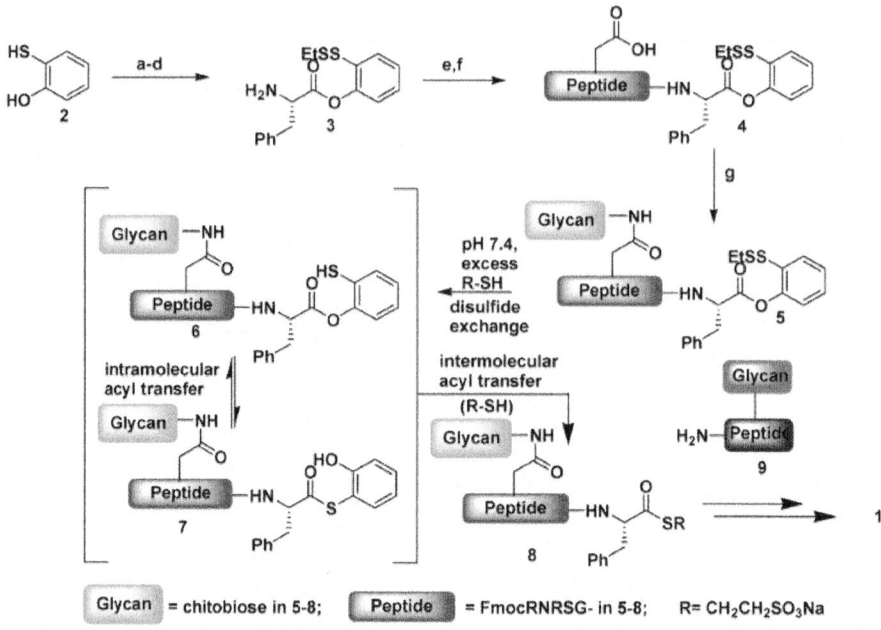

Figure 12. Oxo-ester mediated native chemical ligation, reagents and conditions: (a) I2, MeOH, H2O; (b) BF3-OEt2, EtSSEt, CH2Cl2, 99%, 2 steps; (c) Boc-Phe-OH, EDCI, DMAP, CH2Cl2/THF, 93%; (d) 4 N HCl/dioxane, 94%; (e) Fmoc-Arg(Pbf)-Asp(tBu)-Arg(Pbf)-Ser(tBu)-Gly-OH, HATU, DIEA, DMF, 61%; (f) TFA/phenol/Et-3SiH/H2O, 35:2:1:1, 60%; (g) GlcNAcβ1→4GlcNAcβ1-NH2, HATU, DIEA, DMSO, 52%, (adapted from reference [105] with permission from American Chemical Society).

Synthesis of *p*-nitrophenol ester is easy and less problematic. Weissenborn *et al.* described oxo-ester mediated NCL on oxo-ester activated surfaces and found that 2,3,4,5,6-pentafluorophenyl (PFP) is a more efficient acyl donor than *p*-nitrophenol and *N*-hydroxysuccinimide (NHS) activating agents [110]. Liu *et al.* reported a simple and less activated phenyl oxo-ester of peptide for chemoselective NCL reactions [56]. Borner *et al.* investigated peptide-guided assembly of poly(ethyleneoxide)-peptide conjugate via intramolecular O→N acyl transfer to restore native amide bonds [57]. Our group investigated the role of active *p*-nitrophenyl esters in peptide self-assembly via native chemical ligation [58]. We synthesized Nmoc-protected amino acids/peptides having *C*-terminal *p*-NP esters. The modified ligation precursors undergo self-assembly via NCL with *N*-terminal cysteine residues. Self-assembly via NCL was studied with Nmoc-protected amino acid and peptide *p*-NP esters. Five compounds 1–5 (Figure 13) were synthesized by conventional solution phase methodology.

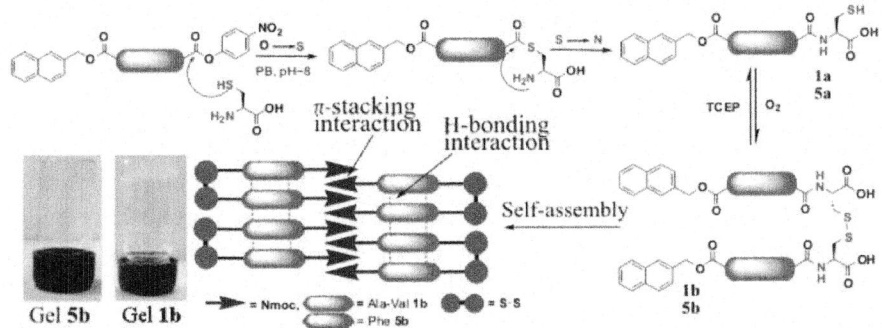

Figure 13. Native chemical ligation at Nmoc-protected-p-NP esters. The cysteine amino acid induces O-S exchange with Nmoc-protected-*p*-NP 1 or 5 (step I) to form a thioester intermediate. Subsequent S-N acyl transfer furnishes the peptide bond 1a or 5a (step II). Air oxidation provides the formation of ligated disulfide 1b or 5b (step III) resulting in supramolecular peptide gels, (adapted from reference [58] with permission from Royal Society of Chemistry).

These kinds of soft biomaterials can be used for cell culture, tissue engineering and supramolecular electronics applications. Messersmith *et al.* described polymer hydrogel formation via oxo-ester mediated NCL between branched polymer precursors containing NHS activated ester and *N*-terminal cysteine group and showed cytocompatibility and *in vivo* acute inflammatory response [111] (Figure 14).

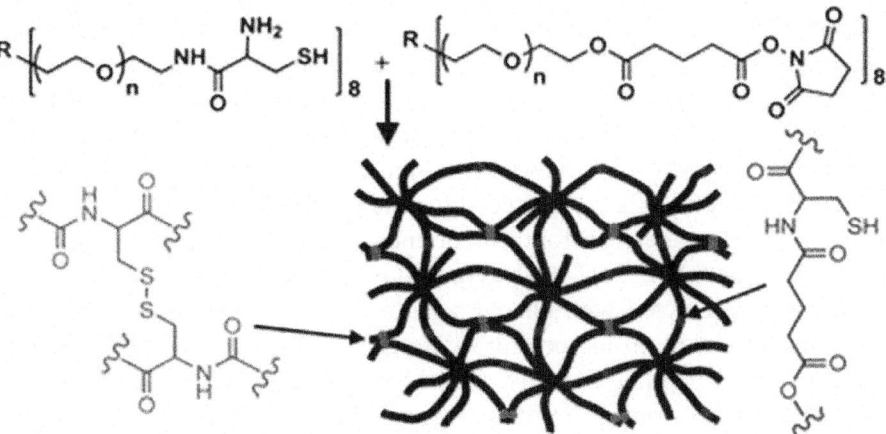

Figure 14. Peptide self-assembly via oxo-ester mediated native chemical ligation, (adapted from reference [111] with permission from Royal Society of Chemistry).

PEPTIDE SELF-ASSEMBLY DRIVEN BY SELENOESTER MEDIATED NATIVE CHEMICAL LIGATION

Over the past decades, selenium and organoselenium compounds are gaining increasing attention due to their properties as antioxidant and antitumor agents, as apoptosis inducers, and in the effective chemoprevention of cancer in a variety of organs [112,113,114,115]. Selenoesters are important intermediates in several organic transformations. The compounds in this class have been used as precursors of acyl radicals [116,117,118] and anions [119] and have attracted attention for the synthesis of new molecular materials, especially superconducting materials, liquid crystals and self-assembled biomaterials. The hydrolysis of selenoesters subsequently generates ionic species of selenium (such as selenols), which can readily participate in redox processes. These charged seleno compounds may possess inherent biological activity, and it would be beneficial if they could enhance the cytotoxic impact on cancer cells. In the case of aryl selenoester, the magnitude of selenoesters hydrolysis as well as its expected biological activity can be tuned by the replacement of different substituent on the aryl ring. The synthesis of selenoesters can be achieved by solution phase methodology as well as Boc solid phase peptide synthesis in peptide chemistry. The reactivity of selenoesters was found higher than the comparable thioesters towards thiol nucleophiles in the first transesterification step of NCL. Therefore, selenoesters have been used for the synthesis of proteins by chemical ligation [120,121], the synthesis of substrates that undergo facile and efficient radical decarbonylation, and the synthesis of natural alkaloid (+)-geissoschizine [122]. Beside the useful application of selenoesters in the above area, there is significant opportunity to explore the selenoesters in the self-assembly process. We have efficiently explored a new method for peptide self-assembly via selenoester-mediated native chemical ligation [59] (Figure 15). In this work, our objective was to develop a simple and efficient method that can direct dynamic peptide self-assembly. To achieve this goal, we synthesized four compounds with an N-terminal capped with an aromatic naphthalene-2-methoxycarbonyl (Nmoc) group. The C-terminals of 1–4 were protected with phenyl selenoester, which could readily undergo NCL reaction at room temperature with N-terminal cysteine and N-terminal cysteine based peptide Cys-Gly. Considering the active role of Cys-Gly in increased risk of women breast cancer, we have used Cys-Gly to ligate with Nmoc-protected selenoesters.

The NCL reaction proceeds through the thioester-linked intermediate where acyl transfers from Se→S [123] are followed by intramolecular S→N acyl transfer to give a peptide bond monitored by reverse phase high performance liquid chromatography. Native chemical ligation reaction with selenoesters was very fast and occurred within five minutes. Self-assembly was studied upon native chemical ligation reactions.

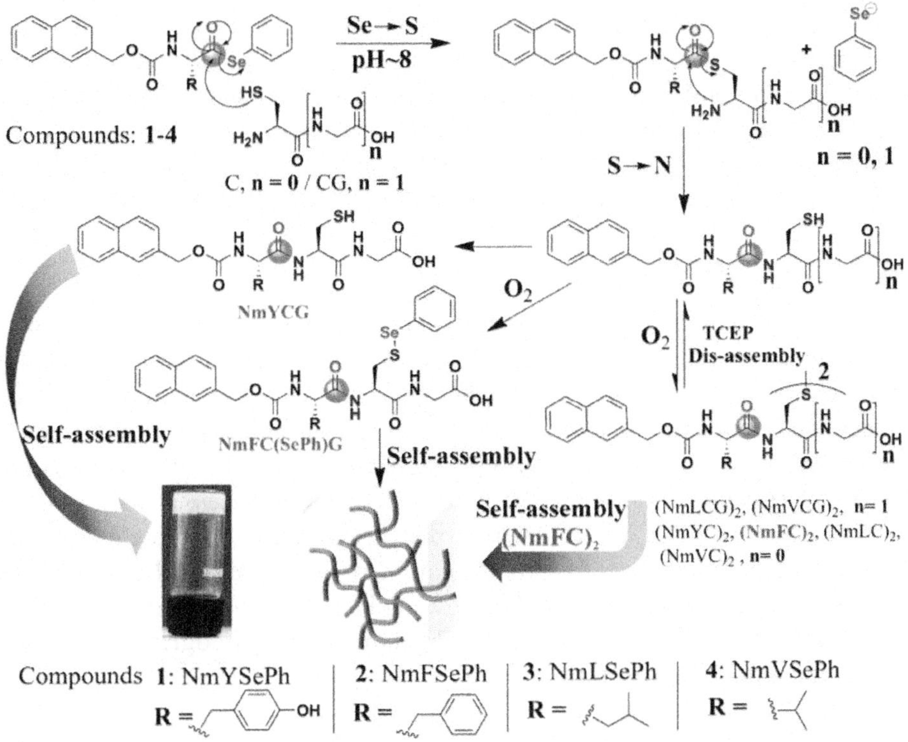

Figure 15. Selenoester mediated native chemical ligation. Ligated products were formed upon the NCL of selenoesters 1-4 with Cys-Gly and cysteine at pH~8. NmYCG self-assembled in its reduced form while oxidized NmFC(SePh)G (sulfur linked with selenophenol) and (NmFC)$_2$ self-assembled to form self-supporting soft materials, (adapted from reference [59] with permission from Royal Society of Chemistry).

APPLICATION OF SELF-ASSEMBLED MATERIALS

There are many applications where self-assembled peptide nanostructures could play an important role as part of biosensing platforms, as efficient drug-delivery systems, cell cultures or as a hydrogels for tissue reparation. Like some of the traditional antibiotics, the short cationic antimicrobial peptides can kill the microbes by interacting and disrupting bacterial cell membranes. Effectiveness of antimicrobial activity depends on the cationic charges and the hydrophobicity of peptides [124]. The self-assembled architectures in associated gel networks and antimicrobial activities of peptide amphiphiles make them potential candidates as cell culture matrices or scaffolds in tissue engineering and regenerative medicine. Extensive study of these self-assembled biomaterials has already proved their biocompatibility [125,126]. Several groups have reported that molecular hydrogels have been widely used as carriers for the delivery of therapeutic agents [127,128]. Self-assembling peptide amphiphiles also have great potential as templates for nanofabrication such as in biomineralization [129,130,131], nucleation, nanowires, and nanocircuits [132]. The self-assembled nanofibers have also been used as templates for the nucleation and growth of CdS nanocrystals [133].

CONCLUSIONS

This review has aimed to provide readers with comprehensive details about peptide self-assembly via mild chemical reactions and various chemical reactions utilized by different research groups for the construction of self-assembled nanostructures in aqueous medium were discussed. Self-assembly is an important process in bottom-up nanotechnology and various covalent and non-covalent interactions govern the self-assembly process. Chemoselective native chemical ligation offers a novel approach for fabrication of self-assembled architectures. *In situ* formation of self-assembled soft materials has significant importance in biology, and peptide based self-assembly is biocompatible and rapid, which allows gelation at physiological conditions to make it an ideal biomaterial.

ACKNOWLEDGMENTS

Apurba K. Das is thankful to CSIR, New Delhi, India for financial support. Dnyaneshwar B. Rasale thanks to MHRD for providing his Postdoctoral Fellowship.

REFERENCES

1.   Whitesides, G.M.; Mathias, J.P.; Seto, C.T. Molecular self-assembly and nanochemistry: A chemical strategy for the synthesis of nanostructures. *Science* 1991, *254*, 1312–1319.

2.   Yanagawa, H.; Kojima, K.; Ito, M.; Handa, N. Synthesis of polypeptides by microwave heating I. Formation of polypeptides during repeated hydration-dehydration cycles and their characterization. *J. Mol. Evol.* 1990, *31*, 180–186.

3.   Lee, D.H.; Granja, J.R.; Martinez, J.A.; Severin, K.; Ghadiri, M.R. A self-replicating peptide. *Nature* 1996, *382*, 525–528.

4. Timpl, R.; Brown, J.C. Supramolecular assembly of basement membranes. *Bioessays* 1996, *18*, 123–132.

5. Schnur, J.M. Lipid tubules: A paradigm for molecularly engineered structures. *Science* 1993, *262*, 1669–1676.

6. Qi, S.Y.; Groves, J.T.; Chakraborty, A.K. Synaptic pattern formation during cellular recognition. *Proc. Natl. Acad. Sci. USA* 2001, *98*, 6548–6553.

7. Conde, C.; Caceres, A. Microtubule assembly, organization and dynamics in axons and dendrites. *Nat. Rev. Neurol.* 2009, *10*, 319–332.

8. Jones, M.R.; Osberg, K.D.; Macfarlane, R.J.; Langille, M.R.; Mirkin, C.A. Templated techniques for the synthesis and assembly of plasmonic nanostructures. *Chem. Rev.* 2011, *111*, 3736–3827.

9. Nogales, E.; Wolf, S.; Downing, K.H. Structure of α,β-tubulin dimer by electron crystallography. *Nature* 1998, *391*, 199–203.

10. Desai, A.; Mitchison, T. Microtubule polymerization dynamics. *Annu. Rev. Cell Dev. Biol.* 1997, *13*, 83–117.

11. Uhlenheuer, D.A.; Petkau, K.; Brunsveld, L. Combining supramolecular chemistry with biology. *Chem. Soc. Rev.* 2010, *39*, 2817–2826.

12. Collier, J.H. Modular self-assembling biomaterials for directing cellular responses. *Soft Matter* 2008, *4*, 2310–2315.

13. Xu, H.; Song, J.; Tian, T.; Feng, R. Estimation of organogel formation and influence of solvent viscosity and molecular size on gel properties and aggregate structures. *Soft Matter* 2012, *8*, 3478–3486.

14. Dunna, B.; Zink, J. Optical properties of sol-gel glasses doped with organic molecules. *J. Mater. Chem.* 1991, *1*, 903–913.

15. Chen, Q.; Lv, Y.; Zhang, D.; Zhang, G.; Liu, C.; Zhu, D. Cysteine and pH-responsive hydrogel based on a saccharide derivative with an aldehyde group. *Langmuir* 2010, *26*, 3165–3168.

16. Ajayaghosh, A.; Praveen, V.K.; Vijayakumar, C. Organogels as scaffolds for excitation energy transfer and light harvesting. *Chem. Soc. Rev.* 2008, *37*, 109–122.

17. George, M.; Weiss, R.G. Molecular organogels soft matter comprised of low-molecular-mass organic gelators and organic liquids. *Acc. Chem. Res.* 2006, *39*, 489–497.

18. Koley, P.; Pramanik, A. Multilayer vesicles, tubes, various porous structures and organo gels through the solvent-assisted self-assembly of two modified tripeptides and their different applications. *Soft Matter* 2012, *8*, 5364–5374.

19. Roy, S.; Banerjee, A. Amino acid based smart hydrogel: Formation, characterization and fluorescence properties of silver nanoclusters within the hydrogel matrix. *Soft Matter* 2011, *7*, 5300–5308.

20. Xu, H.; Das, A.K.; Horie, M.; Shaik, M.S.; Smith, A.M.; Luo, Y.; Lu, X.; Collins, R.; Liem, S.Y.; Song, A.; *et al.* An investigation of the conductivity of peptide nanotube networks prepared by enzyme-triggered self-assembly.*Nanoscale* 2010, *2*, 960–966.

21. Ma, D.; Tu, K.; Zhang, L.M. Bioactive supramolecular hydrogel with controlled dual drug release characteristics.*Biomacromolecules* 2010, *11*, 2204–2212.

22. Naskar, J.; Palui, G.; Banerjee, A. Tetrapeptide-based hydrogels: For incapsulation and slow release of an anticancer drug at physiological pH. *J. Phys. Chem. B* 2009, *113*, 11787–11792.

23. Koley, P.; Gayen, A.; Drew, M.G.B.; Mukhopadhyay, C.; Pramanik, A. Design and self-assembly of a leucine-enkephalin analogue in different nanostructures: Application of nanovesicles. *Small* 2012, *8*, 984–990.

24. Wada, A.; Tamaru, S.; Ikeda, M.; Hamachi, I. MCM-enzyme-supramolecular hydrogel hybrid as a fluorescence sensing material for polyanions of biological significance. *J. Am. Chem. Soc.* 2009, *131*, 5321–5330.

25. Yemini, M.; Reches, M.; Rishpon, J.; Gazit, E. Novel electrochemical biosensing platform using self-assembled peptide nanotubes. *Nano Lett.* 2005, *5*, 183–186.

26. Lee, K.Y.; Mooney, D.J. Hydrogels for tissue engineering. *Chem. Rev.* 2001, *101*, 1869–1880.

27. Tsitsilianis, C. Responsive reversible hydrogels from associative "smart" macromolecules. *Soft Matter* 2010, *6*, 2372–2388.

28. Zhao, X.B.; Pan, F.; Xu, H.; Yaseen, M.; Shan, H.; Hauser, C.A.E.; Zhang, S.; Lu, J.R. Molecular self-assembly and applications of designer peptide amphiphiles. *Chem. Soc. Rev.* 2010, *39*, 3480–3498.

29. Li, J.; Fan, C.; Pei, H.; Shi, J.; Huang, Q. Smart drug delivery nanocarriers with self-assembled DNA nanostructures.*Adv. Mater.* 2013, *25*, 4386–4396.

30. Rahmawan, Y.; Xu, L.; Yang, S. Self-assembly of nanostructures towards transparent, superhydrophobic surfaces. *J. Mater. Chem. A* 2013, *1*, 2955–2969.

31. Berl, V.; Huc, I.; Khoury, R.G.; Krische, M.J.; Lehn, J.M. Interconversion of single and double helices formed from synthetic molecular strands. *Nature* 2000, *407*, 720–723.

32. Ramstrom, O.; Bunyapaiboonsri, T.; Lohmann, S.; Lehn, J.M. Chemical biology of dynamic combinatorial libraries.*Biochim. Biophys. Acta* 2002, *1572*, 178–186.

33. Mercuri, F.; Baldoniab, M.; Sgamellottia, A. Towards nano-organic chemistry: Perspectives for a bottom-up approach to the synthesis of low-dimensional carbon nanostructures. *Nanoscale* 2012, *4*, 369–379.

34. Zhang, S. Fabrication of novel biomaterials through molecular self-assembly. *Nat. Biotechnol.* 2003, *21*, 1171–1178.

35. Wang, Y.; Xia, Y. Bottom-up and top-down approaches to the synthesis of monodispersed spherical colloids of low melting-point metals. *Nano Lett.* 2004, *4*, 2047–2050.

36. Velev, O.D.; Lenhoff, A.M. Colloidal crystals as templates for porous materials. *Curr. Opin. Colloid Interface Sci.* 2000, *5*, 56–63.

37. Xia, Y.; Gates, B.; Yin, Y.; Lu, Y. Monodispersed colloidal spheres: Old materials with new applications. *Adv. Mater.*2000, *12*, 693–713.

38. Ghadiri, M.R.; Granja, J.R.; Buehler, L.K. Artificial transmembrane ion channels from self-assembling peptide nanotubes. *Nature* 1994, *369*, 301–304.

39. Ikeda, M.; Tanida, T.; Yoshii, T.; Hamachi, I. Rational molecular design of stimulus-responsive supramolecular hydrogels based on dipeptides. *Adv. Mater.* 2011, *23*, 2819–2822.

40. Jewett, J.C.; Sletten, E.M.; Bertozzi, C.R. Rapid Cu-free click chemistry with readily synthesized biarylazacyclooctynones. *J. Am. Chem. Soc.* 2010, *132*, 3688–3690.

41. Changa, P.V.; Preschera, J.A.; Slettena, E.M.; Baskina, J.M.; Millera, I.A.; Agarda, N.J.; Loa, A.; Bertozzia, C.R. Copper-free click chemistry in living animals. *Proc. Natl. Acad. Sci. USA* 2010, *107*, 1821–1826.

42. Segarra-Maset, M.D.; Nebot, V.J.; Miravet, J.F.; Escuder, B. Control of molecular gelation by chemical stimuli. *Chem. Soc. Rev.* 2013, *42*, 7086–7098.

43. Bowerman, C.J.; Nilsson, B.L. A reductive trigger for peptide self-assembly and hydrogelation. *J. Am. Chem. Soc.* 2010,*132*, 9526–9527.

44. Chen, C.-S.; Xu, X.-D.; Li, R.-X.; Zhuo, S.-Y.; Zhang, X.-Z. Photo-switched self-assembly of a gemini α-helical peptide into supramolecular architectures. *Nanoscale* 2013, *5*, 6270–6274.

45. Kim, J.H.; Lee, M.; Lee, J.S.; Park, C.B. Self-assembled light-harvesting peptide nanotubes for mimicking natural photosynthesis. *Angew. Chem. Int. Ed.* 2012, *51*, 517–520.

46. Frkanec, L.; Jokic, M.; Makarevic, J.; Wolsperger, K.; Zinic, M. Bis(PheOH) Maleic acid amide-fumaric acid amide photoizomerization induces microsphere-to-gel fiber morphological transition: The photoinduced gelation system. *J. Am. Chem. Soc.* 2002, *124*, 9716–9717.

47. Walsh, C. Correction: Enabling the chemistry of life. *Nature* 2001, *409*, 226–231.

48. Um, S.H.; Lee, J.B.; Park, N.; Kwon, S.Y.; Umbach, C.C.; Luo, D. Enzyme-catalysed assembly of DNA hydrogel. *Nat. Mater.* 2006, *5*, 797–801.

49. Yang, Z.; Xu, B. Using enzymes to control molecular hydrogelation. *Adv. Mater.* 2006, *18*, 3043–3046.

50. Zhang, J.; Hao, R.; Huang, L.; Yao, J.; Chen, X.; Shao, Z. Self-assembly of a peptide amphiphile based on hydrolysed Bombyx mori silk fibroin. *Chem. Commun.* 2011, *47*, 10296–10298.

51. Chronopoulou, L.; Sennato, S.; Bordi, F.; Giannella, D.; Nitto, A.D.; Barbetta, A.; Dentini, M.; Togna, A.R.; Togna, G.I.; Moschinic, S.; *et al.* Designing unconventional Fmoc-peptide-based biomaterials: Structure and related properties. *Soft Matter* 2014, *10*, 1944–1952.

52. John, G.; Zhu, G.; Li, J.; Dordick, J.S. Enzymatically derived sugar-containing self-Assembled organogels with nanostructured morphologies. *Angew. Chem. Int. Ed.* 2006, *45*, 4772–4775.

53. Dirksen, A.; Meijer, E.W.; Adriaens, W.; Hackeng, T.M. Strategy for the synthesis of multivalent peptide-based nonsymmetric dendrimers by native chemical ligation. *Chem. Commun.* 2006, 1667–1669.

54. Ryadnov, M.G.; Woolfson, D.N. Self-assembled templates for polypeptide synthesis. *J. Am. Chem. Soc.* 2007, *129*, 14074–14081.

55. Jung, J.P.; Jones, J.L.; Cronier, S.A.; Collier, J.H. Modulating the mechanical properties of self-assembled peptide hydrogels via native chemical ligation. *Biomaterials* 2008, *29*, 2143–215.

56. Fang, G.-M.; Cui, H.-K.; Zheng, J.-S.; Liu, L. Chemoselective ligation of peptide phenyl esters with *N*-terminal cysteines. *ChemBioChem* 2010, *11*, 1061–1065.

57. Hentschel, J.; Krause, E.; Borner, H.G. Switch-peptides to trigger the peptide guided assembly of poly(ethylene oxide)-peptide conjugates into tape structures. *J. Am. Chem. Soc.* 2006, *128*, 7722–7723.

58. Rasale, D.B.; Maity, I.; Konda, M.; Das, A.K. Peptide self-assembly driven by oxo-ester mediated native chemical ligation. *Chem. Commun.* 2013, *49*, 4815–4817.

59. Rasale, D.B.; Maity, I.; Das, A.K. *In situ* generation of redox active peptides driven by selenoester mediated native chemical ligation. *Chem. Commun.* 2014, *50*, 11397–11400.

60. Boekhoven, J.; Brizard, A.M.; Kowlgi, K.N.K.; Koper, G.J.M.; Eelkema, R.; van Esch, J.H. Dissipative self-assembly of a molecular gelator by using a chemical fuel. *Angew. Chem. Int. Ed.* 2010, *49*, 4825–4828.

61. Miao, X.; Cao, W.; Zheng, W.; Wang, J.; Zhang, X.; Gao, J.; Yang, C.; Kong, D.; Xu, H.; Wang, L.; Yang, Z. Switchable catalytic activity: Selenium-containing peptides with redox-controllable self-assembly properties. *Angew. Chem. Int. Ed.* 2013, *52*, 7781–7785.

62. Sreenivasachary, N.; Lehn, J.-M. Gelation-driven component selection in the generation of constitutional dynamic hydrogels based on guanine-quartet formation. *Proc. Natl. Acad. Sci. USA* 2005, *102*, 5938–5943.

63. Liang, G.; Ren, H.; Rao, J. A biocompatible condensation reaction for controlled assembly of nanostructures in living cells. *Nat. Chem.* 2010, *2*, 54–60.

64. Sayyad, A.S.; Balakrishnan, K.; Ajayan, P.M. Chemical reaction mediated self-assembly of PTCDA into nanofibers.*Nanoscale* 2011, *3*, 3605–3608.

65. Maity, I.; Rasale, D.B.; Das, A.K. Exploiting a self-assembly driven dynamic nanostructured library. *RSC Adv.* 2013, *3*, 6395–6400.

66. Carnall, J.M.A.; Waudby, C.A.; Belenguer, A.M.; Stuart, M.C.A.; Peyralans, J.J.P.; Otto, S. Mechanosensitive self-replication driven by self-organization. *Science* 2010, *327*, 1502–1506.

67. Na, N.; Mu, X.; Liu, Q.; Wen, J.; Wang, F.; Ouyang, J. Self-assembly of diphenylalanine peptides into microtubes with "turn on" fluorescence using an aggregation-induced emission molecule. *Chem. Commun.* 2013, *49*, 10076–10078.

68. Kumaraswamy, P.; Lakshmanan, R.; Sethuraman, S.; Krishnan, U.M. Self-assembly of peptides: Influence of substrate, pH and medium on the formation of supramolecular assemblies. *Soft Matter* 2011, *7*, 2744–2745.

69. Guo, C.; Luo, Y.; Zhou, R.H.; Wei, G.H. Probing the self-assembly mechanism of diphenylalanine-based peptide nanovesicles and nanotubes. *ACS Nano* 2012, *6*, 3907–3918.

70. Yan, X.; Su, Y.; Li, J.; Fruh, J.; Mçhwald, H. Uniaxially oriented peptide crystals for active optical waveguiding. *Angew. Chem. Int. Ed.* 2011, *50*, 11186–11191.

71. Kim, J.H.; Lim, S.Y.; Nam, D.H.; Ryu, J.; Ku, S.H.; Park, C.B. Self-assembled, photoluminescent peptide hydrogel as a versatile platform

for enzyme-based optical biosensors. *Biosens. Bioelectron.* 2011, *26*, 1860–1865.

72. Zhu, P.; Yan, X.; Su, Y.; Yang, Y.; Li, J. Solvent-induced structural transition of self-assembled dipeptide: From organogels to microcrystals. *Chem. Eur. J.* 2010, *16*, 3176–3183.

73. Yan, X.H.; He, Q.; Wang, K.; Duan, L.; Cui, Y.; Li, J.B. Transition of cationic dipeptide nanotubes into vesicles and oligonucleotide delivery. *Angew. Chem. Int. Ed.* 2007, *46*, 2431–2434.

74. Mehler, A.; Reches, M.; Rechter, M.; Cohen, S.; Gazit, E. Rigid self-assembled hydrogel composed of a modified aromatic dipeptide. *Adv. Mater.* 2006, *18*, 1365–1370.

75. Purich, D.L.; Scaife, R.M. Enzymatic modulation of cytoskeletal self-assembly: ADP ribosylation of microtubule protein components. In *Enzyme Dynamics and Regulation*; Springer-Verlag: New York, NY, USA, 1988; pp. 217–223. []

76. Gao, Y.; Shi, J.; Yuan, D.; Xu, B. Imaging enzyme-triggered self-assembly of small molecules inside live cells. *Nat. Commun.* 2012, *3*, 1033.

77. Vemula, P.K.; Li, J.; John, G. Enzyme catalysis: Tool to make and break amygdalin hydrogelators from renewable resources: A delivery model for hydrophobic drugs. *J. Am. Chem. Soc.* 2006, *128*, 8932–8938.

78. Yang, Z.; Liang, G.; Xu, B. Enzymatic hydrogelation of small molecules. *Acc. Chem. Res.* 2008, *41*, 315–326.

79. Das, A.K.; Collins, R.; Ulijn, R.V. Exploiting enzymatic (Reversed) hydrolysis in directed self-assembly of peptide nanostructures. *Small* 2008, *4*, 279–287.

80. Das, A.K.; Hirstb, A.R.; Ulijn, R. V. Evolving nanomaterials using enzyme-driven dynamic peptide libraries (eDPL).*Faraday Discuss.* 2009, *143*, 293–303.

81. Hirst, A.R.; Roy, S.; Arora, M.; Das, A.K.; Hodson, N.; Murray, P.; Marshall, S.; Javid, N.; Sefcik, J.; Boekhoven, J.; *et al.* Biocatalytic induction of supramolecular order. *Nat. Chem.* 2010, *2*, 1089–1094.

82. Guilbaud, J.-B.; Rochas, C.; Miller, A.F.; Saiani, A. Effect of enzyme concentration of the morphology and properties of enzymatically triggered peptide hydrogel. *Biomacromolecules* 2013, *14*, 1403–1411.

83. Robinson, B.A.; Tester, J.W. Kinetics of alkaline hydrolysis of organic esters and amides in neutrally-buffered solution.*Int. J. Chem. Kinet.* 1990, *22*, 431–448.

84. Tsugita, A.; Scheffler, J.-J. A raid method for acid hydrolysis of protein

with a mixture of trifluoroacetic acid and hydrochloric acid. *Eur. J. Biochem.* 1982, *124*, 585–588.

85. Rebecchi, K.R.; Go, E.P.; Xu, L.; Woodin, C.L.; Mure, M.; Desaire, H. A general protease digestion procedure for optimal protein sequence coverage and post-translational modifications analysis of recombinant glycoproteins: Application to the characterization of human lysyl oxidase-like 2 glycosylation. *Anal. Chem.* 2011, *83*, 8484–8491.

86. Rasale, D.B.; Maity, I.; Das, A.K. Emerging p-stacked dynamic nanostructured library. *RSC Adv.* 2012, *2*, 9791–9794.

87. Plunkett, K.N.; Berkowski, K.L.; Moore, J.S. Chymotrypsin responsive hydrogel: Application of a disulfide exchange protocol for the preparation of methacrylamide containing peptides. *Biomacromolecules* 2005, *6*, 632–637.

88. Toledano, S.; Williams, R.J.; Jayawarna, V.; Ulijn, R.V. Enzyme-triggered self-assembly of peptide hydrogels via reversed hydrolysis. *J. Am. Chem. Soc.* 2006, *128*, 1070–1071.

89. Williams, R.J.; Smith, A.M.; Collins, R.; Hodson, N.; Das, A.K.; Ulijn, R.V. Enzyme-assisted self-assembly under thermodynamic control. *Nat. Nanotechnol.* 2009, *4*, 19–24.

90. Rasale, D.B.; Biswas, S.; Konda, M.; Das, A.K. Exploring thermodynamically downhill nanostructured peptide libraries: From structural to morphological insight. *RSC Adv.* 2015, *5*, 1529–1537.

91. Wu, Q.; Soni, P.; Reetz, M.T. Laboratory evolution of enantiocomplementary candida antarctica lipase B mutants with broad substrate scope. *J. Am. Chem. Soc.* 2013, *135*, 1872–1881.

92. Schmid, R.D.; Verger, R. Lipases: Interfacial enzymes with attractive applications. *Angew. Chem. Int. Ed.* 1998, *37*, 1608–1633.

93. Naik, S.; Basu, A.; Saikia, R.; Madan, B.; Paul, P.; Chaterjee, R.; Brask, J.; Svendsen, A. Lipases for use in industrial biocatalysis: Specificity of selected structural groups of lipases. *J. Mol. Catal. B Enzym.* 2010, *65*, 18–23.

94. Gardossi, L.; Bianchi, D.; Klibanov, A.M. Selective acylation of peptides catalyzed by lipases in organic solvents. *J. Am. Chem. Soc.* 1991, *113*, 6328–6329.

95. Chronopoulou, L.; Lorenzoni, S.; Masci, G.; Dentini, M.; Togna, A.R.; Togna, G.; Bordic, F.; Palocci, C. Lipase-supported synthesis of peptidic hydrogels. *Soft Matter* 2010, *6*, 2525–2532.

96. Yu, L.; Banerjee, I.A.; Gao, X.; Matsui, H. Fabrication of enzyme-

incorporated and magnetic peptide nanotubes. *Polym. Prepr.* 2005, *46*, 36. []

97. Rasale, D.B.; Maity, I.; Das, A.K. Lipase catalyzed inclusion of gastrodigenin for the evolution of blue light emitting peptide nanofibers. *Chem. Commun.* 2014, *50*, 8685–8688.

98. Das, A.K.; Maity, I.; Parmar, H.S.; McDonald, T.O.; Konda, M. Lipase-catalyzed dissipative self-assembly of a thixotropic peptide bolaamphiphile hydrogel for human umbilical cord stem-cell proliferation. *Biomacromolecules*2015, *16*, 1157–1168.

99. Dawson, P.E.; Muir, T.W.; Clark-Lewis, I.; Kent, S.B.H. Synthesis of proteins by native chemical ligation. *Science* 1994,*266*, 776–779.

100. Wieland, T.; Bokelmann, E.; Bauer, L.; Lang, H.U.; Lau, H. Über peptidsynthesen. 8. Mitteilung bildung von S-haltigen peptiden durch intramolekulare wanderung von aminoacylresten. *Justus Liebigs Ann. Chem.* 1953, *583*, 129–149.

101. De Duve, C. The beginnings of life on earth. *Am. Sci.* 1995, *83*, 428–437. []

102. Hojo, H.; Aimoto, S. Polypeptide synthesis using the *S*-alkyl thioester of a partially protected peptide segment. Synthesis of the DNA-binding domain of *c*-Myb protein (142-193)-NH$_2$. *Bull. Soc. Chem. Jpn.* 1991, *64*, 111.

103. Dheur, J.; Ollivier, N.; Melnyk, O. Synthesis of thiazolidine thioester peptides and acceleration of native chemical ligation. *Org. Lett.* 2011, *13*, 1560–1563.

104. Hu, B.-H.; Su, J.; Messersmith, P.B. Hydrogels cross-linked by native chemical ligation. *Biomacromolecules* 2009, *10*, 2194–2000.

105. Warren, J.D.; Miller, J.S.; Keding, S.J.; Danishefsky, S.J. Toward fully synthetic glycoproteins by ultimately convergent routes: A solution to a long-standing problem. *J. Am. Chem. Soc.* 2004, *126*, 6576–6578.

106. Hackeng, T.M.; Griffin, J.H.; Dawson, P.E. Protein synthesis by native chemical ligation: Expanded scope by using straightforward methodology. *Proc. Natl. Acad. Sci. USA* 1999, *96*, 10068–10073.

107. Metanis, N.; Keinan, E.; Dawson, P.E. Traceless ligation of cysteine peptides using selective deselenization. *Angew. Chem. Int. Ed.* 2010, *49*, 7049–7053.

108. Pollock, S.B.; Kent, S.B. An investigation into the origin of the dramatically reduced reactivity of peptide-prolyl-thioesters in native chemical ligation. *Chem. Commun.* 2011, 2342–2344.

109. Townsend, S.D.; Tan, Z.P.; Dong, S.W.; Shang, S.Y.; Brailsford, J.A.; Danishefsky, S.J. Advances in proline ligation. *J. Am. Chem. Soc.* 2012, *134*, 3912–3916.

110. Weissenborn, M.J.; Castangia, R.; Wehner, J.W.; Sardzik, R.; Lindhorst, T.K.; Flitsch, S.L. Oxo-ester mediated native chemical ligation on microarrays: An efficient and chemoselective coupling methodology. *Chem. Commun.* 2012, *48*, 4444–4446.

111. Strehin, I.; Gourevitch, D.; Zhang, Y.; Katzb, E.H.; Messersmith, P.B. Hydrogels formed by oxo-ester mediated native chemical ligation. *Biomater. Sci.* 2013, *1*, 603–613.

112. Jacob, C.; Giles, G.I.; Giles, N.M.; Sies, H. Sulfur and selenium: The role of oxidation state in protein structure and function. *Angew. Chem. Int. Ed.* 2003, *42*, 4742–4758.

113. Mugesh, G.; Singh, H.B. Synthetic organoselenium compounds as antioxidants: Glutathione peroxidase activity. *Chem. Soc. Rev.* 2000, *29*, 347–357.

114. Nogueira, C.W.; Zeni, G.; Rocha, J.B.T. Organoselenium and organotellurium compounds: Toxicology and pharmacology. *Chem. Rev.* 2004, *104*, 6255–6268.

115. Nishino, T.; Okada, M.; Kuroki, T.; Watanabe, T.; Nishiyama, Y.; Sonoda, N. One-pot synthetic method of unsymmetrical diorganyl selenides: Reaction of diphenyl diselenide with alkyl halides in the presence of lanthanum metal. *J. Org. Chem.* 2002, *67*, 8696–8698.

116. Keck, G.E.; Grier, M.C. Generation and reactivity of oxazolidinone derived *N*-acyl radicals. *Synlett* 1999, *1999*, 1657–1659.

117. Boger, D.L.; Mathvink, R.J. Acyl radicals: Intermolecular and intramolecular alkene addition reactions. *J. Org. Chem.* 1992, *57*, 1429–1443.

118. Chen, C.; Crich, D.; Papadatos, A. The chemistry of acyl tellurides: Generation and trapping of acyl radicals, including aryltellurium group transfer. *J. Am. Chem. Soc.* 1992, *114*, 8313–8314.

119. Hiiro, T.; Morita, Y.; Inoue, T.; Kambe, N.; Ogawa, A.; Ryu, I.; Sonoda, N. A new access to acyl- and aroyllithiums via lithium-tellurium exchange. *J. Am. Chem. Soc.* 1990, *112*, 455–457.

120. Baca, M.; Muir, T.; Schonolzer, M.; Kent, S.B.H. Chemical ligation of cysteine-containing peptides: Synthesis of a 22 kDa tethered dimer of HIV-1 protease. *J. Am. Chem. Soc.* 1995, *117*, 1881–1887.

121. Inoue, M.; Yamahita, S.; Ishihara, Y.; Hirama, M. Two convergent routes

to the left-wing fragment of ciguatoxin CTX3C using O,S-acetals as key intermediates. *Org. Lett.* 2006, *8*, 5805–5808.

122. Martin, S.F.; Chen, K.X.; Eary, C.T. An enantioselective total synthesis of (+)-geissoschizine. *Org. Lett.* 1999, *1*, 79–82.

123. Durek, T.; Alewood, P.F. Preformed selenoesters enable rapid native chemical ligation at intractable sites. *Angew. Chem. Int. Ed.* 2011, *50*, 12042–12045.

124. Chu-Kung, A.F.; Bozzelli, K.N.; Lockwood, N.A.; Haseman, J.R.; Mayo, K.H.; Tirrell, M.V. Promotion of peptide antimicrobial activity by fatty acid conjugation. *Bioconj. Chem.* 2004, *15*, 530–535.

125. Maity, I.; Parmar, H.S.; Rasale, D.B.; Das, A.K. Self-programmed nanovesicle to nanofiber transformation of a dipeptide appended bolaamphiphile and its dose dependent cytotoxic behaviour. *J. Mater. Chem. B* 2014, *2*, 5272–5279.

126. Ghanaati, S.; Webber, M.J.; Unger, R.E.; Orth, C.; Hulvat, J.F.; Kiehna, S.E.; Barbeck, M.; Rasic, A.; Stupp, S.I.; Kirkpatrick, C.J. Dynamic *in vivo* biocompatibility of angiogenic peptide amphiphile nanofibers. *Biomaterials* 2009, *30*, 6202–6212.

127. Gao, Y.; Kuang, Y.; Guo, Z.-F.; Guo, Z.; Krauss, I.J.; Xu, B. Enzyme-instructed molecular self-assembly confers nanofibers and a supramolecular hydrogel of taxol derivative. *J. Am. Chem. Soc.* 2009, *131*, 13576–13577.

128. Altunbas, A.; Lee, S.J.; Rajasekaran, S.A.; Schneider, J.P.; Pochan, D.J. Encapsulation of curcumin in self-assembling peptide hydrogels as injectable drug delivery vehicles. *Biomaterials* 2011, *32*, 5906–5914.

129. Maity, I.; Manna, M.K.; Rasale, D.B.; Das, A.K. Peptide-nanofiber-supported palladium nanoparticles as an efficient catalyst for the removal of *N*-Terminus protecting groups. *ChemPlusChem* 2014, *79*, 413–420.

130. Maity, I.; Rasale, D.B.; Das, A.K. Peptide nanofibers decorated with Pd nanoparticles to enhance the catalytic activity for C–C coupling reactions in aerobic conditions. *RSC Adv.* 2014, *4*, 2984–2988.

131. Maity, I.; Rasale, D.B.; Das, A.K. Sonication induced peptide-appended bolaamphiphile hydrogels for *in situ* generation and catalytic activity of Pt nanoparticles. *Soft Matter* 2012, *8*, 5301–5308.

132. Dickerson, M.B.; Sandhage, K.H.; Naik, R.R. Protein- and peptide-directed syntheses of inorganic materials. *Chem. Rev.* 2008, *108*, 4935–4978.

133. Sone, E.D.; Stupp, S.I. Semiconductor-encapsulated peptide-amphiphile nanofibers. *J. Am. Chem. Soc.* 2004, *126*, 12756–12757.

Chapter 5

CHEMICAL REACTION OF SOYBEAN FLAVONOIDS WITH DNA: A COMPUTATIONAL STUDY USING THE IMPLICIT SOLVENT MODEL

Hassan H. Abdallah [1], Janez Mavri [2,3], Matej Repič [2], Vannajan Sanghi-ran Lee [4,] and Habibah A. Wahab [5,]

[1]School of Chemical Sciences, University Sains Malaysia, Penang 11800, Malaysia

[2]National Institute of Chemistry, Hajdrihova 19, SI-1001 Ljubljana, Slovenia

[3]EN-FIST Centre of Excellence, Dunajska 156, SI-1000 Ljubljana, Slovenia

[4]Department of Chemistry, Faculty of Science, University of Malaya, Kuala Lumpur 50603, Malaysia

[5]School of Pharmaceutical Sciences, University Sains Malaysia, Penang 11800, Malaysia

ABSTRACT

Genistein, daidzein, glycitein and quercetin are flavonoids present in soybean and other vegetables in high amounts. These flavonoids can be metabolically converted to more active forms, which may react with guanine in the DNA to form complexes and can lead to DNA depurination. We assumed two ultimate carcinogen forms of each of these flavonoids, diol epoxide form and diketone form. Density functional theory (DFT) and Hartree-Fock (HF) methods were used to study the reaction thermodynamics between active forms of flavonoids and DNA guanine. Solvent reaction field method of Tomasi and co-workers and the Langevin dipoles method of Florian and Warshel were used to calculate the hydration free energies. Activation free energy for each reaction was estimated using the linear free energy relation. Our calculations show that diol epoxide forms of flavonoids are more reactive than the corresponding diketone forms and are hence more likely flavonoid ultimate carcinogens. Genistein, daidzein and glycitein show comparable reactivity while quercetin is less reactive toward DNA.

INTRODUCTION

Carcinogenesis is a complex pathological process, where normal cells become neoplastic. It is mainly the process associated with chemical modification of DNA. Chemical modification of DNA could be caused by viruses, photochemical reactions or reactive substances, called carcinogens [1–4]. Carcinogens can either be of endogenous origin [5,6] (*i.e.*, hormones or their metabolites) or of exogenous origin [7] (*i.e.*, those chemicals that originate from the environment). The first step for chemically induced carcinogenicity is the alteration of the cellular DNA by the reactive form of carcinogen. Failure to repair DNA adducts can lead to depurination, as in the case of polyaromatic hydrocarbons [8], or to errors in DNA replication. Both introduce a mutation in DNA which can cause translocation and amplification of specific genes (proto-oncogenes), which translate into transformation from normal to altered cell. Indeed, multiple cumulative mutational events are invariably required for the progression from normal to fully malignant phenotype [9]. The altered cell may remain dormant or under specific circumstances may proliferate into paraneoplastic and ultimately progress to neoplastic cell. It is clear that repair mechanisms and the immune system play very important roles in carcinogenesis. The repair mechanisms in cells are very proficient, but there is evidence that they can be overwhelmed and consequently fail as well. For a recent review see [10].

Genistein, daidzein, glycitein and quercetin are flavonoids present in soybean [11] and other vegetables in high amounts, while quercetin is also sold as a dietary supplement. The majority of publications have focused on the inhibition effect of flavonoids on cell proliferation, although, experiments on the oral squamous cell line SCC-25 have shown a biphasic effect of quercetin on cell proliferation [12]. It was found that quercetin stimulated cell proliferation at concentrations up to 10 µM, whereas at higher concentrations, cell proliferation was inhibited and it is unclear under what conditions either of the effects prevail. Dihal and his group, has shown the ability of quercetin to inhibit the differentiation of Caco-2 cells in a behavior opposite to the anti-carcinogenic adducts with cellular protein and DNA in which, under certain conditions, cancer incidences are increased in experimental animals [13–19]. Some of these flavonoids exhibit estrogenic activity [20–22], but they are usually used for their powerful antioxidant [23] and chemopreventive properties [24–26]; their bioactivated forms are to a certain extent capable of reacting with DNA. Similar to polyaromatic hydrocarbons (PAHs), which are known carcinogens [27,28], flavonoids are not carcinogenic *per se*, but can become carcinogenic after bioactivation. The unmetabolized form is called procarcinogen, while the bioactivated metabolized form is called the ultimate carcinogen. PAHs [29,30], estrogens [31] and flavonoids are all metabolized by the family of cytochromes P450 to a wide variety of primary metabolites, including epoxides, dihydrodiols, quinones, and phenols [32–34], many of which can be reoxidized and recycled through the same metabolic pathways

[29,30,35,36] giving rise to metabolites that are able to chemically react with DNA [27,30,37]. Studies on PAHs and estrogens have led us to conclude that the ultimate carcinogens of studied polyphenols can be either of diol epoxide or quinone type. This assumption is strengthened by studies that showed that the formation of quercetin's quinone type metabolites is fast. The oxidations are performed by peroxidases (one electron oxidation) to form quercetin semiquinone radical and by tyrosinases (two electron oxidation) to form quercetin o-quinone [38,39].

The reaction responsible for the initial phase of carcinogenesis is alkylation of DNA by an ultimate carcinogen, typically at position N7 of guanine, although other alkylation sites are also possible [5]. It is well established, that guanine at position N7 is the most nucleophilic site of DNA and hence most likely to react [40]. Moreover, it was proven experimentally that large majority of the DNA chemical damages are associated with this site. Currently, quantum chemical calculations cannot treat the entire solvated DNA plus the ultimate carcinogen molecule into electronic details. Therefore we truncated the system to methylated guanine and the ultimate carcinogen, while the rest of the system was treated on the level of mean field theory on the Langevin dipoles and solvent reaction field levels. The approach proved to be successful and became a standard way of computational treatment for a large number of systems of biological interest. This approach represents an acceptable trade-off between physical relevance of the model and computational feasibility.

One strategy of preventing chemical damages of DNA is introduction of substances that react with the ultimate carcinogen faster than the ultimate carcinogen reacts with DNA [41]. Ascorbic acid and glutathione are well-known endogenous scavengers of ultimate carcinogens. The increased incidence of prostate cancer induced by complexation of glutathione by heavy metals is well documented in clinical practice [9]. Ellagic acid and polyphenols, such as flavonoids considered in this study, are also typically very efficient scavengers, but their bioactivated diol epoxide or quinone forms are nevertheless able to react with DNA. Thus it is not surprising that there have been studies concerning the safety of quercetin [42–45]. While many long-term studies showed that this compound is not carcinogenic [46,47] there are also contradicting results [48,49]. Moreover, it has been shown experimentally that genistein can mediate DNA strand breaks by H_2O_2/Cu(II) thus exerting influence on oxidative DNA damage [50]. It is worth to give comment on the accessibility of the DNA to the ultimate carcinogens. The ultimate carcinogens have no problems to reach densely packed genetic material. Biological macromolecules are soft and flexible and transport of the substrate to the reactive site is never the rate-limiting step.

The aim of this study was to critically examine the reactivity toward N7 position of guanine of four flavonoids and, for each of them, the both reactive forms were considered. Calculations were performed on Hartree-Fock and DFT level in conjunction with 6-31G(d) basis set. This double zeta basis set augmented with polarization functions on heavy atoms, is flexible enough to reliably describe the thermochemistry of the studied reactions. We considered the density functional B3LYP, where electron exchange is described by the method of Becke [51–53] and electron correlation by the methodology of Lee, Yang and Parr [54]. Structures were first optimized at the PM3 semiempirical MO, followed by HF and B3LYP calculations. Geometry optimizations were performed in the gas phase. For the reactive step we considered linear free energy relationship since the rate-limiting step involves proton transfer via solvent and locating the corresponding transition state is unfeasible. The effects of solvation were included by using solvent reaction field method of Tomasi and coworkers [55] and Langevin dipoles method of Florian and Warshel [56,57].

RESULTS AND DISCUSSION

In this article, we considered two mechanisms for the DNA-adduct formation reaction according to the active form of the flavonoids. Since the selected flavonoids are structurally closely related, the linear free energy relation method was selected for our study. Moreover, the studied reactions are electrophilic substitutions with a complex mechanism involving proton transfer via several solvent molecules. Hence, locating the transition state and calculating the activation free energy for such a complex reaction is not practical. The application of linear free energy relationships to estimate the free energy of activation is well established and is described in reference [58]. In a series of chemical reactions involving similar reactants, which follow the same mechanism, the reaction with the most favorable reaction free energy will have the lowest free energy of activation. Detailed information on the two different proposed mechanisms and the type of adducts formed with guanine, as a model for DNA, using HF and DFT calculations, the chemical reaction will be discussed below.

Diol Epoxide Mechanism

The covalent binding of reactive intermediates to cellular DNA leading to adduct formation is considered to be a critical event in the initiation of carcinogenesis. Flavonoids may interfere with this process, by blocking the formation of reactive intermediates which can potentially prevent the initiation of carcinogenesis inhibiting the metabolic activation of the carcinogen to its reactive intermediates [59]. The effects of flavonoids on procarcinogen-activating enzymes, notably the cytochrome P450 CYP1 family, have been the focus of attention in cancer prevention during the last decade. For flavonoids to bind to DNA, these chemicals need bioactivation to a more reactive diol epoxide form that later attacks the N7 site of the DNA guanine to form DNA-adduct by enzymes such as cytochrome P450 (CYP) and epoxide hydrolase (EH) [60–62].

According to the diol epoxide mechanism the reaction proceeds in the following possible pathway. The N7 site of the DNA guanine approaches the epoxy ring carbon atom. During the reaction the strained epoxy ring opens, and formation of the chemical bond between the epoxy ring carbon atom and guanine is the rate-limiting step, followed by a fast proton transfer involving solvent molecules.

In this reaction, stereoisomeric epoxide reactant gives stereoisomeric product. These are soft degrees of freedom and the energy difference of several conformations is not large (<1 kcal/mol) and it is often difficult to decide simply by inspection which conformation will be adopted. Therefore, we have considered only one stereoisomer for comparison as shown in Scheme 1 for the entire calculation series. Besides, from previous linear free energy relationship study of benzo[a]pyrene diol epoxide stereoisomers [63], different conformations do not have much effect on reactivity.

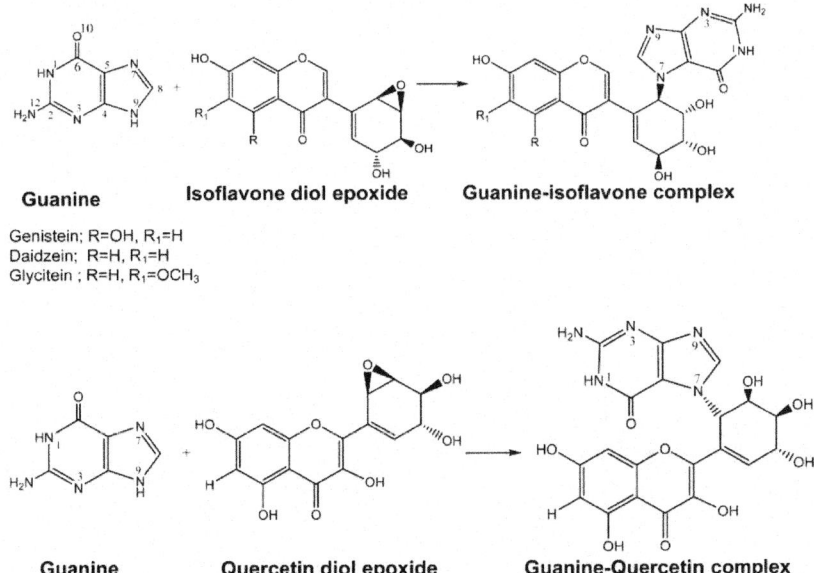

Scheme 1. The diol epoxide mechanism for the DNA-adduct formation.

The calculated reaction energy (ΔE), zero point energy correction (ZPE) and the reaction free energy for the flavonoids are listed in Table 1. The calculated reaction free energy for the chemical reaction calculated as the difference in energy between the reactants (flavonoids + guanine) and the product, which is the flavonoid-guanine complex (DNA adduct) from PCM was found to be lower than LD using both DFT and HF calculation for diol epoxide mechanism. The slightly lower values obtained from HF were found to be −27.98 (quercetin), −27.77 (genistein), −27.32 (daidzein), and −26.91 (glycitein) kcal/mol in similar trend agreed with DFT method among the calculated flavonoids. The negative sign indicates that the reaction is energetically favored. These values include the HF/6-31G(d) calculated classical barrier, 2.40–2.58 kcal/mol of calculated zero point energy correction and 1.48–2.23 kcal/mol of difference in free energy of hydration calculated using Tomasi's approach. In comparison, the Langevin dipoles method for hydration free energy difference gave a value of 13.90–18.60 kcal/mol. It is clear that there is no significant difference between reaction free energies for the reactions of all active flavonoids with guanine. Neither *in vacuo* values of energies differ from each other nor the contributions from hydration free energies for both methods of calculation with the PCM solvent reaction field or the Langevin dipoles method.

Table 1. Free energy components for reactions via diol epoxide and diketone mechanism with density functional theory (DFT) and Hartree-Fock (HF) calculation.

DFT Method (B3LYP/6-31G(d))						
	ΔE a	ΔZPE b	PCM method		LD method	
			ΔG_{hydr} c	ΔG_{react} d	ΔG_{hydr}	ΔG_{react}
Diol epoxide mechanism						
Genistein	−30.04	2.19	1.34	−26.51	6.50	−21.35
Daidzein	−29.74	2.21	1.20	−26.33	5.80	−21.73
Glycitein	−29.66	2.24	1.85	−25.57	5.90	−21.52
Quercetin	−31.95	2.55	1.52	−27.88	11.10	−18.30
Diketone mechanism						
Genistein	−26.83	3.73	11.31	−11.79	5.82	−17.28
Daidzein	−22.79	3.50	7.85	−11.44	3.28	−16.01
Glycitein	−22.94	3.50	7.65	−11.79	1.32	−18.12
Quercetin	−20.44	1.84	0.42	−18.18	12.40	−6.20

HF Method (HF/6-31G(d))						
	ΔE a	ΔZPE b	PCM method		LD method	
			ΔG_{hydr} c	ΔG_{react} d	ΔG_{hydr}	ΔG_{react}
Diol epoxide mechanism						
Genistein	−31.68	2.43	1.48	−27.77	18.60	−10.65
Daidzein	−31.26	2.40	1.54	−27.32	17.00	−11.86
Glycitein	−31.16	2.40	1.85	−26.91	15.10	−13.66
Quercetin	−32.79	2.58	2.23	−27.98	13.90	−16.31
Diketone mechanism						
Genistein	−23.30	1.57	1.39	−20.34	23.30	1.57
Daidzein	−17.58	1.59	0.93	−15.06	17.70	1.71
Glycitein	−17.49	1.60	1.06	−14.83	21.70	5.81
Quercetin	−14.45	1.52	1.06	−11.87	14.50	1.57

Experimental and theoretical results show that the polyaromatic hydrocarbons diol epoxide form is biologically active in the Fjord and Bay region and this activity is attributed to the high affinity for protonation and the high thermodynamic stability of the resulting cation [30,50]. According to our calculations, the diol epoxide forms of the flavonoids are highly active for attack on DNA, which may be attributed to the presence of the epoxide group. Figure 1 shows the geometry of the flavonoid-guanine complex according to the diol epoxide mechanism in which the guanine molecule has a nearly perpendicular position to the plane of the flavonoid molecule.

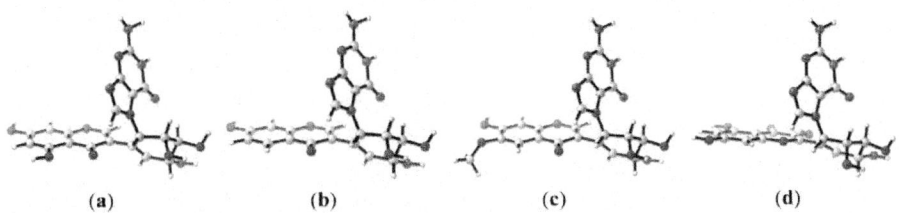

(a)   (b)   (c)   (d)

Figure 1. DFT optimized structures of the complex of the flavonoids diol epoxide-N7-guanine for (**a**) genistein-guanine, (**b**) daidzein-guanine, (**c**) glycitein-guanine, and (**d**) quercetin-guanine.

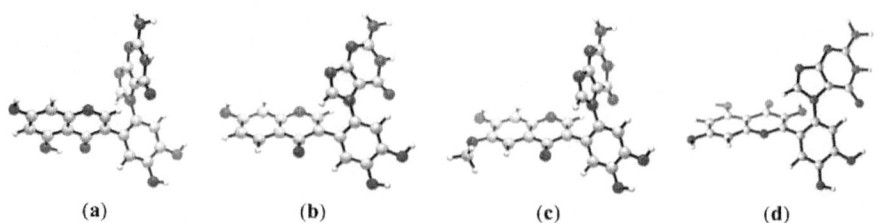

    (a)   (b)   (c)   (d)

Figure 2. DFT optimized structures of the diketone flavonoid-*N*7-guanine complexes for **(a)** genistein-guanine, **(b)** daidzein-guanine, **(c)** glycitein-guanine, and **(d)** quercetin-guanine.

Diketone Mechanism

The active form of the flavonoids in this reaction mechanism is in the diketone form and the reaction proceeds as in Scheme 2. The N7 site of the DNA guanine approaches the diene ring carbon atom, which is followed by the reduction of diketone to form the hydroxyl group. The calculated reaction free energy for the chemical reaction for DFT and HF are quite different. As a result in general, three isoflavones have a lower reaction free energy or the lower activation energy barrier in comparison to quercetin from diketone mechanism. However, with LD method using HF calculation, the diketone mechanism of all flavonoids is not favorable with positive reaction free energy. Also, quercetin appears with the most negative reaction free energy (−18.18 kcal/mol) on DFT with PCM solvation model. Calculations of hydration free energy on the HF/6-31g(d) level of theory exaggerates with the dipole moment and it corresponds to the polarized case as is the situation in aqueous solution. Langevine dipoles model was also parametrized for this level and hydration free energies are quite reliable. B3LYP functional systematically underestimates the reaction barrier due to absence of parametrization set for diketone species in the calculation as it is observed in PCM model for isoflavones with −11.44 to −11.79 kcal/mol (DFT) and −14.83 to −20.34 kcal/mol (HF). These values in B3LYP/6-31G(d) level also include the calculated classical barrier, 1.84–3.73 kcal/mol of B3LYP/6-31G(d) calculated zero point energy correction and 1.32–12.40 kcal/mol of difference in free energy of hydration calculated using the Langevin dipoles method as implemented in program ChemSol 2.1. In comparison, Tomasi's approach for hydration free energy difference gave a value of 0.42–11.31 kcal/mol. It is also clear that there is no significant difference between reaction free energies for the reactions of all active flavonoids with guanine under this mechanism either*in vacuo* or with solvation contribution.

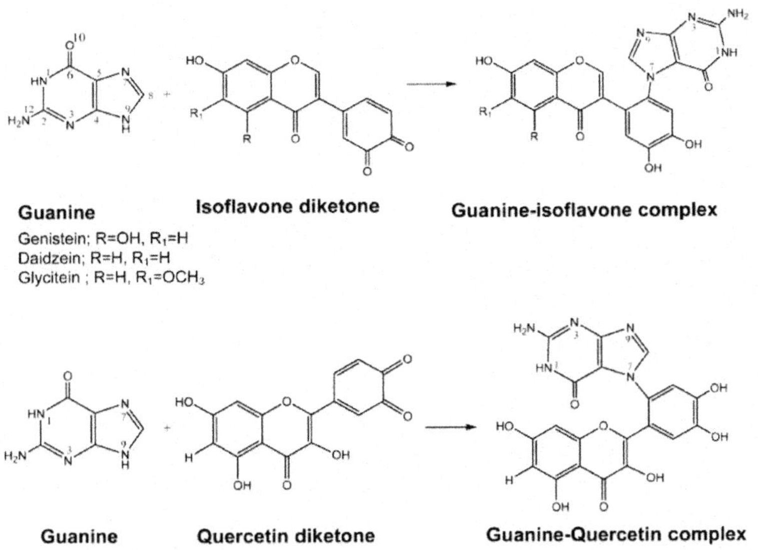

Guanine   **Isoflavone diketone**   **Guanine-isoflavone complex**

Genistein; R=OH, R₁=H
Daidzein; R=H, R₁=H
Glycitein ; R=H, R₁=OCH₃

Guanine   **Quercetin diketone**   **Guanine-Quercetin complex**

Scheme 2. The diketone mechanism for the DNA-adducts formation.

According to the values of the reaction energy and the reaction free energy diketone metabolites are less reactive than the diol epoxide. The diol epoxide mechanism shows no activity difference among flavonoids either in gas or solvent calculation. The calculated zero point contributions to reaction free energies for the three isoflavones are essentially identical, whereas a slightly higher value found for quercetin, as free energy of hydration was modeled on two levels of theory, PCM and LD. As parametrization of the Langevin dipole (LD) model is developed for quantum calculations of chemical processes in aqueous solution, the extension of the model to other polar solvents would require some simple reparameterization. The implementation and parametrization for aqueous solution was described with the training set that encompassed solvation free energies of 44 neutral and 39 ionic solutes of the C, O, N, P, S, F, and Cl atoms and was found to be comparable or slightly better than the PCM continuum model of Tomasi and co-workers to predict the acid dissociation constants (pK_a) of the chemicals involved in proton transfer. Simplified explicit representation of solvent molecules of the LD model gives better insight into the molecular origin of different solvent effects than that obtained by continuum models. Solvent in LD model is approximated by polarizable discrete dipoles fixed on a cubic grid whereas the Polarizable Continuum Model calculates the molecular free energy in solution as the sum over the electrostatic, the dispersion-repulsion contributions to the free energy, and the cavitation using a cavity defined through interlocking *van der Waals* spheres centered at atomic positions. The three flavonoids (genistein, glycitein, and daidzein) according to the first mechanism show no substantial difference in hydration free energy. Interestingly, we noticed that quercetin according to both mechanisms on DFT with LD method needed more energy for hydration in comparison with genistein, glycitein, and daidzein. This behavior can be attributed to the extra hydroxyl group with higher polarization and electron correlation effect. The comparison of the values of the total energy in the gas phase by using HF and DFT theory and the values of the reaction free energy between the two mechanisms clearly showed that the first mechanism gives better results.

Comparing the two different calculation methods from Table 1, we achieved similar barrier heights in gas phase in diol epoxide mechanism, 29.66–31.95 kcal/mole for DFT and 31.16–32.79 kcal/mol for HF, however, application of HF significantly reduced the barrier (14.45–23.30 kcal/mol) in diketone mechanism. In previous work on alkylation of guanine by styrene-7,8-oxide, the combination of Hartree-Fock calculation using flexible basis sets and Langevin dipoles calculation of hydration free energies gives very reasonable agreement with the experimental free energy (26.52 kcal/mol) whereas B3LYP calculations predict a lower activation free energy [64]. We have the impression that the applied DFT method outperforms the HF calculations. Strong evidence indicates that for diketone mechanism, with hydration contributions included, the free energies are positive. The DFT calculated reaction free energies are lower for epoxide mechanism than for diketone mechanism for all the species. Therefore it is very likely, that the carcinogenic reaction proceeds via epoxide mechanism unless flavonoids activation by P450 is the rate-limiting step. Our calculations give strong evidence that genistein, daidzein and glycitein are basically equally reactive toward DNA, while quercetin has different reactivity, less reactive on DFT or more reactive on HF with LD solvation model. This is a possible reason why isoflavones may interfere with the promotion stage of carcinogenesis by blocking the formation of the metabolic activation of the carcinogen to its reactive intermediates. Thus, it remains a major challenge to fine tune the reactivity of quercetin analogs to widen the therapeutic window.

Finally, we can conclude that chemical reactions for four selected flavonoids either with diol epoxide or diketone for this, since free reaction energies are quite similar for all four flavonoid diol epoxides. One possible explanation is that different metabolic transformations at different rates may be the source of discrepancies in carcinogenicity. Other metabolic transformations at different rates could be found in particular reactions catalyzed by enzymes such as catechol-O-methyl transferase (COMT), glutathione-S-transferase, P450 (CYP families), aromatase and various peroxidases [63,65]. In addition, different reaction rates may occur in DNA depurination step, which would also result in different carcinogenicity profiles of studied compounds. Moreover, not only

DNA, but proteins, as well as cellular lipids and some metallic ions (iron and copper) are targets for reactions with quinines [66,67]. It is therefore clear that reaction rates of metabolic transformations, adduct formation, depurination and reactivity with other cellular components all contribute to the overall carcinogenicity of studied compounds. The molecular modeling of chemical reactivity of each of these steps will play an important role in cancer research and will ultimately contribute to the prevention and improvement of cancer treatment.

COMPUTATIONAL DETAIL

Initial structures of the flavonoids and the flavonoid-guanine adducts were built using Molden program. PM3 semiempirical MO method was used to refine the starting structures, which were submitted to the HF and DFT optimization with 6-31G(d) basis set. No constraints were applied. The double-zeta basis set augmented with polarization functions on heavy atoms is flexible enough to describe chemical processes and is also computationally tractable. To ensure that the optimized structure is the real minima rather than saddle points, vibrational analysis was performed in the harmonic approximation and the absence of imaginary frequencies proves that the stationary points are real minima rather that the saddle points. To consider the solvent effect on the reaction mechanism, the polarizable continuum model of Tomasi and coworkers as implemented in Gaussian03 was applied. Moreover, Langevin dipoles (LD) method in which the solute is described by a set of point charges on the grid of the Langevin dipoles, together with a proper parameterization, has received attention during the past years to model solvation effects. Langevin dipoles calculations were performed using CHEMSOL versions 2.1 package kindly provided by Jan Florian. The Merz-Kollman charges were calculated at B3LYP/6-31G(d) level using Tomasi's PCM SCRF.

The reaction free energy was calculated as the free energy difference between products and the reactants corrected for zero point energy contribution calculated in the harmonic approximation and the hydration free energy.

$$\Delta Greact = \Delta E + \Delta ZPE + \Delta Ghydr$$

Langevin dipoles and Gaussian03 [68] calculations were performed on a dual-CPU PC/Linux cluster.

CONCLUSIONS

Reaction free energy and the free energy of hydration were calculated for four flavonoids present in soybean: genistein, daidzein, glycitein, and quercetin, for their reaction with guanine according to two kinds of mechanisms; diol epoxide and diketone mechanism. HF and DFT levels of theory in conjunction with 6-31G(d) basis set were employed for calculating the reaction energies *in vacuo*. LD and PCM solvation models were used for calculation of hydration free energies thus enabling the calculation of reaction free energies. This work sheds light on the mechanism of action of flavonoids. The results show that the diol epoxide mechanism describes the carcinogenic reaction better than the diketone mechanism. Our study shows that metabolites of soy food products are slightly reactive toward DNA but flavonoids are even more reactive toward the ultimate carcinogens of the epoxy type. In terms of cancer prevention, it is still beneficial to consume soy products. Quantum chemical calculations are valuable for the design of flavonoids derivatives with even lower activation energy and increased reactivity toward ultimate carcinogens as well as controlled reactivity toward DNA.

ACKNOWLEDGMENTS

This work is funded by the Malaysian Ministry of Higher Education through the Fundamental Grant Scheme, No. 203/PFARMASI/671001 and Slovenian Research Agency through the grant P1-0012. Matej Repič benefited through the Slovenian Young Investigators Program.

REFERENCES

1. Underwood, J.C.E. *General and Systematic Pathology*, 3rd ed; Churchill Livingstone: Edinburgh, UK, 2000; pp. 374–375.

2. Brookes, P.; Lawley, P.D. Evidence for the binding of polynuclear aromatic hydrocarbons to the nucleic acids of mouse skin: Relation between carcinogenic power of hydrocarbons and their binding to deoxyribonucleic acid. *Nature* 1964,*202*, 781–784.

3. Volk, D.E.; Rice, J.S.; Luxon, B.A.; Yeh, H.J.C.; Liang, C.; Xie, G.; Sayer, J.M.; Jerina, D.M.; Gorenstein, D.G. NMR evidence for *syn*-anti interconversion of a trans opened (10*R*)-dA adduct of benzo[*a*]pyrene (7*S*,8*R*)-diol (9*R*,10*S*)-epoxide in a DNA duplex. *Biochemistry* 2000, *39*, 14040–14053.

4. Smela, M.E.; Hamm, M.L.; Henderson, P.T.; Harris, C.M.; Harris, T.M.; Essigmann, J.M. The aflatoxin B1 formamidopyrimidine adduct plays a major role in causing the types of mutations observed in human hepatocellular carcinoma. *Proc. Natl. Acad. Sci. USA* 2002, *99*, 6655–6660.

5. Stack, D.E.; Byun, J.; Gross, M.L.; Rogan, E.G.; Cavalieri, E.L. Molecular characteristics of catechol estrogen quinones in reactions with deoxyribonucleosides. *Chem. Res. Toxicol* 1996, *9*, 851–859.

6. Huetz, P.; Kamarulzaman, E.E.; Wahab, H.A.; Mavri, J. Chemical reactivity as a tool to study carcinogenicity: Reaction between estradiol and estrone 3,4-quinones ultimate carcinogens and guanine. *J. Chem. Inf. Comput. Sci* 2004, *44*, 310–314.

7. Watanabe, S.; Kobayashi, Y. Exogenous hormones and human cancer. *Jpn. J. Clin. Oncol* 1993, *23*, 1–13.

8. Cavalieri, E.; Rogan, E. Mechanisms of Tumor Initiation by Polycyclic Aromatic Hydrocarbons in Mammals. In *The Handbook of Environmental Chemistry, PAHs and Related Compounds*; Neilson, A.H., Ed.; Springer-Verlag: Berlin, Germany, 1998; Volume 3, pp. 81–117.

9. Fauci, A.S.; Braunwald, E.; Kasper, D.L.; Hauser, S.L.; Longo, D.L.; Jameson, J.L.; Loscalzo, J. *Harrison's Principles of Internal Medicine*, 17th ed; McGraw-Hill: New York, NY, USA, 2008.

10. Mellon, I. DNA Repair. In *Molecular and Biochemical Toxicology*; Smart, R.C., Hodgson, E., Eds.; Wiley: Hoboken, NJ, USA, 2008; pp. 493–535.

11. Murphy, P.A.; Farmakalidis, E.; Johnson, L.D. Isoflavone content of soya-based laboratory animal diets. *Food Chem. Toxicol* 1982, *20*, 315–317.

12. Elattar, T.M.; Virji, A.S. The inhibitory effect of curcumin, genistein, quercetin and cisplatin on the growth of oral cancer cells *in vitro*. *Anticancer Res* 2000, *20*, 1733–1738.

13. Dihal, A.A.; Woutersen, R.A.; Ommen, B.V.; Rietjens, I.M.; Stierum, R.H. Modulatory effects of quercetin on proliferation and differentiation of the human colorectal cell line Caco-2. *Cancer Lett* 2006, *238*, 248–259.

14. Dunnick, J.K.; Hailey, J.R. Toxicity and carcinogenicity studies of quercetin, a natural component of foods. *Fundam. Appl. Toxicol* 1992, *19*, 423–431.

15. Pamukcu, A.M.; Yalciner, S.; Hatcher, J.F.; Bryan, G.T. Quercetin, a rat intestinal and bladder carcinogen present in bracken fern (*Pteridium aquilinum*). *Cancer Res* 1980, *40*, 3468–3472.

16. Erturk, E.; Hatcher, J.F.; Nunoya, T.; Pamukcu, A.M.; Bryan, G.T. Hepatic tumors in Sprague-Dawley (SD) and Fischer 344 (F) female rats chronically exposed to quercetin (Q) or its glycoside rutin (R). *Proc. Am. Assoc. Cancer Res* 1984, *25*, 25–95.

17. Morino, K.; Matsukara, N.; Kawachi, T.; Ohgaki, H.; Sugimura, T.; Hirono, I. Carcinogenicity test of quercetin and rutin in golden hamsters by oral administration. *Carcinogenesis* 1982, *3*, 93–97.

18. Pereira, M.A.; Grubbs, C.J.; Barnes, L.H.; Li, H.; Olson, G.R.; Eto, I.; Juliana, M.; Whitaker, L.M.; Kelloff, G.J.; Steele, V.E.; Lubet, R.A. Effects of the phytochemicals, curcumin and quercetin, upon azoxymethane-induced colon cancer and 7,12-dimethylbenz[*a*]anthracene-induced mammary cancer in rats. *Carcinogenesis* 1996, *17*, 1305–1311.

19. Ishikawa, M.; Oikawa, T.; Hosokawa, M.; Hamada, J.; Morikawa, K.; Kobayashi, H. Enhancing effect of quercetin on 3-methylcholanthrene carcinogenesis in C57Bl/6 mice. *Neoplasma* 1985, *32*, 435–441.

20. Lamartiniere, C.A.; Murrill, W.B.; Manzolillo, P.A.; Zhang, J.X.; Barnes, S.; Zhang, X.; Wei, H.; Brown, N.M. Genistein alters the ontogeny of mammary gland development and protects against chemically-induced mammary cancer in rats. *Proc. Soc. Exp. Biol. Med* 1998, *217*, 358–364.

21. Song, T.T.; Hendrich, S.; Murphy, P.A. Estrogenic activity of glycitein, a soy isoflavone. *J. Agric. Food Chem* 1999, *47*, 1607–1610.

22. Shelby, M.D.; Newbold, R.R.; Tully, D.B.; Chae, K.; Davis, V.L. Assessing environmental chemicals for estrogenicity using a combination of *in vitro* and *in vivo* assays. *Environ. Health Perspect* 1996, *104*, 1296–1300.

23. Kalaiselvan, V.; Kalaivani, M.; Vijayakumar, A.; Sureshkumar, K.; Venkateskumar, K. Current knowledge and future direction of research on soy isoflavones as a therapeutic agents. *Pharmacogn. Rev* 2010, *4*, 111–117.

24. Zhao, L.; Chen, Q.; Brinton, R.D. Neuroprotective and neurotrophic efficacy of phytoestrogens in cultured hippocampal neurons. *Exp. Biol. Med* 2002, *227*, 509–519.

25. Anthony, M.S.; Clarkson, T.B.; Williams, J.K. Effects of soy isoflavones on atherosclerosis: Potential mechanisms. *Am. J. Clin. Nutr* 1998, *68*, 1390S–1393S.

26. Carroll, K.K. Review of clinical studies on cholesterol-lowering response to soy protein. *J. Am. Diet. Assoc* 1991, *91*, 820–827.

27. Cavalieri, E.L.; Rogan, E.G. The approach to understanding aromatic hydrocarbon carcinogenesis. The central role of radical cations in metabolic activation. *Pharmacol. Ther* 1992, *55*, 183–199.

28. Chakravarti, D.; Felling, J.C.; Cavalieri, E.L.; Rogan, E.G. Relating aromatic hydrocarbon-induced DNA adducts and c-H-ras mutations in mouse skin papillomas: The role of apurinic sites. *Proc. Natl. Acad. Sci. USA* 1995, *92*, 10422–10426.

29. Gelboin, H.V. Benzo[alpha]pyrene metabolism, activation and carcinogenesis: Role and regulation of mixed-function oxidases and related enzymes. *Physiol. Rev* 1980, *60*, 1107–1166.

30. Hall, M.; Grover, P.L. Policyclic Aromatic Hydrocarbons: Metabolism, Activation and Tumor Initiation. In *Chemical Carcinogenesis and Mutagenesis I*; Cooper, C.S., Grover, P.L., Eds.; Springer-Verlag: Berlin, Germany, 1990; pp. 327–372.

31. Cavalieri, E.L.; Stack, D.E.; Devanesan, P.D.; Todorovic, R.; Dwivedy, I.; Higginbotham, S.; Johansson, S.L.; Patil, K.D.; Gross, M.L.; Gooden, J.K.; *et al*. Molecular origin of cancer: Catechol estrogen-3,4-quinones as endogenous tumor initiators. *Proc. Natl. Acad. Sci. USA* 1997, *94*, 10937–10942.

32. Pauwels, W.; Veulemans, H. Comparison of ethylene, propylene and styrene 7,8-oxide *in vitro* adduct formation on *N*-terminal valine in human haemoglobin and on *N*-7-guanine in human DNA. *Mutat. Res* 1998, *418*, 21–33.

33. Goldstein, J.A.; Faletto, M.B. Advances in mechanisms of activation and deactivation of environmental chemicals.*Environ. Health Perspect* 1993, *100*, 169–176.

34. Rodeiro, I.; Donato, M.T.; Lahoz, A.; Garrido, G.; Delgado, R.; Gomez-Lechon, M.J. Interactions of polyphenols with the P450 system: Possible implications on human therapeutics. *Mini Rev. Med. Chem* 2008, *8*, 97–106.

35. Pelkonen, O.; Nebert, D.W. Metabolism of polycyclic aromatic hydrocarbons: Etiologic role in carcinogenesis.*Pharmacol. Rev* 1982, *34*, 189–222.

36. Dipple, A.; Moschel, R.C.; Bigger, C.A.H. Polynuclear aromatic carcinogens. *Chem. Carcinog* 1984, *1*, 41–163.

37. Russo, J.; Hu, Y.F.; Tahin, Q.; Mihaila, D.; Slater, C.; Lareef, M.H.; Russo, I.H. Carcinogenicity of estrogens in human breast epithelial cells. *Acta Pathol. Microbiol. Immunol. Scand* 2001, *109*, 39–52.

38. Galati, G.; Chan, T.; Wu, B.; O'Brien, P.J. Glutathione-dependent generation of reactive oxygen species by the peroxidase-catalyzed redox cycling of flavonoids. *Chem. Res. Toxicol* 1999, *12*, 521–525.

39. Awad, H.M.; Boersma, M.G.; Boeren, S.; van Bladeren, P.J.; Vervoort, J.; Rietjens, I.M.C.M. Structure—Activity study on the quinone/quinone methide chemistry of flavonoids. *Chem. Res. Toxicol* 2001, *14*, 398–408.

40. Friedberg, E.C.; Walker, G.C.; Siede, W.; Wood, R.D.; Schultz, R.A.; Ellenberger, T. *DNA Repair and Mutagenesis*, 2nd ed; American Society

for Microbiology: Washington, DC, USA, 1995.

41. Lagerqvist, A.; Håkansson, D.; Frank, H.; Seidel, A.; Jenssen, D. Structural requirements for mutation formation from polycyclic aromatic hydrocarbon dihydrodiol epoxides in their interaction with food chemopreventive compounds.*Food Chem. Toxicol* 2011, *49*, 879–886.

42. MacGregor, J.T.; Jurd, L. Mutagenicity of plant flavonoids: Structural requirements for mutagenic activity in Salmonella typhimurium. *Mutat. Res* 1978, *54*, 297–309.

43. Nagao, M.; Morita, N.; Yahagi, T.; Shimizu, M.; Kuroyanagi, M.; Fukuoka, M.; Yoshihira, K.; Natori, S.; Fujino, T.; Sugimura, T. Mutagenicities of 61 flavonoids and 11 related-compounds. *Environ. Mutagen* 1981, *3*, 401–419.

44. Vanderhoeven, J.C.M.; Bruggeman, I.M.; Debets, F.M.H. Genotoxicity of quercetin in cultured mammalian-cells. *Mutat. Res* 1984, *136*, 9–21.

45. Rietjens, I.M.C.M.; Boersma, M.G.; van der Woude, H.; Jeurissen, S.M.F.; Schutte, M.E.; Alink, G.M. Flavonoids and alkenylbenzenes: Mechanisms of mutagenic action and carcinogenic risk. *Mutat. Res* 2005, *574*, 124–138.

46. Stoewsand, G.S.; Anderson, J.L.; Boyd, J.N. Quercetin: A mutagen, not a carcinogen, in Fischer rats. *J. Toxicol. Environ. Health* 1984, *14*, 105–114.

47. Okamoto, T. Safety of quercetin for clinical application (Review). *Int. J. Mol. Med* 2005, *16*, 275–278.

48. Pamukcu, A. M.; Yalcener, S.; Hatcher, J. F. Quercetin as an intestinal and bladder carcinogen present in bracken fern.*Teratog. Carcinog. Mutagen* 1980, *1*, 213–221.

49. Pamukcu, A.M.; Yalciner, S.; Hatcher, J.F.; Bryan, G.T. Quercetin, a rat intestinal and bladder carcinogen present in bracken fern (*Pteridium aquilinum*). *Cancer Res* 1980, *40*, 3468–3472.

50. Win, W.; Cao, Z.; Peng, X.; Trush, M.A.; Li, Y. Different effects of genistein and resveratrol on oxidative DNA damage*in vitro*. *Mutat. Res. Genet. Toxicol. Environ. Mutagen* 2002, *513*, 113–120.

51. Becke, A.D. Density-functional thermochemistry. I. The effect of the exchange-only gradient correction. *J. Chem. Phys*1992, *96*, 2115–2160.

52. Becke, A.D. Density-functional thermochemistry. II. The effect of the Perdew-Wang generalized-gradient correlation correction. *J. Chem. Phys* 1992, *97*, 9173–9177.

53. Becke, A.D. Density-functional thermochemistry. III. The role of exact exchange. *J. Chem. Phys* 1993, *98*, 5648–5652.

54. Yang, W.; Parr, R.G.; Lee, C. Various functionals for the kinetic energy density of an atom or molecule. *Phys. Rev. A* 1986, *34*, 4586–4590.

55. Tomasi, J.; Persico, M. Molecular interactions in solution: An overview of methods based on continuous distributions of the solvent. *Chem. Rev* 1994, *94*, 2027–2094.

56. Florián, J.; Warshel, A. Calculations of hydration entropies of hydrophobic, polar, and ionic solutes in the framework of the langevin dipoles solvation model. *J. Phys. Chem. B* 1999, *103*, 10282–10288.

57. Florián, J.; Warshel, A. Langevin dipoles model for *ab initio* calculations of chemical processes in solution: Parametrization and application to hydration free energies of neutral and ionic solutes and conformational analysis in aqueous solution. *J. Phys. Chem. B* 1997, *101*, 5583–5595.

58. Warhsell, A. *Computer Modeling of Chemical Reactions in Enzymes and Solutions*; Wiley-Interscience: New York, NY, USA, 1997.

59. Wattenberg, L.W. Chemoprevention of cancer. *Cancer Res* 1985, *45*, 1–8.

60. Murray, G.I.; Taylor, V.E.; McKay, J.A.; Weaver, R.J.; Ewen, S.W.B.; Melvin, W.T.; Burke, M.D. The immunohistochemical localization of drug-metabolizing enzymes in prostate cancer. *J. Pathol* 1995, *177*, 147–152.

61. Murray, G.I.; Taylor, V.E.; McFadyen, M.C.; McKay, J.A.; Greenlee, W.F.; Burke, M.D.; Melvin, W.T. Tumor-speci.c expression of cytochrome P450 CYP1B1. *Cancer Res* 1997, *57*, 3026–3031.

62. Shimada, T.; Fujii-Kuriyama, Y. Metabolic activation of polycyclic aromatic hydrocarbons to carcinogens by cytochromes P450 1A1 and 1B1. *Cancer Sci* 2004, *95*, 1–6.

63. Dunning, A.M.; Healey, C.S.; Pharoah, P.D.P.; Teare, M.D.; Ponder, B.A.J.; Easton, D.F. A systematic review of genetic polymorphisms and breast cancer risk. *Cancer Epidemiol. Biomark. Prev* 1999, *8*, 843–854.

64. Kržan, M.; Mavri, J. Carcinogenicity of styrene oxide: Calculation of chemical reactivity. *Croat. Chem. Acta* 2009, *82*, 317–322.

65. Matsui, A.; Ikeda, T.; Enomoto, K.; Nakashima, H.; Omae, K.; Watanabe, M.; Hibi, T.; Kitajima, M. Progression of human breast cancers to the metastatic state is linked to genotypes of catechol-*O*-methyltransferase. *Cancer Lett* 2000, *150*, 23–31.

66. Yager, J.D.; Liehr, J.G. Molecular mechanisms of estrogen carcinogenesis. *Annu. Rev. Pharmacol. Toxicol* 1996, *36*, 203–232.

67. Wang, M.Y.; Liehr, J.G. Identification of fatty acid hydroperoxide cofactors in the cytochrome P450-mediated oxidation of estrogens to quinone metabolites. Role and balance of lipid peroxides during estrogen-induced carcinogenesis. *J. Biol. Chem* 1994, *269*, 284–291.

68. Frisch, M.J.; Trucks, G.W.; Schlegel, H.B.; Scuseria, G.E.; Robb, M.A.; Cheeseman, J.R.; Montgomery, J.A.; Vreven, T.; Kudin, K.N.; Burant, J.C.; *et al.* Gaussian 03*, Revision C.02*; Gaussian, Inc: Wallingford, CT, USA, 2008.

Chapter 6

MEMORY SWITCHES IN CHEMICAL REACTION SPACE

Naren Ramakrishnan[1, 2], Upinder S. Bhalla[3]

[1] Department of Computer Science, Virginia Tech, Blacksburg, Virginia, United States of America

[2] Institute of Bioinformatics and Applied Biotechnology, Bangalore, India

[3] National Centre for Biological Sciences, Tata Institute of Fundamental Research, Bangalore, India

ABSTRACT

Just as complex electronic circuits are built from simple Boolean gates, diverse biological functions, including signal transduction, differentiation, and stress response, frequently use biochemical switches as a functional module. A relatively small number of such switches have been described in the literature, and these exhibit considerable diversity in chemical topology. We asked if biochemical switches are indeed rare and if there are common chemical motifs and family relationships among such switches. We performed a systematic exploration of chemical reaction space by generating all possible stoichiometrically valid chemical configurations up to 3 molecules and 6 reactions and up to 4 molecules and 3 reactions. We used Monte Carlo sampling of parameter space for each such configuration to generate specific models and checked each model for switching properties. We found nearly 4,500 reaction topologies, or about 10% of our tested configurations, that demonstrate switching behavior. Commonly accepted topological features such as feedback were poor predictors of bistability, and we identified new reaction motifs that were likely to be found in switches. Furthermore, the discovered switches were related in that most of the larger configurations were derived from smaller ones by addition of one or more reactions. To explore even larger configurations, we developed two tools: the "bistabilizer," which converts almost-bistable systems into bistable ones, and frequent motif mining, which helps rank untested configurations. Both of these tools increased the coverage of our library of bistable systems. Thus, our systematic exploration of chemical reaction space has produced a valuable resource for investigating the key signaling motif of bistability.

INTRODUCTION

Most chemical reaction systems have a single steady state, but a few interesting cases are known to oscillate [1], form spatial patterns [2], or have multiple stable states [3],[4]. Aside from their intrinsic mathematical and chemical significance, systems with multiple stable states are of particular biological interest because they can retain a "memory" of past inputs and cellular decisions [3],[4]. Bistability is a particularly interesting case of multi-stability, as it leads to switch-like behavior. Chemical stimuli can trigger a state change from one stable state to another. The current state of the chemical system is therefore a "memory" of this earlier stimulus.

A few biochemical switches have been extensively analyzed, including complex enzyme mechanisms [5],[6], kinase feedback [7],[8], dual phosphorylation [9], the cell cycle [10], triggering of caspases [11], and synaptic memory switches [12]–[14]. Two observations emerge from this set of known switches. First, relatively few switches are known. A recent computational exploration yielded only about 2% bistable models among those tested [15]. Furthermore, no entries are annotated as bistable in either KEGG (331 pathways) or BIOCARTA (355 pathways). Somewhat at odds with this absence of bistable pathways in pathway databases, kinetic models of bistable pathways are more common. There are several bistable models in the signaling model databases DOQCS (10/69; [16]) and BioModels.net (12/147; [17]), coming to about 10% of recorded models. This may be an overrepresentation, due to modeling interest in bistability. In particular, there are several signaling models that explore bistability as a basis for synaptic memory [12]–[14].

A second observation about the known bistable switches is that they are quite different in their chemical topologies. While feedback loops are a recurring motif [3],[18], there are some cases where enzyme saturation appears to play a role [13], and others where the balance between competing reactions itself generates bistability [9].

While signaling models tend to result in rather complex reaction systems, a distinct approach to the study of chemical bistability is driven from theoretical analyses of enzyme kinetics and flux reaction systems [3],[5]. These studies show that very few reactions are needed to achieve bistability. This raises the interesting question of whether there are core sets of reactions, or motifs, that are embedded in all bistable chemical reaction systems, despite their diversity. A corollary is whether such a set of motifs may help to detect bistable sub-systems in complex biological signaling networks.

Necessary conditions for bistability, such as positive loops in the system Jacobian, have been well characterized [18]. Earlier work by Clarke [19] parametrically defines all steady states of a given reaction system, but does not yield specific solutions when concentrations and rate constants are given. Recent studies detect chemical switches by testing for correlates of bistability [3],[5] or by looking for properties that frequently co-occur with bistability and, optionally, engineering bistability by minor modifications to such networks [15],[20],[21]. We sought to identify bistable systems without placing any "top-down" requirement on the mechanistic details, and use this unbiased search to reconstruct relationships between the switches.

Here we systematically explore chemical reaction space to show that bistable chemical switches are remarkably common. We show that all small bistable systems are related, and that larger ones frequently share motifs that may be predictive of bistability.

RESULTS

Bistables Are Common

In our first phase of analysis, we began with a basis set of 12 reactions (Figure 1A) and systematically tested all reaction configurations involving 2 molecules, 3 molecules from 1 to 6 reactions, and 4 molecules from 1 to 3 reactions. In our second phase of analysis, we sampled a subset of possible reaction configurations involving 3 molecules from 7 to 15 reactions, 4 molecules from 4 to 5 reactions, and 5 molecules from 1 to 4 reactions. The number of possible configurations rose rapidly with the number of molecules and reactions (Figure 1D), and it took longer to test each configuration for bistability, hence we sampled a small subset of configurations for the second phase of analysis. For each configuration we generated ~100 models using Monte Carlo assignment of concentration and rate parameters (see Methodssection) and tested each for bistability. The propensity of a configuration for bistability was defined as the fraction of tested models for that configuration that exhibited two or more stable steady states.

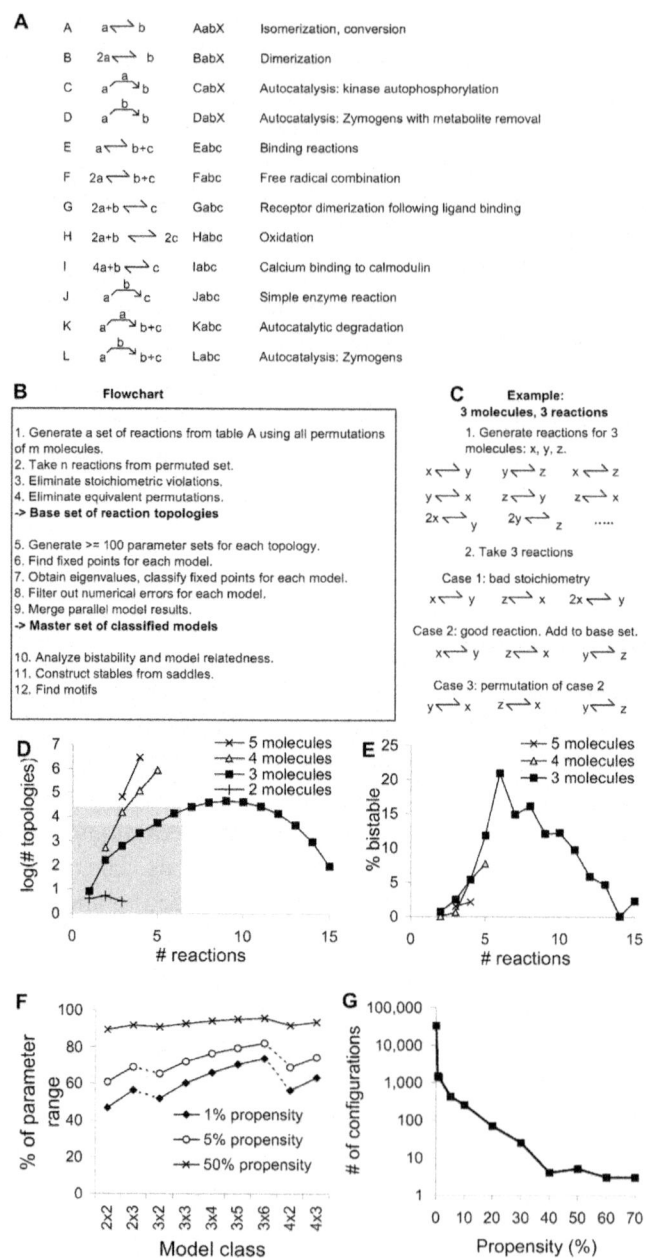

Figure 1: Finding Bistability.

(A) Basis set of reactions with signatures and examples. Reactions C, D, J, K, and L are enzymatic and the enzyme name is at the bend of the arrow. (B) Flowchart for finding bistability. (C) Example of steps 1–4 in the flowchart.

(D) Number of possible configurations rises steeply with increasing number of molecules and initially also with the number of reactions. Shaded region indicates configurations fully sampled in this study, the remainder were subsampled. (E) Bistable configurations initially become more common with increasing numbers of reactions, and for 3 molecules the percentage declines for more than 6 reactions. (F) Bistability must persist over a wide parameter range to be detected. Fraction of parameter range$=^{\#parameters}\sqrt{(Propensity)}$. Model class is expressed as m×n where m is the number of molecules and n is the number of reactions. (G) Frequency of occurrence of bistability as a function of propensity. Some configurations exhibit bistability over 30% of any parameter set in our selected range.

doi:10.1371/journal.pcbi.1000122.g001

We observed a large number of bistable systems even with our very sparse sampling of reaction parameter space (Figure 1E). 3,562 of the fully sampled configurations (~10%) had at least one bistable model, and 918 of the larger reaction systems (~5%) did. This large percentage was surprising for two reasons. First, known bistable configurations from biology are rare, as discussed above. Second, our sampling of parameter space was very sparse, so we would be likely to detect bistable configurations only if they remained bistable in a substantial portion of parameter space. Most known chemical bistable switches exhibit bistability in a relatively narrow range of parameters rarely exceeding a factor of two [9],[22]. While a factor of two may be substantial from a biological viewpoint, we required a 30-fold range to detect bistability. This was because even small models have a large number of parameters. For instance, a 3 molecule, 3 reaction system has 7 parameters. In order to obtain bistability in 1% of tested models of this configuration, bistability would have to be present over approximately half the sampling range (Figure 1F) for each parameter: $(0.5)^7 \sim 0.01$. Our logarithmic sampling spanned 3 orders of magnitude, so half this range is about 30-fold for each parameter. A few configurations had a propensity of over 50% (Figure 1G). This suggests that bistability in these systems is very robust.

Admittedly, due to our sparse sampling of parameter space, there could be undetected bistables in the space of systems sampled here. While a single configuration is sufficient to prove that a network has the capability for exhibiting bistability, our analysis methods do not support an impossibility proof for bistability. The range in which a system exhibits bistability can depend intricately on how the phase space is structured in terms of the system parameters such as molecule concentrations and rate constants. Bifurcation analysis can shed insight into parameter ranges feasible for realizing bistability. Nevertheless, even with the possibility of false negatives, it is significant that nearly 10% of explored systems are bistable and this percentage can only improve with greater analysis and exploration.

Bistables Are Diverse

The simplest bistable system (3×2M101) involved 3 molecules and 2 reactions (Figure 2A). We tested its switch-like behavior by introducing perturbations from its stable states (Figure 2C). Small perturbations in 3×2M101 (small arrows in Figure 2C) caused transients which return to the originating stable state whereas large perturbations (large arrows) caused state flips. An intuitively appealing simpler system with only 2 molecules (Figure 2B) turned out not to be bistable with our mass-action formulation for enzymes.

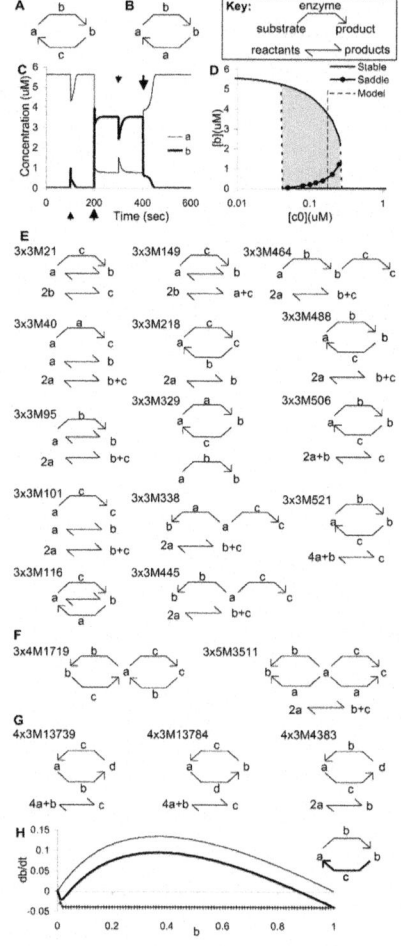

Figure 2: Example Bistable Models.

(A) The simplest configuration, 3×2M101, i.e., having 3 molecules, 2 reactions, and being the 101st configuration in this class. (B) A simpler model that is bistable with Michaelis–Menten/Briggs–Haldane kinetics, but not with a mass action explicit representation of the enzyme-substrate complex. (C) Time-course of response of 3×2M101 to perturbations. (D) Stability diagram of 3×2M101, as a function of c0 (initial concentration of c). The bistable region is shaded. The specific model in panel A has c0 indicated by the dashed line. (E) All the 3×3 bistable configurations. (F) Two configurations with bistability propensity >70%. Model 3×3M445 in (E) is also over 70%. Note the model similarities and symmetry. (G) Three non-autocatalytic configurations with propensity 26%, 23%, and 16%, respectively. (H) Schematic of stability curve for an autocatalytic reaction (inset, thin arrow). There is a stable point at 0, and a saddle at 1. Addition of a rapidly saturating fast back-reaction (inset, thick arrow, and graph, crosses) converts this to a bistable model with the same configuration as (A). Now the stability curve (thick line) has a stable at 0, a saddle at ~0.05, and another stable at ~0.9.

doi:10.1371/journal.pcbi.1000122.g002

Positive feedback loops, such as autocatalysis and catalytic loops, have been implicated as a common motif leading to bistability in signaling [3],[18],[23]. In our study, autocatalysis (reactions D and L) was frequently present in bistable models, but it was not necessary. When we excluded autocatalysis from small reaction systems with only 3 molecules, there were fewer possible reaction configurations and a ten-fold reduction in percentage of bistable configurations. However, autocatalysis had little effect on the percentage of bistable configurations for 4 and 5 molecules (Figure 3).

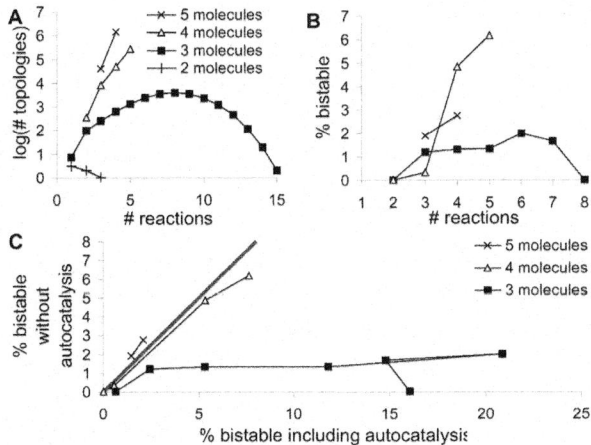

Figure 3: Effects of Autocatalysis.

(A) Number of nonautocatalytic configurations as a function of number of reactions. (B) Percentage of bistable configurations for 3, 4, and 5 molecules as a function of number of reactions. Same symbols as in (A). (C) Comparison of bistability percentage for entire dataset (x axis) and set without autocatalysis (y axis). The thick line indicates equal percentages. 3-molecule systems have fewer bistables without autocatalysis, but 4 and 5 molecule systems have nearly equal bistables if autocatalytic reactions are excluded. The different points on each plot are for numbers of reactions. As bistability declines for large reaction numbers for 3 molecules, this curve folds back on itself.

doi:10.1371/journal.pcbi.1000122.g003

In addition to autocatalysis, we found several cases where bistability arose from more subtle chemical interactions (e.g., Figure 2G and 2F and 3×3M40 in Figure 2E). Such reaction sets would have been difficult to identify as bistable by searching for similarities to published networks [24],[25]. Interestingly, all our switches had exactly two stable states; the lack of higher levels of multi-stability may simply be due to our sparse sampling of parameter combinations.

Uniqueness of Bistables

Are all discovered bistables distinct? Because isomorphisms were removed at the time of generating possible reaction signatures, we ensured that each discovered bistable mapped to a unique signature composed of the 12 basic reaction types. A remaining concern was that there might be equivalences in terms of the underlying dynamical system when the chemical systems were converted to mathematical models. We investigated this possibility by reducing all the composite reactions to approximate equivalences in the form of either a single reactant-single product reaction (type A) or a double reactant-single product reaction (type E). We emphasize that these are "approximate" equivalences, for the following three reasons. First, many higher-order reactions required the inclusion of intermediate molecular species which were not present in the mathematical formulation of the original basis reactions. Second, the expanded reactions treated enzyme-substrate complexes as distinct molecular species having their own trajectories beginning from non-zero concentrations, whereas E-S complexes in the original reactions were initialized to zero in our modeling (see Methods). Third, backward rates from the E-S complexes to the reactants were assumed to be zero in our original modeling (e.g., for reactions C, D, J) whereas in the expanded modeling all reactions (forward and backward) have non-zero reaction rates. With these caveats in mind, we found situations where configurations were isomorphic according to our approximate mappings and both were bistable, and also cases where the configurations were approximately equivalent but one was bistable, and the others were not. These examples reveal that composite reactions such as are commonly used in biochemistry and in our study, complicate stability analysis in two ways. First, they may hide mechanistic similarities between systems. This can be addressed by expanding composite reactions into more basic steps, as we have done. Second, they may hide key assumptions such as intermediate species and fundamental reaction steps, which may cause major differences in the dynamical behavior of the reaction system. While this issue is important from a rigorous mathematical viewpoint, we point out that such approximations are inevitable when translating cellular biochemistry into idealized mathematical forms. We suggest that in many cases bistability is indeed preserved across approximations. Our study provides a framework for further systematic analysis of this question.

Bistables Are Related

Does bistability "run" in families of related reaction topologies? To test this hypothesis, we constructed a directed acyclic graph (DAG) of configurations where each bistable configuration was a node, and each addition/removal of a reaction between nodes was an edge. We found that almost all bistable configurations from the first phase (3,415/3,562=95.9%) formed a single, highly interconnected set, i.e., a giant component. Most of the 147 "orphans" occurred at the boundaries of our sampling (98 at 3×6 and 47 at 4×3). These may simply represent novel 'roots' that connect further up in the reaction hierarchy. In Figure 4A, the DAG is represented as a multiply rooted "banyan-tree" like diagram where there is one main root (3×2M101) and multiple higher-order roots linked to the primary root through more complex, bistable "branches". We may have missed low-propensity bistable configurations, so it is possible that isolated islands of lower propensity may be present. Conversely, it is also possible that finer sampling may uncover intermediate bistable systems that link orphan configurations into the DAG. We constructed a radial diagram restricted to those configurations that were derived from the simplest bistable configuration (3×2M101, Figure 4B). In Figure 4A and 4B, there was an apparent clustering of high-propensity nodes. We investigated this further by comparing projections of Figure 4A onto high-propensity nodes. A giant component persisted even when we increased the threshold for bistability propensity from >0 to ≥0.3. This showed that highly bistable systems form a connected subgraph in the graph of all bistable systems. The much smaller non-autocatalytic subset of bistables was also multiply rooted with a giant component and a few separated nodes.

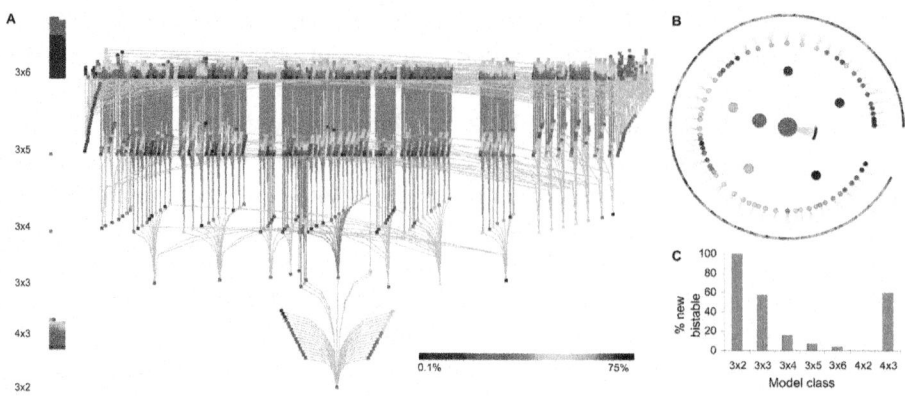

Figure 4: Bistability Runs in Families.

(A,B) Representations of relationships between bistable configurations. Node color represents bistability propensity. Color scales for (A) and (B) are the same. (A) "Banyan tree" diagram showing multiple "root" bistable configurations that cannot be generated by addition of a single reaction to a smaller bistable configuration but are connected through larger configurations. Model classes are labeled on the left. Nodes are staggered vertically within bands for visualization. "Root" edges are in sky blue and deeper edges are in green. On the left are orphan models. (B) Minimum spanning tree rooted at $3 \times 2M101$. Inner nodes with smaller reaction sizes are drawn as larger circles. A few 3×3 bistables and the primarily low propensity systems they derive are not shown to minimize crowding. The "pie" denotes restriction of exploration of 4 molecule systems to only 3 reactions in this study. (C) The number of novel bistables drops sharply in larger reaction systems.

doi:10.1371/journal.pcbi.1000122.g004

These graphs suggested that most bistable systems were derived from smaller ones. As reactions were added (Figure 4C), we encountered a decreasing number of novel bistables (i.e., cases that could not be derived by addition of a reaction to a smaller bistable configuration). This suggests that bistable systems involving small numbers of molecules may form the architectural core of more complex reaction networks that are also bistable.

Relationship to Published Bistables

We tested two implications of the "bistables are related" observation. First, we asked if we could take a large published bistable system and remove one reaction at a time without losing bistability. If we could continue this process till we ended up at a bistable configuration present in our dataset, then we had a continuous trajectory from our known DAG of bistables to the published model. Second, we asked if the large bistable system was a superset of a known bistable configuration, without requiring that there were intermediate bistables between the two. We performed this analysis on several known bistable reaction systems from published work (Figure 5). We found that in four of these cases, the published bistables were either already among our catalog, or had a subset of reactions that was bistable. In the remaining three configurations, there was neither a connection between the published models to the tree, nor was there a subset of reactions that was bistable in the DAG.

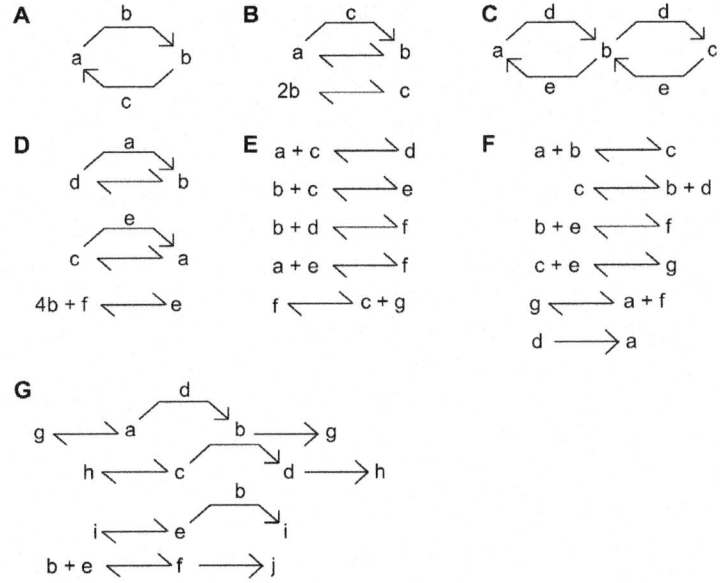

Figure 5: Known Bistables.

(A,B) Published bistables from [20]. These are already present in our DAG. (C,D) Published bistables from [6] and [9]. These can be reduced to smaller models and have some bistable motifs from our dataset. (E,F) Bistables from [5]. (G) Bistable from [11]. None of (E), (F), and (G) had motifs from our dataset.

doi:10.1371/journal.pcbi.1000122.g005

We therefore hypothesized that the DAG of bistables may be nearly complete for small systems, but the increasing degrees of freedom afforded by greater numbers of molecules and reactions helped realize bistability in new, unseen, ways. We developed two analysis tools that work in complementary ways to explore such larger configurations.

The "Bistabilizer"

A suggestive observation from our first phase was that a large fraction of configurations (~60%) contained saddle points and line solutions. Most of these saddles (80.8% of the non-bistable set) occurred when the concentration of all but one molecule in the system was zero. A simple example of this is in Figure 2H, where molecule **b** catalyzes its own formation from **a**. When **b** is at zero, **a** does not change – it is metastable. However, the addition of a small amount of **b** causes the system to 'fall' into a truly stable state where all molecules have been converted into **b**. As has been previously analyzed [20] a rapid but saturating back-reaction is one way to convert this into a true stable system. This can be done using an enzyme to remove **b** more rapidly than it builds up, at low levels of **b** (Figure 2H). In this case, we reconstruct our original simplest bistable system (compare Figure 2A and 2H). We developed an algorithm that introduced several reactions to achieve bistability using this approach, in a general but not necessarily minimal manner. Due to the added reaction complexity, we generated bistables from non-bistable systems involving only 3 molecules and up to 5 reactions. The bistabilizer added at least 2 molecules, 3 enzymes and a reaction for each converted saddle point. We were able to generate a 5-fold higher proportion of bistables than were present in the source configurations in a sample of 70,000 models. This construction may usefully complement bifurcation discovery tools [26] to generate and refine bistable configurations.

Frequent Motif Mining

Our second tool used motif matching. We analyzed the configurations of smaller bistables plus the sparsely sampled larger bistables to find frequently occurring groups of reactions, and then searched for these motifs in unexplored configurations. We analyzed bistables in each configuration class (3 molecules, 4 molecules, 5 molecules) separately for frequent motifs. A motif must occur with sufficient frequency in the given class to be detected. Because motifs are subsets of chemical reaction systems, they may not quite have the same number of molecules as the class of systems from which they are mined; for instance, a motif mined from 5-molecule bistable systems may not itself have 5 molecules. Furthermore, observe that while a motif is a subset of reactions that is well-represented in bistable systems, it may not be bistable. We found the greatest number of motifs (1615) from 3-molecule systems, and smaller numbers for 4 and 5 molecules (143 and 28, respectively), probably because our initial harvesting of bistables in the larger systems had yielded fewer confirmed bistables to scan for motifs.

The motifs were mostly independent and only one motif occurred in all three reaction classes. Coincidentally, this common, two-reaction, motif (composed of reactions DabX and Jbca) was identical to a bistable found both in our analysis and in previous work ([20]; Figure 2A and 5A).

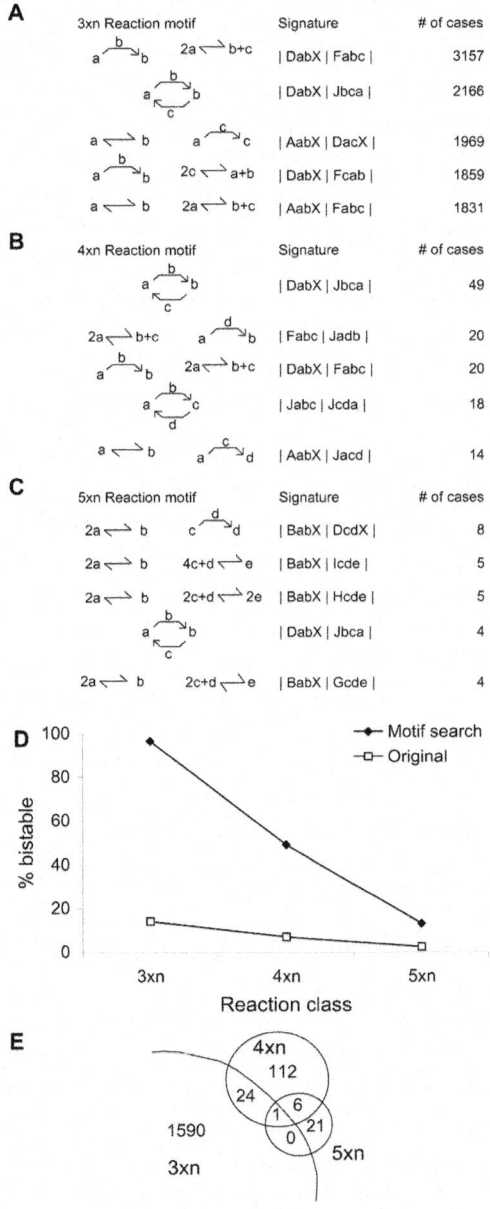

Figure 6: Motifs in Bistable Switches.

(A–C) Top 5 motifs mined for 3-molecule, 4-molecule, and 5-molecule bistable systems. (D) Motifs improve search for bistables by about 7-fold, for each of the reaction classes tested. (E) Venn diagram of motifs among different classes of reaction configurations. The one motif shared by all classes of configurations is the same as the smallest bistable, model M101.

doi:10.1371/journal.pcbi.1000122.g006

This motif/bistable also occurred in the top-5 motifs mined for each class when the motifs were ranked by their frequencies (Figure 6A–C). Interestingly, four of the top-5 motifs in the 3 molecule case contained at least one of reaction D or reaction J and all of the top-5 motifs in the 4 molecule case contained at least one of these two reactions. In the case of 5 molecules, only two of the top-5 motifs involved either D or J, though our much smaller sample set may have led to skewed results in this case. The remaining top-5 motif among the 3 molecule systems and the three other top-5 motifs in the 5-molecule systems utilized basic reversible reaction types such as A, B, F, G, H, and I that did not involve autocatalysis or even basic enzyme catalysis.

Just as a motif occurred in multiple bistable systems, a given bistable system could exhibit many distinct motifs. We used this property to advantage to help rank untested configurations for their potential to exhibit bistability. In each configuration class (3-molecule, 4-molecule, or 5-molecule systems), we searched for motifs specific to that class, and ranked the (untested) configurations in terms of the number of motifs they exhibited. We evaluated ≥ 100 of the top configurations for each class, exploring 120 parameter sets for each. We found bistability in 96% of the 3 molecule systems (214/222), 49% of the 4 molecule systems (49/100), and 13% of the 5 molecule systems (82/641). These numbers significantly improved upon the random sampling results of the second phase of analysis (Figure 6D).

Finally, we compared the motifs across the three classes (3, 4, and 5 molecule systems) to investigate whether there were any overlaps in mechanisms by which 3-molecule, 4-molecule, and 5-molecule systems exhibit bistability. On face value, there was little overlap between the motif sets taken pairwise (Figure 6E). To understand the distinctions better, we searched for all three sets of motifs in all three classes of bistable systems. For each motif, we counted its frequency in each class and identified the absolute value of the difference in frequencies across classes. The median difference in frequency is cataloged in Table 1 which shows that the motifs in 4-molecule systems were more similar to those in 5-molecule systems than either to 3-molecule systems. This suggests that there are qualitatively different mechanisms for bistability vis-à-vis 3 molecule systems and higher molecular systems.

Table 1: Median Difference in Frequency of Motif Sets Searched across Different Classes of Bistable Systems.

doi:10.1371/journal.pcbi.1000122.t001

	3xn	4xn	5xn
3xn	–	15%	29%
4xn	21.5%	–	1.6%
5xn	45%	3.5%	–

Discussion

Our study draws the first stability map of chemical reaction space. We find that bistables are common, especially in smaller reaction systems. They are also very robust, i.e., we find many configurations that are bistable over a very wide parameter range. Smaller bistables are all related to each other in a tree-like manner. While the overall configurations that support bistability are very diverse, there are frequently recurring motifs of reaction groupings in such configurations. These motifs serve to identify promising candidates in higher order systems.

A Resource for Studying Bistability

Signaling motifs have been regarded as a good way to abstract out the chemical complexity of signaling [4],[24]. Specifically, positive feedback loops have long been considered good indicators for bistability. Our study shows that such broad network features are inadequate. The simplest form of positive feedback, that is, autocatalysis, is a good predictor for bistability only in very small reaction sets. In reactions with 4 or 5 molecules the proportion of bistables does not seem to depend on the presence or absence of autocatalysis (Figure 3). Instead we propose that our library of bistable configurations is a more complete and stronger approach. Our catalog, available from the DOCSS (Database of Chemical Stability Space) website athttp://docss.ncbs.res.in, provides complete model descriptions in chemical reaction signature format, as well as selected bistables in SBML format. Together with recent methods to reduce chemical networks to their core reactions [27],[28], our catalog may open up chemical and bioinformatics approaches to searching for bistability in biochemical signaling pathways.

Evolutionary Implications

Bistable switches are important in biology in maintaining cellular history and decisions. Our study shows that there is a large repertoire of such switches for natural selection to draw upon, including many very simple switches. Furthermore, several of these switches are highly robust with respect to parameter variations. This has two implications for evolution. First, it is easy for evolving biochemical networks to stumble upon parameters that will give a switch. Second, such switches themselves will work effectively over a wide range of parameter conditions. The relatedness of the switches through addition or removal of individual reactions is also a good substrate for evolutionary modification. For example, a mutation that adds another enzyme regulator to a bistable switch is, by this argument, quite likely to retain the original bistability, along with the new regulatory properties. Overall, our survey of chemical topologies hints at an interconnected and rather well-populated terrain of bistability in a biologically biased region of chemical space.

METHODS

Modeling Chemical Reaction Space

We selected a set of 12 primary chemical reaction steps (Figure 1A) as the "alphabet" from which we performed our exhaustive search of chemical reaction space. In principle a small set of reactions may suffice to build up to arbitrary reaction schemes. For instance, using two reactions of type $EA + B \leftrightarrow C$ we can realize the higher order enzymatic reaction $A - B \rightarrow C$ which is reaction J in our system, by modeling it as a composite of $A + B \leftrightarrow A.B, \ A.B \leftrightarrow B + C$

Our choice of primary chemical steps was biologically-inspired. In other words, we found a different proportion of bistables than we would see if we used, say, only the most elementary reactions such as type A and type E. Instead we reduced the parameter space by using biologically-inspired composite reactions, and hence sampled more completely in biological chemistry space. From this set of 12 primary chemical reaction steps, we constructed all possible reaction configurations involving 2 or 3 molecules, 4 molecules up to 5 reactions, and 5 molecules up to 4 reactions. This was a total of ~2,800,000 configurations. These reaction configurations were topological: they defined the molecules and chemical steps, but did not specify concentrations or kinetic parameters.

We constructed reaction architectures involving m molecules as follows. We first selected one reaction out of the set of 12 reactions. We then assigned molecules to the slots of the reaction. For example, the reaction J has three slots, so we could assign molecules a, b, and c to this reaction. Having set up the first reaction, we then repeated the process $n - 1$ times to obtain a configuration of m molecules and n reactions (Figure 1). We eliminated all stoichiometrically invalid configurations by row-reducing the augmented stoichiometric matrix and checking for conserved moieties [29]. We repeated this entire process for all possible permutations of the m molecules. Similar approaches have been employed at a more elementary level of chemical reactive species to computationally analyze reaction systems[30],[31].

Signature Typing

We signed each reaction with a terse unique 4-character string that completely specified all reactants and products, so that the first character of a reaction signature denotes one of the 12 reaction types (A–L), and the remaining two or three characters denote the molecular species participating in various roles in the reaction. The signature for a reaction architecture was obtained by concatenating the signatures for the constitutent reactions. We checked for isomorphic signatures and only one signature per unique system was retained. The number of such unique, stoichiometrically valid reaction architectures was combinatorially large (Figure 1D). As further reactions were added, the number of possible configurations peaked and then declined because of stoichiometric constraints and symmetry (Figure 1D).

Our set of configurations did not deal with two cases that have previously been analyzed for bistability: continuous flux and buffered systems [21]. Instead our reactions required that there was mass conservation among the named molecules, but did permit the presence of 'hidden' molecules that were folded into the rate terms. Thus we represented a kinase reaction as an elementary enzymatic step, by "hiding" the ATP and ADP exchange: Substrate–Kinase → Product.

In this manner our reaction systems also accommodated steady state cases where continuous metabolic input was necessary to sustain stability. We stipulated that these 'hidden' molecules were stoichiometrically balanced within individual reactions, such as the enzymatic step above. We were able to approximate many cases of buffering simply by having a high concentration of the 'buffered' species.

Parameterization

In order to assess bistability we needed to work with specific models, with all parameters specified. We generated at least 100 models for each configuration we tested. Each model was generated from one of the configurations using Monte Carlo sampling to assign rate constants and concentrations. We chose concentrations using logarithmic sampling in the range 10 nM to 10 µM. This spans the concentration range of most biochemical reagents. We chose rate constants using logarithmic sampling in the range 0.01 $\mu M^{-N} s^{-1}$ to 10 $\mu M^{-N} s^{-1}$, where N was the order of the reaction from 0 to 4. Again, these rates were chosen to span the common range of biochemical reaction parameters.

Due to computational limitations we sampled only the smaller reaction sets completely for bistability. We completely sampled all configurations with 2 molecules, 3 molecules up to 6 reactions, and 4 molecules up to 3 reactions. This amounted to a total of 100,000 configurations and ~20e6 models, sampling at least 100 models per configuration. We sampled the remaining reaction sets more sparsely, mostly 1 in 100, but we used 1 in 1,000 sampling for the very large 4×5 and 5×4 reaction sets.

Identifying Steady States

We found steady states for each model using two distinct methods: homotopy continuation[32] and time course analysis. Briefly, homotopy continuation finds steady states by tracking solution paths of systems of simultaneous equations. Time-course analysis simulates models from a number of distinct initial conditions toward steady states. Neither of these methods determined the global bifurcation behavior of the reaction system: they only identified fixed points of the fully parameterized model. We classified solutions as *stable*, *saddle*, or*other* using eigenvalue calculation and simulation around steady states.

ACKNOWLEDGMENTS

We thank Sanjay Jain, Dennis Bray, Charles Wampler II, and Layne Watson for many ideas on the project. Vishal Surana and Evan Maxwell helped verify the results obtained from frequent motif mining. We acknowledge the use of the Anantham and System X clusters at Virginia Tech on which our simulations were conducted. We thank Harsha Rani for setting up the DOCSS web page and SBML conversions of models.

AUTHOR CONTRIBUTIONS

Conceived and designed the experiments: NR UB. Performed the experiments: NR UB. Analyzed the data: NR UB. Contributed reagents/materials/analysis tools: NR UB. Wrote the paper: NR UB.

REFERENCES

1. Kuramoto Y (2003) Chemical Oscillations, Waves and Turbulence. Mineola (New York): Dover Publications.

2. Shoji H, Iwasa Y, Kondo S (2003) Stripes, spots, or reversed spots in two-dimensional Turing systems. J Theor Biol 224 (3): 339–350.

3. Angeli D, Ferrell JE, Sontag ED (2004) Detection of multistability, bifurcations, and hysteresis in a large class of biological positive-feedback systems. Proc Natl Acad Sci USA 101 (7): 1822–1827.

4. Tyson JJ, Chen KC, Novak B (2003) Sniffers, buzzers, toggles and blinkers: dynamics of regulatory and signaling pathways in the cell. Curr Opin Cell Biol 15 (2): 221–231.

5. Craciun G, Tang Y, Feinberg M (2006) Understanding bistability in complex enzyme-driven reaction networks. Proc Natl Acad Sci USA 103 (23): 8697–8702.

6. Brandman O, Ferrell JE Jr, Li R, Meyer T (2005) Positive feedback loops drive reliable cell decisions. Science 310 (5747): 496–498.

7. Ferrell JE, Machleder EM (1998) The biochemical basis of an all-or-none cell fate switch in Xenopus oocytes. Science 280 (5365): 895–898.

8. Bhalla US, Ram PT, Iyengar R (2002) MAP kinase phosphotase as a locus of flexibility in a mitogen-activated protein kinase signaling network. Science 297 (5583): 1018–1023.

9. Markevich NI, Hoek JB, Kholodenko BN (2004) Signaling switches and bistability arising from multisite phosphorylation in protein kinase cascades. J Cell Biol 164 (3): 353–359.

10. Novak B, Tyson JJ (2004) A model for restriction point control of the mammalian cell cycle. J Theor Biol 230 (4): 563–579.

11. Eissing T, Conzelmann H, Gilles ED, Allgower F, Bullinger E, et al. (2005) Bistability analyses of a caspase activation model for receptor-induced apoptosis. J Biol Chem 279 (35): 36892–36897.

12. Miller P, Zhabotinsky AM, Lisman JE, Wang X-J (2005) The stability of a stochastic CaMKII switch: dependence on the number of enzyme molecules and protein turnover. PLoS Biol 3 (4): e107. doi:10.1371/journal.pbio.0030107.

13. Hayer A, Bhalla US (2005) Molecular switches at the synapse emerge from receptor and kinase traffic. PLoS Comput Biol 1 (2): 137–154.

14. Kuroda S, Schweighofer N, Kawato M (2001) Exploration of signal transduction pathways in cerebellar long-term depression by kinetic

simulation. J Neurosci 21 (15): 5693–5702.

15. Paladugu SR, Chickarmane V, Deckard A, Frumkin JP, McCormack M, et al. (2006) In silico evolution of functional modules in biochemical networks. IEE Proc Syst Biol 153 (4): 223–235.

16. Sivakumaran S, Hariharaputran S, Mishra J, Bhalla US (2003) The database of quantitative cellular signaling: management and analysis of chemical kinetic models of signaling networks. Bioinformatics 19 (3): 408–415.

17. Le Novère N, Bornstein B, Broicher A, Courtot M, Donizelli M, et al. (2006) BioModels database: a free, centralized database of curated, published, quantitative kinetic models of biochemical and cellular systems. Nucleic Acids Res 34: D689–D691.

18. Thomas R (1994) The role of feedback circuits: Positive feedback circuits are a necessary condition for positive real eigenvalues of the Jacobian matrix. Ber Bunsenges Phys Chem 98 (9): 1148–1151.

19. Clarke BL (1981) Complete set of steady states for the general stoichiometric dynamical system. J Chem Phys 75 (10): 4970–4979.

20. Ferrell JE, Xiong W (2001) Bistability in cell signaling: how to make continuous processes discontinuous, and reversible processes irreversible. Chaos 11 (1): 227–236.

21. Guidi GM, Goldbeter A (1997) Bistability without hysteresis in chemical reaction systems: a theoretical analysis of irreversible transitions between multiple steady states. J Phys Chem A 101 (49): 9367–9376.

22. Bhalla US, Iyengar R (2001) Robustness of the bistable behavior of a biological signaling feedback loop. Chaos 11 (1): 221–226.

23. Kholodenko BN (2006) Cell-signaling dynamics in time and space. Nat Rev Mol Cell Biol 7 (3): 165–176.

24. Milo R, Shen-Orr S, Itzkovitz S, Kashtan N, Chklovskii D, et al. (2002) Network motifs: simple building blocks of complex networks. Science 298 (5594): 824–827.

25. Prill RJ, Iglesias PA, Levchenko A (2005) Dynamic properties of network motifs contribute to biological network organization. PLoS Biol 3 (11): e343. doi:10.1371/journal.pbio.0030343.

26. Chickarmane V, Paladugu SR, Bergmann F, Sauro HM (2005) Bifurcation discovery tool. Bioinformatics 21 (18): 3688–3690.

27. Liu G, Swihart MT, Neelamegham S (2005) Sensitivity, principal component and flux analysis applied to signal transduction: the case of epidermal growth factor mediated signaling. Bioinformatics 21 (7):

1194–1202.

28. Maurya MR, Bornheimer SJ, Venkatasubramanian V, Subramaniam S (2005) Reduced-order modelling of biochemical networks: application to the GTPase-cycle signaling module. IEE Proc Syst Biol 152 (4): 229–242.

29. Sauro HM, Ingalls B (2004) Conservation analysis in biochemical networks: computational issues for software writers. Biophys Chem 109 (1): 1–15.

30. Németh A, Vidóczy T, Héberger H, Kúti Z, Wágner J (2002) MECHGEN: computer aided generation and reduction of reaction mechanisms. J Chem Inf Comput Sci 42 (2): 208–214.

31. Fic G, Nowak G (2001) Implementation of similarity model in the CSB system for chemical reaction predictions. Comput Chem 25 (2): 177–186.

32. Sommese AJ, Wampler CW II (2005) The Numerical Solution of Systems of Polynomials Arising in Engineering and Science. Singapore: World Scientific.

Chapter 7

DISTINGUISHABILITY IN ENTROPY CALCULATIONS: CHEMICAL REACTIONS, CONFORMATIONAL AND RESIDUAL ENTROPY

Ernesto Suárez

Departamento de Química Física y Analítica, Universidad de Oviedo, Julián Clavería 8, 33006, Oviedo, Spain; Tel.: +34-985103492; Fax: +34-985103125

ABSTRACT

BY analyzing different examples of practical entropy calculations and using concepts such as conformational and residual entropies, I show herein that experimental calorimetric entropies of single molecules can be theoretically reproduced considering chemically identical atoms either as distinguishable or indistinguishable particles. The broadly used correction in entropy calculations due to the symmetry number and particle indistinguishability is not mandatory, as an *ad hoc* correction, to obtain accurate values of absolute and relative entropies. It is shown that, for *any* chemical reaction of *any* kind, considering distinguishability or indistinguishability among identical atoms is irrelevant as long as we act consistently in the calculation of all the required entropy contributions.

INTRODUCTION

In the statistical treatment of a system of N identical particles, it is customary to divide the partition function by N! in order to avoid the overcounting of states due to particle indistinguishability [1]. Analogously, the single molecule partition function is divided by the external symmetry number σ_{ext} that corresponds to the number of indistinguishable molecular orientations, and by the internal symmetry number σ_{int} that accounts for the number of indistinguishable conformers (see [2] for a convincing discussion on symmetry numbers). In all cases, the reduction in microstates is related to the concept of indistinguishability: since chemically identical atoms are considered as indistinguishable, any permutation among them would lead to the same state.

The solution of the so-called *Gibbs paradox* [3,4,5,6,7,8], is probably the most famous example where the same kind of correction has been applied. Gibbs proposed an *ad hoc* reduction in the entropy of an *N*-particle system by the amount $-kB\ln N!$, where k_B is Boltzmann's constant. This entropy diminution, which corresponds to a reduction in the number of microstates accessible to the system by the permutation symmetry number N!, was able to correct what Gibbs considered as an unphysical situation, that is, the fact that entropy increases after mixing two (identical) ideal gases both being initially at the same temperature and pressure.

The unphysical situation that Gibbs tried to avoid in mixing processes is consistent with the concept of entropy as extensive property as held by Gibbs himself. However, it may be interesting to remember that the thermodynamic definition of entropy proposed by Clausius in 1865 does not reveal anything about how the entropy behaves as the number of particles *N* changes. The Clausius definition only allows us to compute the difference in entropy between two thermodynamics states of a *closed system*. Pauli noticed this incompleteness and showed what additional condition must be imposed in order to define an extensive entropy, suggesting that entropy, as defined by Clausius, is not intrinsically an extensive property [4,9]. In any case, the extensivity and the indistinguishability are concepts closely connected to each other in the context of the Gibbs arguments, which have been supported and rejected more than once in an ongoing debate [3,4,7,10,11].

From the point of view of classical statistics, whenever identical particles are distinguishable (Maxwell–Boltzmann statistics), it turns out that the entropy reduction by a term of $-k_B N \ln \sigma$ is in contrast with the idea of a magnitude that grows with the number of microscopic complexions compatible with the macroscopic state of the system [6], because ultimately, the result of any symmetry operation including any permutation is another microscopic complexion. Nevertheless, it is well known that the experimental 3rd law entropies of small molecules can be reproduced with extreme accuracy if the entropy reduction is employed. It therefore *appears* that the experimental values can be reproduced only by using this correction, or equivalently, by adopting truly quantum statistics from which the entropy reduction emerges naturally due to the symmetry of the wave function [12].

The residual entropy [13,14,15] is another concept that can be linked to the indistinguishability. When we assert that a perfect crystal at 0K has null entropy, we are implicitly assuming from the statistical standpoint that any permutation of two identical particles does not lead to a new microstate. Although the residual entropy is only relevant when is empirically detectable [13,14] and can be related to a potentially measurable latent head [15], the concept will be helpful when we analyse absolute entropies in the context of distinguishable particles. Because in this scenario, the residual entropy would be present even if a reversible path to the solid state at 0K were available.

Herein, through the careful analysis of various practical cases, I support the idea that considering identical atoms as indistinguishable particles is not mandatory in order to compute entropy values that are in agreement with experiment [10,16,17]. I show that classical treatment (distinguishable particles) can reproduce experimental entropy values without the need for any adjustment due to *weaknesses* of the classical model. All that is required is to be consistent with all the implications arising from distinguishability, including the consequences for the residual and conformational entropy (if any) of the involved molecules. I also show with two examples the innocuous effect of the distinguishability on the entropy change in chemical reactions obtaining for all cases the same result as that obtained by considering identical atoms as indistinguishable.

The entropy of mixing and the Gibbs paradox, however, is out of the scope of this work because the problem has recently been solved for distinguishable particles without any *ad hoc* correction [8,17]. In this respect, the present work tries to generalize the idea to any other chemical transformation of any kind, where symmetry changes might take place. The implications of these ideas could be particularly relevant for approximate calculations of absolute entropies in which quantum mechanical and classical statistics are mixed in order to estimate different entropic contributions.

DISCUSSION

Indistinguishable Particles and Third Law Entropies

Nowadays, 3rd law entropies of small molecules in the gas-phase can be computed easily and with remarkable precision by feeding thermodynamic statistical formulae with molecular properties computed with quantum chemical methods [18]. For example, Table 1 shows both theoretical and experimental entropy values reported in the literature for small alkanes [19]. For simplicity, the examples in Table 1 are selected so that the ω possible conformers for each molecule, if any, are not only isoenergetic, but also chemically identical. Thus, the conformational entropy would be zero, depending whether or not we are considering indistinguishability or distinguishability among identical conformers.

Table 1. B3LYP/cc-pVTZ theoretical and experimental entropies in $JK^{-1}mol^{-1}$ [19]

Molecule	σ_{ext}	Theory	Experiment	Abs. Error
methane	12	186.20	186.37	0.17
ethane	6	228.50	229.16	0.66
propane	2	270.20	270.31	0.11
methylpropane	3	295.50	295.70	0.20
dimethylpropane	12	306.74	306.00	0.74
2,2-dimethylbutane	1	358.70	358.40	0.30

Since molar entropies are being dealing with, the number of particles is chosen to be the Avogadro Number (Na), expressing the entropy corrections preferably in terms of the gas constant R=kBNa. The theoretical values in Table 1 are Rigid-Rotor Harmonic-Oscillator (RRHO) entropies obtained from standard statistical thermodynamic formulae at the B3LYP/cc-pVTZ level of theory, where B3LYP is a hybrid density functional and cc-pVTZ denotes the correlation consistent basis set used [20]. Standard formulae refers to the fact that in all cases the reported theoretical entropies are reduced due to the symmetry including the permutation symmetry (*i.e.*, reduced by the terms $-R \ln \sigma_{ext}$ and $-k_B \ln N_a!$) [12]. In principle, we should also correct the entropy due to the internal symmetry number σint=ω by adding $-R_{\ln \omega}$. However, it is well known that RRHO entropies do not capture all the intramolecular entropy, as they lack the purely conformational part of the entropy [21,22,23], which is in our case exactly $Rl_{\ln \omega}$ and, therefore, the last correction is automatically done due to the deficiencies of the RRHO method. As can be seen in Table 1, the theoretical results are, without any doubt, in good agreement with the experimental values.

Distinguishable Particles and Third Law Entropies

If the particles are distinguishable, the entropy correction is not justified and there are new entropy terms that should be taken into account. The conformational entropy, for instance, is now not canceled and consequently the corresponding term $R_{\ln \omega}$, as well as the one due to the external symmetry $R\ln \sigma_{ext}$, must be added to each of the theoretical values in Table 1. By doing so, the agreement of the theoretical data with the experimental values *apparently* worsens. However, we realise that standard experimental calorimetric entropies are ultimately an entropy change from T=0K to T=298K. This change is equal to the absolute entropy if the 3rd law holds, *i.e.*,

$$S_{T=298K} = S_{T=0K} + \int_{T=0K}^{T=298K} \frac{\delta Q}{T}$$

(1)

To interpret the experimental results assuming distinguishable particles, it can be noted first that, in the examples, any *formal* conformational change of a single molecule near 0K, as well as any rotational symmetry operation, will lead to a different microscopic complexion compatible with the macroscopic state [13]. Therefore, a residual entropy should be considered for these molecules having a value of $R_{ln}(\omega\sigma_{ext})$. This quantity must be added to the original experimental values and the resulting entropy values, which assume particle distinguishability, maintain the agreement between theory and experiment (see Table 2).

Table 2. Theoretical (B3LYP/cc-pVTZ) and experimental entropy values in $JK^{-1}mol^{-1}$ augmented by $R\ln(\omega\sigma_{ext})$ due to the distinguishability.

Molecule	ω	σ_{ext}	Theory	Experiment	Abs. Error
methane	1	12	206.86	207.03	0.17
ethane	3	6	252.53	253.19	0.66
propane	3^2	2	294.23	294.34	0.11
methylpropane	3^3	3	332.03	332.23	0.20
dimethylpropane	3^4	12	363.93	363.19	0.74
2,2-dimethylbutane	3^5	1	404.37	404.07	0.30

At this point the reader might wonder why, if all particles are taken as distinguishable, the uncertainty due to the permutation symmetry in the solid state at 0K has not been considered. After all, any permutation would give a new *different* microstate. In fact, it could have been done, but it would have changed nothing, because in such a case the term kBlnNa! needs to also be added to the theoretical value because the translational part of the entropy is computed in its corrected form $S_t = R(\ln q_t + \frac{5}{2})$, where qt is the translational partition function [1]. Note that the corrected form is conceptually equivalent to the "reduced" entropy used by Cheng [17]. In general any other intra- or extra-molecular permutation between identical but distinguishable atoms can both be considered in the gas phase and in the solid state nearby 0K, and the agreement between theory and experiment would be unaffected.

Entropy Changes in Chemical Reactions: Is There Any Difference?

The statement that the absolute entropy of a system depends on a subjective decision, to consider or not that identical atoms or particles are indistinguishable, most likely seems awkward. It is no less subjective, however, than setting an arbitrary reference in order to transform a relative magnitude into an absolute one. There are an infinite number of possible functions that would give the correct experimental entropy change, and therefore they all meet the original Clausius thermodynamic definition. The entropy change is the magnitude that must be invariant regardless of any considerations. Through two simple examples, both points of view discussed above (considering identical particles distinguishable or not) will be shown as totally equivalent.

Let us first consider the following equilibrium reactions in the gas phase:

$$CH_3Cl + Cl_2 \rightleftharpoons CH_2Cl_2 + HCl \tag{R-1}$$

If a quantum chemical program is used to optimize the molecular geometries, carry out the corresponding frequency calculations and compute the RRHO entropies without considering any symmetry operation except the identity, the entropy values, say, SRRHOCH3Cl, SRRHOCl2, SRRHOCH2Cl2, and SRRHOHCl would be obtained. For convenience the required entropy corrections are introduced explicitly, then the estimated entropy change in (R-1) is

$$\Delta S = \Delta S_{nosym}^{RRHO} + \Delta(-R \ln \sigma_{ext})$$
$$= \Delta S_{nosym}^{RRHO} - R \ln \frac{\sigma_{ext}(CH_2Cl_2)\sigma_{ext}(HCl)}{\sigma_{ext}(CH_3Cl)\sigma_{ext}(Cl_2)}$$
$$= \Delta S_{nosym}^{RRHO} + R \ln 3 \tag{2}$$

where $\Delta S_{nosym}^{RRHO} = S_{CH_2Cl_2}^{RRHO} + S_{HCl}^{RRHO} - S_{CH_3Cl}^{RRHO} - S_{Cl_2}^{RRHO}$. In principle, ΔS would reproduce the experimental entropy change provided that the level of theory in the calculations is adequate.

Considering distinguishability in the same reaction (R-1), it is now obvious that the CH$_3$Cl molecule can be formed by any of the 3 distinguishable Cl atoms involved, furthermore, the three numerable H atoms can be reordered in two different forms not superimposable by rotations, being the atoms in Cl$_2$ completely determined by our first selection. Hence, the reactants would have an additional uncertainty that contributes to the entropy in Rln(2×3). On the other hand, the CH$_2$Cl$_2$ molecule can be formed by any two of the three Cl and any two of the three H atoms, and once selected, there are two possible arrangements to be chosen between. Note that the atoms are numerable and we could obtain two enantiomeric configurations. The atoms in HCl will be determined once again by the previous selection and finally the entropy estimation under this new formalism is

$$\Delta S = \Delta S_{nosym}^{RRHO} + R \ln \left\{ 2 \binom{3}{2} \binom{3}{2} \right\} - R \ln (2 \times 3)$$

$$= \Delta S_{nosym}^{RRHO} + R \ln 18 - R \ln 6$$

$$= \Delta S_{nosym}^{RRHO} + R \ln 3 \tag{3}$$

which is exactly the same result obtained above.

Let us consider a more complex example where the conformational entropy is also involved. In (R-2), the symmetry number is three for the methylpropane, one in the methylcyclopropane and two for the H, being $\Delta(-R \ln \sigma_{ext}) = R \ln (3/2)$.

$$\tag{R-2}$$

Considering identical atoms as indistinguishable, there is no conformational entropy either in the methylpropane or in the methylcyclopropane molecules. The entropy change involved is $\Delta S = \Delta S_{nosym}^{RRHO} + R \ln (3/2)$.

If, on the contrary, identical atoms are distinguishable, the H atoms can be arranged in multiple different ways and the entropy value is not lower due to symmetry, but higher. Additionally, the conformational entropy must be taken into account since any conformational change in any methyl group would give a new different conformer. The uncertainty due to the arrangements of the carbon atoms (excluding the connectivity) is the same in reactant and products and will not be considered.

In order to build the reactant molecule (methylpropane), 10 H atoms need to be distributed into 4 "boxes" of capacities 3, 3, 3 and 1, where, in the boxes of capacity 3 (methyl groups), there are two possible enantiomeric arrangements. Also, each methyl group will contribute to the conformational entropy with 3 conformers, being the total number of complexions

$$\frac{10!}{3!3!3!1!} \times 2^3 \times 3^3 = 10!$$

For the products (methylcyclopropane and H_2) the carbon atoms which will close the cycle are selected first (there are $\binom{3}{2}$ possibilities), then we have 10 H atoms for 5 boxes of capacities 3, 2, 2, 1 and 2, where we included the H_2 molecule as the last box. Once again the methyl groups as well as the $-CH_2-$ groups have two possible arrangements and each methyl group generates three different conformers. As a consequence, the total number of complexions is

$$\binom{3}{2} \times \frac{10!}{3!2!2!2!1!} \times 2^3 \times 3 = \frac{3}{2} \times 10!$$

and therefore the computed entropy is

$$\Delta S = \Delta S_{nosym}^{RRHO} + R\ln\left(10! \times (3/2)\right) - R\ln 10!$$
$$= \Delta S_{nosym}^{RRHO} + R\ln\left(3/2\right) \tag{4}$$

obtaining again the same result under both formalisms.

However, two examples do not equate to a formal proof, the idea needs to be extended to *any* chemical reaction. To this end, notice that for distinguishable particles, those permutations that lead to a different arrangement, *i.e.*, not superimposable with the original one by rigid rotations are being considered. For a given system, a systematic way to compute the required number of permutations would be to consider all the possible permutations and then reduce this value taking into account the total symmetry number. For example, it is known that there are only two possible arrangements of the distinguishable atoms in the CH_4 molecule, this quantity is equal to the number of permutations of the H atoms (4!) divided by the symmetry number of a tetrahedral molecule (σ=12).

In general, the number of permutations not superimposable by rotations (internal or external) of a system that have n_1 atoms of type 1, n_2 atoms of type 2, and so on, is equal to

$$\frac{\prod_i n_i!}{\prod_j \sigma_j} \tag{5}$$

where the denominator is the product of all the symmetry numbers of the system (reactants or products). If, for instance, the last expression in the reaction (R-1) is applied, it results in $(3!3!1!)/(2\times3)=6$ and $(3!3!1!)/(2\times3)=18$ complexions for the reactants and products respectively, the same results obtained above (see Equation (3)).

For a general reaction React$\rightleftharpoons$Prod, since the number and type of atoms is conserved, the numerator in (5) always cancels out in the difference $R\ln\frac{\prod_i n_i!}{\prod_j \sigma_{j,prod}} - R\ln\frac{\prod_i n_i!}{\prod_j \sigma_{j,react}}$, where $\sigma_{j,react}$,and $\sigma_{j,prod}$ are respectively the symmetry number of reactants and products. Consequently, the effect is equivalent to correcting the entropy change by $\Delta(-R\ln\sigma)$, being $\sigma = \prod_j \sigma_j$ the total symmetry number on each side of the chemical reaction. Note that the correction is the same as that for indistinguishable particles except for one point; in the above two examples when indistinguishable particles were considered, only external symmetry numbers were used, not because the internal symmetry was not present, but simply because of the flaws of the RRHO approach taken as "reference". In other words, considering whether identical atoms are distinguishable or not has no effect on the entropy change in chemical reactions.

CONCLUSIONS

It has been explicitly shown through practical examples that, for all practical applications, it is irrelevant to consider indistinguishable particles or not in entropy calculations. The classical statistical treatment (distinguishable particles) is equally valid provided that the new degrees of freedom involved are taken into account properly. These arguments could be of particular interest for computing entropies in biochemical reactions where the classical treatment is ubiquitous. Even though in such systems it is quite common to observe chemical reactions like binding processes where no symmetry change takes place [21,22], care must be taken, because as we have seen, even under a classical formalism the symmetry should be considered for a proper entropy estimation.

ACKNOWLEDGEMENTS

The author thanks Dimas Suárez and Ramón López (Universidad de Oviedo) for their careful reading of the manuscript and their suggestions.

REFERENCES

1.   McQuarrie, D. *Statistical Mechanics*; University Science Books: Sausalito, CA, USA, 2000.

2.   Gilson, M.K.; Irikura, K.K. Symmetry numbers for rigid, flexible, and fluxional molecules: Theory and applications. *J. Phys. Chem. B* 2010, *114*, 16304–16317.

3.   Ben-Naim, A. On the so-called Gibbs paradox, and on the real paradox. *Entropy* 2007, *9*, 132–136.

4.   Jaynes, E. The Gibbs paradox. In *Maximum Entropy and Bayesian Methods*; Kluwer Academic Publishers: Dordrecht, The Netherlands, 1992.

5.   Lesk, A. On the Gibbs paradox: What does indistinguishability really mean? *J. Phys. A: Math. Gen.* 1980, *13*, L111–L114.

6.   Lin, S.K. Correlation of entropy with similarity and symmetry. *J. Chem. Inf. Comput. Sci.* 1996, *36*, 367–376.

7.   Lin, S.K. Gibbs paradox and the concepts of information, symmetry, similarity and their relationship. *Entropy* 2008, *10*, 1–5.

8. Versteegh, M.A.M.; Dieks, D. The Gibbs paradox and the distinguishability of identical particles. *Am. J. Phys.* 2011, *79*, 741–746.

9. Pauli, W. Thermodynamics and the Kinetic Theory of Gases. In *Pauli Lectures on Physics*; MIT Press: Cambridge, MA, USA, 1973.

10. Swendsen, R. Statistical mechanics of classical systems with distinguishable particles. *J. Stat. Phys.* 2002, *107*, 1143–1166.

11. Nagle, J. Regarding the entropy of distinguishable particles. *J. Stat. Phys.* 2004, *117*, 1047–1062.

12. Pathria, R. *Statistical Mechanics*; Butterworth-Heinemann: Oxford, UK, 1996.

13. Pauling, L. The structure and entropy of ice and of other crystals with some randomness of atomic arrangement. *J. Am. Chem. Soc.* 1935, *57*, 2680–2684.

14. Kozliak, E. Consistent application of the boltzmann distribution to residual entropy in crystals. *J. Chem. Educ.* 2007, *84*, 493–498.

15. Kozliak, E.; Lambert, F.L. Residual entropy, the third law and latent heat. *Entropy* 2008, *10*, 274–284.

16. Ercolany, G.; Piguet, C.; Borkovec, M.; Hamacek, J. Symmetry numbers and statistical factors in self-assembly and multivalency. *J. Phys. Chem. B* 2007, *111*, 12195–12203.

17. Cheng, C.H. Thermodynamics of the system of distinguishable particles. *Entropy* 2009, *11*, 326–333.

18. DeTar, D.F. Theoretical ab initio calculation of entropy, heat capacity, and heat content. *J. Phys. Chem. A* 1998, *102*, 5128–5141.

19. NIST Computational Chemistry Comparison and Benchmark Database. Available online: http://cccbdb.nist.gov/ (accessed on 10 June 2011).

20. Dunning, T. Gaussian basis sets for use in correlated molecular calculations. I. The atoms boron through neon and hydrogen. *J. Chem. Phys.* 1989, *90*, 1007.

21. Chang, C.; Chen, C.; Gilson, M.K. Ligand configurational entropy and protein binding. *Proc. Nat. Acad. Sci. USA* 2007, *104*, 1534–1539.

22. Suárez, E.; Díaz, N.; Suárez, D. Entropic control of the relative stability of triple-helical collagen peptide models. *J. Phys. Chem. B* 2008, *112*, 15248–15255.

23. Zhou, H.X.; Gilson, M.K. Theory of free energy and entropy in noncovalent binding. *Chem. Rev.* 2009, *109*, 4092–4107.

Chapter 8

CONCORDANT CHEMICAL REACTION NETWORKS AND THE SPECIES-REACTION GRAPH

Guy Shinar[1] and Martin Feinberg[2]

[1]InBrain Therapeutics Ltd., 12 Metzada St., Ramat Gan 52235, Israel

[2]The William G. Lowrie Department of Chemical & Biomolecular Engineering and Department of Mathematics, Ohio State University, 140 W. 19th Avenue, Columbus, OH, USA 43210

ABSTRACT

This article is intended as a supplement to another one [1], in which we defined the large class of *concordant chemical reaction networks*[3] and deduced many of the striking properties common to all members of that class. We argued that, so long as the kinetics is weakly monotonic (§3.2), network concordance enforces behavior of a very circumscribed kind.

Among other things, we showed that the class of concordant networks coincides *precisely* with the class of networks which, when taken with any weakly monotonic kinetics, invariably give rise to kinetic systems that are *injective* — a quality that precludes, for example, the possibility of switch-like transitions between distinct positive stoichiometrically compatible steady states. Moreover, we showed that certain properties of concordant networks taken with weakly monotonic kinetics extend to still broader categories of kinetics — including kinetics that involve product inhibition — provided that the networks considered are not only concordant but also *strongly concordant* [1].

Although reaction network concordance is a subtle structural property, determination of whether or not a network is concordant (or strongly concordant) is readily accomplished with the help of easy-to-use and freely available software [2], at least if the network is of moderate size.[4] In this way, one can easily determine whether the several dynamical consequences of concordance or strong concordance accrue to a particular reaction network of interest.

INTRODUCTION

Background

This article is intended as a supplement to another one [1], in which we defined the large class of *concordant chemical reaction networks*[3] and deduced many of the striking properties common to all members of that class. We argued that, so long as the kinetics is weakly monotonic (§3.2), network concordance enforces behavior of a very circumscribed kind.

Among other things, we showed that the class of concordant networks coincides *precisely* with the class of networks which, when taken with any weakly monotonic kinetics, invariably give rise to kinetic systems that are *injective* — a quality that precludes, for example, the possibility of switch-like transitions between distinct positive stoichiometrically compatible steady states. Moreover, we showed that certain properties of concordant networks taken with weakly monotonic kinetics extend to still broader categories of kinetics — including kinetics that involve product inhibition — provided that the networks considered are not only concordant but also *strongly concordant* [1].

Although reaction network concordance is a subtle structural property, determination of whether or not a network is concordant (or strongly concordant) is readily accomplished with the help of easy-to-use and freely available software [2], at least if the network is of moderate size.[4] In this way, one can easily determine whether the several dynamical consequences of concordance or strong concordance accrue to a particular reaction network of interest.

The role of the Species-Reaction Graph

The *Species-Reaction Graph* (SR Graph), defined in Section 2, is a graphical depiction of a chemical reaction network resembling pathway diagrams often drawn by biochemists. For so-called "fully open" systems (§1.2.1), earlier work [4, 5, 6, 7] indicated that, when the SR Graph satisfies certain structural conditions and when the kinetics is within a specified class, the governing differential equations can only admit behavior of a restricted kind. A survey of some of that earlier work, beginning with the Ph.D. research of Paul Schlosser [8, 9] is provided in [1]. Although most of the initial SR Graph results for fully open networks were focused on mass action kinetics, that changed with the surprising work of Banaji and Craciun [6, 7]. (In contrast to the SR Graph results, however, the network analysis tools in [2] are indifferent to whether the network is fully open.)

In [1] we asserted without proof that attributes of the SR Graph shown earlier to be sufficient for other network properties [6, 4, 5, 7] are, in fact, sufficient to ensure not only concordance but also strong concordance, at least in the "fully open" setting. In turn, those other properties (e.g., the absence of multiple steady states when the kinetics is weakly monotonic) largely *derive* from concordance. *That certain SR Graph attributes might ensure concordance seems, then, to be the fundamental idea, with attendant dynamical consequences of those same SR Graph attributes ultimately* descending *from concordance.*

Although handy computational tools in [2] will generally be more incisive than the SR Graph in determining a network's concordance properties and although those computational tools are indifferent to whether the network is fully open, the SR Graph nevertheless has its strong attractions. Not least of these is the close relationship that the SR Graph bears to reaction network diagrams often drawn by biochemists. For this reason alone, far-from-obvious dynamical consequences that might be inferred from these ubiquitous diagrams take on considerable interest. Moreover, theorems that tie network concordance to properties of the SR Graph point to consequences of reaction network structure that are not easily gleaned from implementation of computational tests on a case-by-case basis.

Thus, our primary purpose here is to prove the assertions made in [1] about connections between concordance of a network and the nature of its SR Graph. In fact, we go further in two directions, directions that would have been somewhat askew to the main thrusts of [1]:

The Species-Reaction Graph and the Concordance of "Fully Open" Networks

First, we show that one can infer concordance of a so-called "fully open" network when its SR graph satisfies conditions, stated in Theorem 2.1, that are substantially weaker than those stated in [1] (or in [4, 5, 6, 7]); Corollary 2.4 invokes the SR Graph conditions stated in[1]. A fully open network is one containing a "degradation reaction" $s \to 0$ for each species s in the network.[5] Proof of concordance when the aforementioned SR Graph conditions are satisfied is by far the main undertaking of this article.

Remark 1.1

In [7] Banaji and Craciun restricted their attention to networks in which no reaction has the *same*species as *both* a reactant and a product. Thus, for example, no reaction would be of the form $A + E \to B + E$, but a reaction sequence such as $A + E \to C \to B + E$ would be deemed admissible. In fact, from a chemical viewpoint, the presence of an intermediate compound is to be expected, so the Banaji-Craciun restriction hardly amounts to a serious loss of generality. (To some extent it serves to enforce more realistic chemistry.) Throughout this article *it will be understood that we too restrict our attention to networks in which no reaction has the same species as both a reactant and a product.*

The Species-Reaction Graph and the Concordance of Networks that are not "Fully Open"

Second, we examine more completely concordance information that the SR Graph gives for reaction networks that are *not* fully open.

For a network that is not fully open, its *fully open extension* is the network obtained from the original one by adding degradation reactions for all species for which such degradation reactions are not already present. In [1] we proved that a *normal* network is concordant (respectively, strongly concordant) if its fully open extension is concordant (strongly concordant). A normal network is one satisfying a very mild condition described in [10], [1], and, here again, in §7.1. *The large class of normal networks includes all weakly reversible networks and, in particular, all reversible networks* [10].

Thus, if inspection of the SR Graph can serve to establish, by means of theorems crafted *for fully open networks*, that a *normal* network's fully open extension is concordant or strongly concordant, *then the original network itself has that same property*. In this way, the seemingly restricted power of "fully open" SR graph theorems extends far beyond the fully open setting.

This has some importance, especially in relation to reversible networks. Consider a reaction network whose SR Graph ensures, by virtue of Corollary 2.4, concordance of the network's fully open extension. Then the original network can *fail* to be concordant only if it is not reversible. But a kinetic system based on a network that is not reversible is usually deemed to be an approximation of a more exact "nearby" kinetic system in which all reactions are reversible, perhaps with some reverse reactions having extremely small rates.

To the extent that this is the case, lack of concordance in the original network is an artifact of the approximation: The reversible network underlying the more exact kinetic system is concordant, so that system inherits all of the dynamical properties in [1] that are consequences of network concordance. (In particular, it inherits those properties listed as items (i) – (iii) in Theorem 2.8.) Thus, any absence of these attributes in a kinetic system based on the original network is, again, an artifact of the approximation.

Even so, kinetic system models based on networks that are not reversible (or, more generally, weakly reversible) have an intrinsic interest. They are ubiquitous, and one would like to understand their inherent properties. Results in [1] about network normality and concordance already go a long way in this direction, for the class of normal networks extends well beyond the weakly reversible class.

Here we argue that these same results extend to the still larger class of *nondegenerate* reaction networks, a class that includes all normal networks and, in particular, all weakly reversible networks. A network is nondegenerate if, taken with *some* differentiably monotonic kinetics [1], the resulting kinetic system admits*even one* positive composition at which the derivative of the species-formation-function is nonsingular[6]; otherwise we say that the network is *degenerate* (Definition 7.6).

In fact, degenerate reaction networks are *never* concordant (§7.1), so for them questions about the possibility of concordance are moot. On the other hand, a *nondegenerate* network is concordant (strongly concordant) if the network's fully open extension is concordant (strongly concordant) (§7.2). *Thus, SR Graph theorems that give information about the concordance or strong concordance of a nondegenerate network's fully open extension give that same information about the network itself.*

Implications of network normality were considered in some depth in [1]. Because corresponding considerations of the broader notion of nondegeneracy are similar and because those considerations are rather different in spirit from the largely graph-theoretical aspects of most of this article, we chose to defer the entire discussion of nondegeneracy to Section 7, which includes computational tests whereby normality and nondegeneracy of a network can be affirmed.

In any case, it should be kept in mind that the computational tools provided in [2] are indifferent to whether the network under study is fully open (or nondegenerate).

The Species-Reaction Graph and Consequences of Concordance

For the most part the theorems in this article have as their objective the drawing of connections between the concordance of a reaction network and properties of the network's Species-Reaction Graph. However, it is important to keep in mind the dynamical consequences of these theorems. When an SR Graph theorem asserts concordance of a particular network, *the same theorem is also asserting that the network inherits all of the properties shown in [1] to accrue to all concordant networks.*

Thus, for example, a theorem that asserts concordance for any reaction network whose SR Graph has Property X is *also* asserting that, for any reaction network whose SR Graph has Property X, *there is no possibility of two distinct stoichiometrically compatible equilibria, at least one of which is positive, no matter what the kinetics might be, so long as it is weakly monotonic.*

A theorem that does make explicit connections between the SR Graph and dynamical consequences is offered at the close of the next section.

MAIN THEOREMS

The *Species-Reaction Graph (SR Graph)* of a chemical reaction network is a bipartite graph constructed in the following way: The vertices are of two kinds — *species vertices* and *reaction vertices*. The species vertices are simply the species of the network. The reaction vertices are the reactions of the network but with the understanding that a reversible reaction pair such as $A + B \rightleftarrows P$ is identified with a single vertex. If a species appears in a particular reaction, then an edge is drawn that connects the species with that reaction, and the edge is labeled with the name of the complex in which the species appears. (The *complexes* of a reaction network are the objects that appear before and after the reaction arrows. Thus, the complexes of reaction $A + B \rightarrow P$ are $A + B$ and P.) The *stoichiometric coefficient of an edge* is the stoichiometric coefficient of the adjacent species in the labeling complex. We show the Species-Reaction Graph for network (1) in Figure 1. The arrows on some of the edges will be explained shortly.

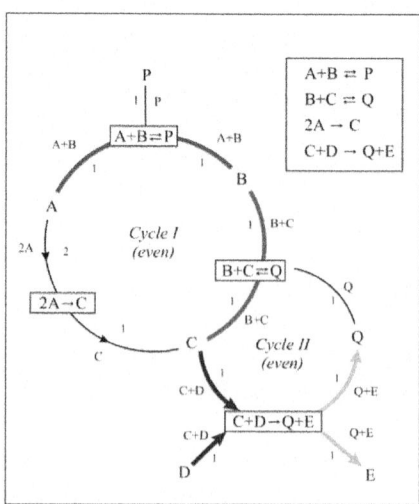

Figure 1: An example of a Species-Reaction Graph.

$$
\begin{aligned}
A + B &\;\rightleftarrows\; P \\
B + C &\;\rightleftarrows\; Q \\
2A &\;\rightarrow\; C \\
C + D &\;\rightarrow\; Q + E
\end{aligned}
$$

$$(1)$$

It will be understood that our focus is on concordance of a "fully open" reaction network $\{S, C, \mathcal{R}\}$. To say that the network is fully open is to say that, for each species s in the network there is a "degradation reaction" of the form $s \rightarrow 0$. (There might also be "synthesis reactions" of the form $0 \rightarrow s$.) Recall too that we suppose throughout that a species can appear on only one side of a reaction. The first of these requirements was invoked in [7, 11, 4], while the second was invoked only in [7].

Even in consideration of fully open reaction networks we will always restrict our attention to the Species-Reaction Graph for the "true chemistry" — that is, the Species-Reaction Graph for the original fully open network but with degradation-synthesis "reactions" such as $A \rightarrow 0$ or $0 \rightarrow A$ omitted. As a reminder we will sometimes refer to the "true-chemistry Species-Reaction Graph," but this understanding will always be implicit even when it is not made explicit.

By the *intersection of two cycles* in the Species-Reaction Graph, we mean the subgraph consisting of all vertices and edges common to the two cycles. We say that two cycles have a *species-to-reaction intersection (S-to-R intersection)* if their intersection is not empty and each of the connected components of the intersection is a path having a species at one end and a reaction at the other. In the phrase "S-to-R intersection" no directionality is implied. In Figure 1, Cycle I and Cycle II have an S-to-R intersection consisting of the edge connecting species C to reactions $B + C \Downarrow Q$. There is a third unlabeled "outer" cycle, hereafter called Cycle III, that traverses species A,C,Q,B and returns to A. It has an S-to-R intersection with Cycle I, in particular, the long outer path connecting reactions $B + C \Downarrow Q$ to species C via species A. Cycle III also has an S-to-R intersection with Cycle II.

An *edge-pair* in the SR Graph is a pair of edges adjacent to a common *reaction* vertex. A *c-pair*(abbreviation for *complex-pair*) is an edge-pair whose edges carry the same complex label. For readers with access to color, the c-pairs in Figure 1 are given distinct colors.

An *even cycle* in the SR Graph is a cycle whose edges contain an even number of c-pairs.[7] Cycle I contains two c-pairs and Cycle II contains none, so both are even. Cycle III, the large outer cycle, contains one c-pair, so it is not even.

A *fixed-direction edge-pair* is an edge-pair that is not a c-pair and for which the reaction of the edge-pair is irreversible. In this case, we assign a *fixed direction* to each of the two edges in the intuitively obvious way: The edge adjacent to the reactant species (i.e., the species appearing in the reactant complex) is directed away from that species, and the edge adjacent to the product species (i.e., the species appearing in the product complex) points toward that species. Thus, for example, a fixed-direction edge-pair such as

$$A \overset{A+B}{\to} \boxed{A + B \to C} \overset{C}{\to} C|$$

has the fixed direction

$$A \overset{A+B}{\to} \boxed{A + B \to C} \overset{C}{\to} C.|$$

In Figure 1 there are fixed-direction edge-pairs centered at the reactions $2A \to C$ and $C + D \to Q + E$; the corresponding fixed-directions for the adjacent edges are shown in the figure.

An *orientation* for a simple cycle $s_1 R_1 s_2 R_2 \ldots s_n R_n s_1$ in the SR Graph is an assignment of one of two directions to the edges, either

$$s_1 \to R_1 \to s_2 \to R_2 \cdots \to s_n \to R_n \to s_1$$

or

$$s_1 \leftarrow R_1 \leftarrow s_2 \leftarrow R_2 \cdots \leftarrow s_n \leftarrow R_n \leftarrow s_1,$$

that is consistent with any fixed directions the cycle might contain. A cycle is *orientable* if it admits at least one orientation. By an *oriented cycle* in the SR Graph we mean an orientable cycle taken with a choice of orientation.

Note that a cycle that has no fixed-direction edge-pair will be orientable and have two orientations. (This happens when every reaction in the cycle is either reversible or else is contained in a c-pair within the cycle.) A cycle that has just one fixed-direction edge-pair is orientable and has a unique orientation. A cycle that has more than one fixed-direction edge-pair might or might not be orientable, depending on whether the fixed-direction edge-pairs all point "clockwise" or "counterclockwise," but if there is an orientation it will be unique. All cycles in Figure 1 are orientable, with each cycle having a unique orientation.

A set of cycles in an SR graph *admits a consistent orientation* if the cycles can all be assigned orientations such that each edge contained in one of the cycles has the same orientation-direction in every cycle of the set in which that edge appears. To see that a set of cycles might not admit a consistent orientation, consider a pair of cycles sharing a common edge, each cycle having the same compulsory orientation (e.g., "clockwise"). In fact, this is precisely the situation for Cycles I and II, so the set consisting of those two cycles does not admit a consistent orientation. On the other hand the set consisting of Cycles I and III (the large outer cycle) does admit a consistent orientation, as does the set consisting of Cycles II and III.

A *critical subgraph* of an SR Graph is a union of a set of *even* cycles taken from the SR Graph that admits a consistent orientation. Because Cycle III is not even, it is easy to see that there are only two critical subgraphs in Figure 1: One consists only of Cycle I and the other consists only of Cycle II.

Consider an oriented cycle in the SR Graph, $s_1 \rightarrow R_1 \rightarrow s_2 \rightarrow R_2 \ldots \rightarrow s_n \rightarrow R_n \rightarrow s_1$. For each directed reaction-to-species edge $R \rightarrow s$ we denote by $f_{R \rightarrow s}$ the stoichiometric coefficient associated with that edge. For each directed species-to-reaction edge, we denote by $e_{s \rightarrow R}$ the stoichiometric coefficient associated with it. The cycle is *stoichiometrically expansive* relative to the given orientation if

$$\left. \frac{f_{R_1 \rightarrow s_2} f_{R_2 \rightarrow s_3} \cdots f_{R_n \rightarrow s_1}}{e_{s_1 \rightarrow R_1} e_{s_2 \rightarrow R_2} \cdots e_{s_n \rightarrow R_n}} > 1 \right. \quad (2)$$

One of the major goals of this article is proof of the following theorem:

Theorem 2.1

A fully open reaction network is concordant if its true-chemistry Species-Reaction Graph has the following properties:

- No even cycle admits a stoichiometrically expansive orientation.
- In no critical subgraph do two even cycles have a species-to-reaction intersection.

Remark 2.2

If there are no critical subgraphs — in particular if there are no orientable even cycles — then the conditions of the theorem are satisfied trivially, and the network is concordant.

Remark 2.3

Determination of which subgraphs of the Species-Reaction Graph are critical requires consideration of orientability. However, it should be understood that condition (*ii*) of Theorem 2.1 is imposed on every *undirected* critical subgraph of the *undirected* Species-Reaction Graph. In particular, account should be taken of all even cycles in the subgraph, whether or not they be orientable.

(Although, a critical subgraph is the union of a set of even cycles that admit a consistent orientation, the resulting subgraph might also contain cycles that are not orientable.) Again, no directionality should be associated with the phrase "species-to-reaction" intersection.

With regard to condition (i), note that a cycle that admits two orientations can be non-expansive relative to both orientations only if, relative to one of the orientations, the number on the left side of (2) is *equal to one*(in which case it will also have that same value with respect to the opposite orientation).

In the language of [12, 5, 7], we say that a (not necessarily orientable) cycle in the SR Graph is an *s-cycle* if relative to an arbitrarily imposed "clockwise" direction, the calculation on the left side of (2) yields the value*one*. With this in mind we can state an "orientation-free" corollary of Theorem 2.1:

Corollary 2.4

A fully open reaction network is concordant if its true-chemistry Species-Reaction Graph has the following properties:

- Every even cycle is an s-cycle.
- No two even cycles have a species-to-reaction intersection.

Especially for networks in which there are several irreversible reactions, Theorem 2.1 is likely to be more incisive than its corollary. Consider, for example, the Species-Reaction Graph for network (1). *Both*conditions of Corollary 2.4 fail. Nevertheless, *both* of the (weaker) conditions of Theorem 2.1 are satisfied, so a fully open network having (1) as its true chemistry is concordant.

Remark 2.5

We shall see very easily in Section 7 that network (1) is normal. Because every normal network with a concordant fully open extension is itself concordant, we can conclude that network (1), like its fully open extension, is concordant.

When a fully open network *does* satisfy the stronger conditions of Corollary 2.4 we can say more, this time about *strong* concordance.

Theorem 2.6

A fully open reaction network is strongly concordant if its true-chemistry Species-Reaction Graph has the following properties:

- Every even cycle is an s-cycle.
- No two even cycles have a species-to-reaction intersection.

Although the fully open extension of network (1) is concordant, it is not strongly concordant. (This can be quickly determined by the freely available *Chemical Reaction Network Toolbox* [2].) The Species-Reaction Graph for network (1) does not satisfy the conditions of Theorem 2.6.

Remark 2.7

Although these theorems nominally speak in terms of concordance or strong concordance of a fully open network, they are even more generous than they seem: Recall that in Section 7 we indicate why the concordance properties ensured by the theorems actually extend beyond the fully open setting to a far wider class of reaction networks. In particular, they extend to the class of *nondegenerate* networks described earlier in §1.2.2 (including *all* weakly reversible networks). For nondegenerate networks, then, the fully open requirement of Theorems 2.1 and 2.6 becomes moot.

In the spirit of §1.3 we close this section with the statement of a theorem that ties the hypothesis of Theorem 2.1 not to concordance itself but, instead, to dynamical consequences of concordance proved in [1]. The following theorem is essentially a corollary of Theorem 2.6, theorems in [1], and material to be discussed in Section 7. Language in the theorem is that used in [1]. Much of it is reviewed in Section 3 of this article.

Theorem 2.8

For any nondegenerate reaction network whose true chemistry Species-Reaction Graph satisfies conditions (i) and (ii) of Theorem 2.1 the following statements hold true:

- For each choice of weakly monotonic kinetics, every positive equilibrium is unique within its stoichiometric compatibility class. That is, no positive equilibrium is stoichiometrically compatible with a different equilibrium, positive or otherwise.

- If the network is weakly reversible then, for each choice of kinetics (not necessarily weakly monotonic), no nontrivial stoichiometric compatibility class has an equilibrium on its boundary. In fact, at each boundary composition in any non-trivial stoichiometric compatibility class the species-formation-rate vector points into the stoichiometric compatibility class in the sense that there is an absent species produced at strictly positive rate. If, in addition, the network is conservative then, for any choice of a continuous weakly monotonic kinetics, there is precisely one equilibrium in each nontrivial stoichiometric compatibility class, and it is positive.

- If the kinetics is differentiably monotonic then every positive equilibrium is nondegenerate. Moreover, every real eigenvalue associated with a positive equilibrium is negative.

As in [1], it is understood that eigenvalues in the theorem statement are those associated with eigenvectors in the stoichiometric subspace. Similarly, when we say that a positive equilibrium in nondegenerate, we mean that, for the equilibrium, zero is not an eigenvalue corresponding to an eigenvector in the stoichiometric subspace. A stoichiometric compatibility class is nontrivial if it contains at least one positive composition.

Remark 2.9

Consider a (not necessarily nondegenerate) reaction network whose true chemistry SR Graph satisfies conditions (i) and (ii) of Theorem 2.1. With respect to the possibility of nondegenerate positive equilibria, Theorem 2.8 describes something of an all or nothing situation: If there is *some* differentiably monotonic kinetics that gives rise to *even one* nondegenerate positive equilibrium, then the network itself is nondegenerate, in which case *every* positive equilibrium arising from *any* differentiably monotonic kinetics is nondegenerate (and unique within its stoichiometric compatibility class).

Remark 2.10

Consider a nondegenerate reaction network that is not necessarily fully open. If its true-chemistry SR Graph satisfies the conditions of Theorem 2.6, then considerations in Section 7 will indicate that the network is strongly concordant. In this case, the dynamical properties of strongly concordant networks given in [1] accrue to the network at hand, and one can again deduce a theorem which, in the spirit of Theorem 2.8, makes statements about general properties of kinetic systems the network engenders, this time including those that derive from so-called "two-way monotonic kinetics."

We conclude this section with a biologically-motivated example that *fails* to satisfy the conditions of Theorem 2.1. Network (3) depicts a model for part of the *E. coli* IDHKP-IDH glyoxylate bypass regulation system, which controls the partitioning of carbon flux between the tricarboxylic acid (TCA) cycle and the glyoxylate bypass. Glyoxylate bypass is controlled by regulation of the phosphorylation level of the TCA cycle enzyme isocitrate dehydrogenase (IDH), denoted I. The phosphorylated form of IDH is denoted I_p. Phosphorylation and dephosphorylation of the protein I is implemented by the bi-functional enzyme isocitrate dehydrogenase kinase phosphatase (IDHKP), denoted E. See [13] for a discussion of this and other models for glyoxylate bypass regulation, along with a discussion of experimental background.

$$EI_p + I \rightleftarrows EI_pI \rightarrow EI_p + I_p \Big|$$
$$E + I_p \rightleftarrows EI_p \rightarrow E + I.$$

$$(3)$$

In Figure 2 we depict network (3) in a diagrammatic form not uncommon in the biochemical literature. Note that the Species-Reaction Graph for network (3), shown in Figure 3, is, in a sense, a more detailed articulation of Figure 2. In Figure 3 there are four c-pairs.

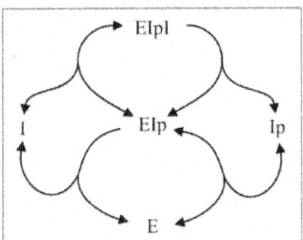

Figure 2: Diagrammatic depiction of network (3).

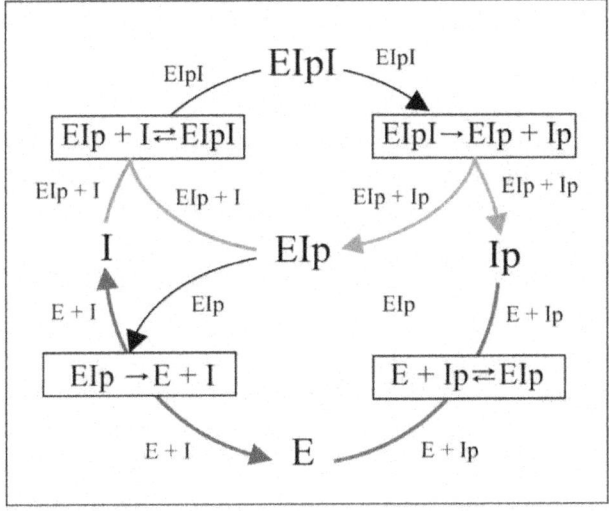

Figure 3: The Species-Reaction Graph for network (3).

In this case, condition (ii) of Theorem 2.1 is not satisfied: There is a critical subgraph in Figure 3 that is the union of two cycles. The first (smaller) cycle begins at species EI_pI, passes through EI_p and then returns to EI_pI. The second cycle also begins at species EI_pI but this time traverses both EI_p and I before returning to EI_pI. Neither cycle contains a c-pair, so both cycles are even.

Moreover, the two cycles intersect in a path having reaction $EI_p + I \downdownarrows EI_pI$ at one end and species EI_p at the other end.

Because condition (ii) is not satisfied, Theorem 2.1 is silent about the concordance of network (3). In fact,*The Chemical Reaction Network Toolbox* [2] indicates quickly that neither the network nor its fully open extension is concordant. This, of course, implies that, for network (3), at least one of the two conditions in Theorem 2.1 *must* be violated. We note in passing that network (3) does not share all of the dynamical properties, described in Theorem 2.8, that are possessed by every concordant network. In particular, with mass action kinetics the network admits two distinct stoichiometrically compatible equilibria, with one strictly positive. For a discussion of robustness properties of network (3) see [14, 15].

REACTION NETWORK THEORY PRELIMINARIES

This section is, for the most part, is a compendium of the definitions and infrastructure used in [1], repeated here for the reader's convenience. Reference [1] has more in the way of discussion, and [16] has still more in the way of motivation. We begin with notation.

Notation

When I is a finite set (for example, a set of species or a set of reactions), we denote the vector space of real-valued functions with domain I by $\mathbb{R}^I$. If N is the number of elements in the set I, then $\mathbb{R}^I$ is, in effect, a copy of the standard vector space $\mathbb{R}^N$, with the components of a vector $x \in \mathbb{R}^I$ indexed by the names of the members of I instead of the integers 1, 2, ..., N. For $x \in \mathbb{R}^I$ and $i \in I$, the symbol x_i denotes the value (component) of x corresponding to the element $i \in I$.

For example, if $S = \{NO, O_2, NO_2\}$ is the set of species in a chemical system and if c_{NO}, c_{O2}, and c_{NO2} are the molar concentrations of the three species in a particular mixture state, then that state can be represented by a "composition vector" c in the vector space $\mathbb{R}^S$. That is, c represents an assignment to each species of a number, the corresponding molar concentration. Readers who wish to do so can, in this case, simply regard c to be the 3-vector $[c_{NO}, c_{O2}, c_{NO2}]$, with the understanding that the arrangement of the three numbers in such an *ordered* array is superfluous to the mathematics at hand.

Vector representations in spaces such as $\mathbb{R}^I$ rather than $\mathbb{R}^N$ have advantages in consideration of graphs and networks, where one wants to avoid nomenclature that imparts an artificial numerical order to vertices or edges. See, for example, [17].

The subset of $\mathbb{R}^I$ consisting of vectors having only positive (nonnegative) components is denoted $\mathbb{R}^I_+$ ($\overline{\mathbb{R}}^I_+$). For each $x \in \mathbb{R}^I$ and for each $z \in \mathbb{R}^{I+}$, the symbol x^z denotes the real number defined by

$$x^z := \prod_{i \in I} (x_i)^{z_i},$$

where it is understood that $0^0 = 1$. For each $x, x' \in \mathbb{R}^I$, the symbol $x \circ x'$ denotes the vector of $\mathbb{R}^I$ defined by

$$(x \circ x')_i := x_i x'_i, \ \forall i \in I.$$

For each $i \in I$, we denote by ω_i the vector of $\mathbb{R}^I$ such that $(\omega_i)_j = 1$ whenever $j = i$ and $(\omega_i)_j = 0$ whenever $j \neq i$.

The *standard basis* for $\mathbb{R}^I$ is the set $\{\omega_i \in \mathbb{R}^I : i \in I\}$. Thus, for each $x \in \mathbb{R}^I$, we can write $x = \Sigma_{i \in I} x_i \omega_i$. The *standard scalar product* in $\mathbb{R}^I$ is defined as follows: If x and x' are elements of $\mathbb{R}^I$, then

$$x \cdot x' = \sum_{i \in I} x_i x'_i.$$

The standard basis of $\mathbb{R}^I$ is orthonormal with respect to the standard scalar product. It will be understood that $\mathbb{R}^I$ carries the standard scalar product and the norm derived from the standard scalar product. It will also be understood that $\mathbb{R}^I$ carries the corresponding norm topology.

If U is a linear subspace of $\mathbb{R}^I$, we denote by $U^\perp$ the orthogonal complement of U in $\mathbb{R}^I$ with respect to the standard scalar product.

By the *support* of $x \in \mathbb{R}^I$, denoted supp x, we mean the set of indices $i \in I$ for which x_i is different from zero. When ξ is a real number, the symbol sgn (ξ) denotes the sign of ξ. When x is a vector of $\mathbb{R}^I$, sgn (x) denotes the function with domain I defined by

$$(\text{sgn}(x))_i := \text{sgn}\, x_i, \ \forall i \in I.$$

Some Definitions

As indicated in Section 2, the objects in a reaction network that appear at the heads and tails of the reaction arrows are the *complexes* of the network. Thus, in network (4), the complexes are 2*A*, *B*, *C*, *C* + *D* and *E*. A reaction network can then be viewed as a directed graph, with complexes playing the role of the vertices and reaction arrows playing the role of the edges.

$$2A \rightleftarrows B$$
$$\diagdown \quad \diagup$$
$$C$$
$$C+D \rightleftarrows E \qquad\qquad\qquad\qquad (4)$$

Remark 3.1

Let S be the set of species in a network. In chemical reaction network theory, it is sometimes the custom to replace symbols for the standard basis of $\mathbb{R}^S$ with the names of the species themselves. For example, in network (4), with $S =$ $\{A, B, C, D, E\}$, a vector such as $_C + _D \in \mathbb{R}^S$ can instead be written as $C + D$, and $2\omega_A$ can be written as $2A$. In this way, $\mathbb{R}^S$ can be identified with the vector space of formal linear combinations of the species. As a result, the complexes of a reaction network with species set S can be identified with vectors in $\mathbb{R}^S$.

Definition 3.2

A *chemical reaction network* consists of three finite sets:

- a set S of distinct *species* of the network;
- a set $C \subset \mathbb{R}^{\overline{}} S+$ of distinct *complexes* of the network;
- a set $\mathcal{R} \subset C \times C$ of distinct *reactions*, with the following properties:
  a. $(y, y) \notin \mathcal{R}$ for any $y \in C$;
  b. for each $y \in C$ there exists $y' \in C$ such that $(y, y') \in \mathcal{R}$ or such that $(y', y) \in \mathcal{R}$.

If (y, y') is a member of the reaction set $\mathcal{R}$, we say that y *reacts to* y', and we write $y \rightarrow y'$ to indicate the reaction whereby complex y reacts to complex y'. The complex situated at the tail of a reaction arrow is the *reactant complex* of the corresponding reaction, and the complex situated at the head is the reaction's *product complex*.

For network (4), $S = \{A, B, C, D, E\}$, $C = \{2A, B, C, C + D, E\}$, and $\mathcal{R} = \{2A \rightarrow B, B \rightarrow 2A, B \rightarrow C, C \rightarrow 2A, C + D \rightarrow E, E \rightarrow C + D\}$. The diagram in (4) is an example of a *standard reaction diagram*: each complex in the network is displayed precisely once, and each reaction in the network is indicated by an arrow in the obvious way.

In the context of the present paper and its predecessor [1] the idea of *weak reversibility* plays an important role. The following definition provides some preparation.

Definition 3.3

A complex $y \in \mathcal{C}$ *ultimately reacts* to a complex $y' \in \mathcal{C}$ if any of the following conditions is satisfied:

- $y \to y' \in \mathcal{R}$;
- There is a sequence of complexes $y(1)$, $y(2)$, ..., $y(k)$ such that $y \to y(1) \to y(2) \to ... \to y(k) \to y'$.

In network (4) the complex $2A$ ultimately reacts to the complex C, but the complex C does not ultimately react to the complex $C + D$.

Definition 3.4

A reaction network $\{\mathcal{S}, \mathcal{C}, \mathcal{R}\}$ is *reversible* if $y' \to y \in \mathcal{R}$ whenever $y \to y' \in \mathcal{R}$. The network is *weakly reversible* if for each y, $y' \in \mathcal{C}$, y' ultimately reacts to y whenever y ultimately reacts to y'.

Network (4) is weakly reversible but not reversible. On the other hand, every reversible network is also weakly reversible. A reaction network is weakly reversible if and only if in its standard reaction diagram every arrow resides in a directed arrow-cycle.

Definition 3.5

The *reaction vectors* for a reaction network $\{\mathcal{S}, \mathcal{C}, \mathcal{R}\}$ are the members of the set

$$\{y' - y \in \mathbb{R}^{\mathcal{S}} : y \to y' \in \mathcal{R}\}.$$

The *rank* of a reaction network is the rank of its set of reaction vectors.

For network (4) the reaction vector corresponding to the reaction $2A \to B$ is $B - 2A$. The reaction vector corresponding to the reaction $C + D \to E$ is $E - C - D$.

Definition 3.6

The *stoichiometric subspace* S of a reaction network $\{\mathcal{S}, \mathcal{C}, \mathcal{R}\}$ is the linear subspace of $\mathbb{R}^{s}$ defined by

$$S := \mathrm{span}\{y' - y \in \mathbb{R}^{\mathcal{S}} : y \to y' \in \mathcal{R}\}. \qquad (5)$$

The dimension of the stoichiometric subspace is identical to the network's rank. The stoichiometric subspace is a proper linear subspace of $\mathbb{R}^{s}$ whenever the network is *conservative*:

Definition 3.7

A reaction network $\{S, C, R\}$ is *conservative* if the orthogonal complement $S^\perp$ of the stoichiometric subspace S contains a strictly positive member of $\mathbb{R}^s$:

$$S^\perp \cap \mathbb{R}_+^{\mathscr{S}} \neq \varnothing.$$

Network (4) is conservative: The strictly positive vector $(A+2B+2C+D+3E)\in \mathbb{R}_+^{\mathscr{S}}$ is orthogonal to each of the reaction vectors of (4).

For a reaction network $\{S, C, R\}$ a mixture state is generally represented by a *composition vector* $c \in \mathbb{R}_+^{\mathscr{S}}$, where, for each $s \in S$, we understand c_s to be the molar concentration of s. By a *positive composition* we mean a strictly positive composition — that is, a composition in $\mathbb{R}_+^{\mathscr{S}}$.

Definition 3.8

A *kinetics* K for a reaction network $\{S, C, R\}$ is an assignment to each reaction $y \to y' \in R$ of a *rate function* $\mathscr{K}_{y\to y'} : \overline{\mathbb{R}}_+^{\mathscr{S}} \to \overline{\mathbb{R}}_+$ such that

Ky→y'(c)>0 if and only if supp y ⊂ supp c.

Definition 3.9

A *kinetic system* $\{S, C, R, K\}$ is a reaction network $\{S, C, R\}$ taken with a kinetics K for the network.

Many of the dynamical consequences of network concordance require that the kinetics be *weakly monotonic* or *differentiably monotonic*. Both are formally defined below:

Definition 3.10

A kinetics K for reaction network $\{S, C, R\}$ is *weakly monotonic* if, for each pair of compositions c^* and c^{**}, the following implications hold for each reaction $y \to y' \in R$ such that supp $y \subset$ supp c^* and supp $y \subset$ supp c^{**}:

- $\mathscr{K}_{y\to y'}(c^{**}) > \mathscr{K}_{y\to y'}(c^*) \Rightarrow$ there is a species $s \in$ supp y with $c^{**}s > c^*s$.

- $\mathscr{K}_{y\to y'}(c^{**}) = \mathscr{K}_{y\to y'}(c^*) \Rightarrow c_{ss}^{**} = c_s^*$ for all $s \in$ s for all s∈supp y or else there are a species s, s'∈supp y with c**s>c*s and c**s'<c*s'.

We say that the kinetic system $\{S, C, R, K\}$ is weakly monotonic when its kinetics K is weakly monotonic.

Definition 3.11

A kinetics $\mathcal{K}$ for a reaction network $\{S, C, R\}$ is *differentiably monotonic* at $c^* \in \mathbb{R}_+^{\mathcal{S}}$ if, for every reaction $y \to y' \in R$, $\mathcal{K}_{y \to y'}(\cdot)$ is differentiable at c^* and, moreover, for each species $s \in S$,

$$\frac{\partial \mathcal{K}_{y \to y'}}{\partial c_s}(c^*) \geq 0,$$

(6)

with inequality holding if and only if $s \in$ supp y. A *differentiably monotonic kinetics* is one that is differentiably monotonic at every positive composition.

When a kinetics $\mathcal{K}$ for a reaction network $\{S, C, R\}$ is differentiably monotonic at $c* \in \mathbb{R}S+$, we denote by the symbol $\nabla \mathcal{K}_{y \to y'}(c^*)$ the member of $\mathbb{R}^S$ defined by

$$\nabla \mathcal{K}_{y \to y'}(c^*) := \sum_{s \in \mathcal{S}} \frac{\partial \mathcal{K}_{y \to y'}}{\partial c_s}(c^*)s.$$

Note that every mass action kinetics is both weakly monotonic and differentiably monotonic. Recall that a*mass action kinetics* is a kinetics in which the rate function corresponding to each reaction $y \to y'$ takes the form $\mathcal{K}_{y \to y'}(c) = k_{y \to y'}c^y, \forall c \in \mathbb{R}_+^{\mathcal{S}}$, is a positive *rate constant* for the reaction $y \to y'$.

Definition 3.12

The *species formation rate function* for a kinetic system $\{S, C, R, \mathcal{K}\}$ with stoichiometric subspace S is the map $f: \mathbb{R}_+^{\mathcal{S}} \to S$ defined by

$$f(c) = \sum_{y \to y' \in \mathcal{R}} \mathcal{K}_{y \to y'}(c)(y' - y).$$

(7)

Definition 3.13

The *differential equation* for a kinetic system with species formation rate function $f(\cdot)$ is given by

$$\dot{c} = f(c). \tag{8}$$

From equations (5), (7), and (8), it follows that, for a kinetic system $\{S, C, R, \mathcal{K}\}$, the vector $\dot{c}$ will invariably lie in the stoichiometric subspace S for the network $\{S, C, R\}$. Thus, two compositions $c \in \overline{\mathbb{R}}_+^{\mathcal{S}}$ and $c' \in \overline{\mathbb{R}}_+^{\mathcal{S}}$ can lie along the same solution of (8) only if their difference $c' - c$ lies in S. This motivates the following definition:

Definition 3.14

Let $\{S,\ C,\ R\}$ be a reaction network with stoichiometric subspace S. Two compositions c c' in $\overline{\mathbb{R}}_+^{\mathscr{S}}$ are *stoichiometrically compatible* if $c'-c$ lies in S.

For a network $\{S,\ C,\ R\}$, the stoichiometric compatibility relation serves to partition $\overline{\mathbb{R}}_+^{\mathscr{S}}$ into equivalence classes called the *stoichiometric compatibility classes* for the network. Thus, the stoichiometric compatibility class containing an arbitrary composition c, denoted $(c+S)\cap\overline{\mathbb{R}}_+^{\mathscr{S}}$, is given by

$$(c+S)\cap\overline{\mathbb{R}}_+^{\mathscr{S}}=\left\{c'\in\overline{\mathbb{R}}_+^{\mathscr{S}}:c'-c\in S\right\}$$

(9)

The notation is intended to suggest that $(c+S)\cap\overline{\mathbb{R}}_+^{\mathscr{S}}$ is the intersection of $\overline{\mathbb{R}}_+^{\mathscr{S}}$ with the parallel of S containing c. A stoichiometric compatibility class is *nontrivial* if it contains at least one (strictly) positive composition.

Definition 3.15

An *equilibrium* of a kinetic system $\{S,\ C,\ R,\ K\}$ is a composition $c\in\overline{\mathbb{R}}_+^{\mathscr{S}}$ for which $f(c)=0$. A *positive equilibrium* is an equilibrium that lies in $\mathbb{R}_+^{\mathscr{S}}$.

Because compositions along solutions of (8) are stoichiometrically compatible, one is typically interested in changes in values of the species formation rate function that result from changes in composition that are stoichiometrically compatible. In particular, if $f(c^*)$ is the value of the species formation rate function at composition c^*, then one might be interested in the value of $f(c)$ for a composition c very close to c^* and stoichiometrically compatible with it. Thus, for a kinetic system $\{S,\ C,\ R,\ K\}$ with stoichiometric subspace S, with smooth reaction rate functions, and with species formation rate function f: $\overline{\mathbb{R}}_+^{\mathscr{S}}\to S$, we will want to work with the derivative $df(c^*):S\to S$, defined by

$$df(c^*)\sigma=\frac{df(c^*+\theta\sigma)}{d\theta}\bigg|_{\theta=0},\qquad\forall\sigma\in S.$$

(10)

We say that $c*\in\mathbb{R}_+^{\mathscr{S}}$ is a *nondegenerate equilibrium* if c^* is an equilibrium and if, moreover, $df(c^*)$ is nonsingular. An *eigenvalue associated with a positive equilibrium* $c*\in\mathbb{R}_+^{\mathscr{S}}$ is an eigenvalue of the derivative $df(c^*)$.

CONCORDANT REACTION NETWORKS

Here we recall the definition of reaction network concordance [1]. We consider a reaction network $\{S, C, R\}$ with stoichiometric subspace $S \subset \mathbb{R}^s$, and we let $L : \mathbb{R}^R \to S$ be the linear map defined by

$$L\alpha = \sum_{y \to y' \in R} \alpha_{y \to y'} (y' - y).$$

$$(11)$$

Definition 4.1

The reaction network $\{S, C, R\}$ is *concordant* if there do not exist an $\alpha \in \ker L$ and a nonzero $\sigma \in S$ having the following properties:

- For each $y \to y'$ such that $\alpha_{y \to y'} \neq 0$, supp y contains a species s for which $\mathrm{sgn}\, \sigma_s = \mathrm{sgn}\, \alpha_{y \to y'}$.

- For each $y \to y'$ such that $\alpha_{y \to y'} = 0$, $\sigma_s = 0$ for all $s \in$ supp y or else supp y contains species s and s' for which $\mathrm{sgn}\, \sigma_s = -\mathrm{sgn}\, \sigma_{s'}$, both not zero.

A network that is not concordant is *discordant*.

Note that for a fully open network with species set S the stoichiometric subspace coincides with $\mathbb{R}^s$. The following lemma will prove useful later on:

Lemma 4.2

If a fully open network is discordant, it is always possible to choose from each reversible pair of non-degradation reactions at least one (and sometimes both) of the reactions for removal such that the resulting fully open subnetwork is again discordant.

Proof

Suppose that a fully open network $\{S, C, R\}$ is discordant. Then there are $\alpha \in \ker L$ and nonzero σ that together satisfy conditions (i) and (ii) in Definition 4.1. In particular, we have

$$\sum_{y \to y' \in R} \alpha_{y \to y'} (y' - y) = 0.$$

$$(12)$$

Let $\bar{y} \rightleftarrows \bar{y}'$ be a pair of reversible non-degradation reactions in R, and suppose that $\alpha_{\bar{y} \to \bar{y}'} \neq \alpha_{\bar{y}' \to \bar{y}}$ with the complexes labeled such that $|\alpha_{\bar{y} \to \bar{y}'}| > |\alpha_{\bar{y}' \to \bar{y}}|$. In this case, let $R^\dagger := R \setminus \{\bar{y}' \to \bar{y}\}$, let $\alpha^\dagger_{\bar{y} \to \bar{y}'} := \alpha_{\bar{y} \to \bar{y}'} - \alpha_{\bar{y}' \to \bar{y}}$, and, for all other $y \to y'$ in $R^\dagger$, let $\alpha^\dagger_{y \to y'} := \alpha_{y \to y'}$. From (12) it follows easily that

$$\sum_{y \to y' \in \mathscr{R}^\dagger} \alpha^\dagger_{y \to y'}(y'-y)=0.$$

(13)

If, on the other hand, $a_{\bar{y} \to \bar{y}'} = a_{\bar{y}' \to \bar{y}}$, then we can choose $\mathscr{R}^\dagger = \mathscr{R} \backslash \{\bar{y} \rightleftarrows \bar{y}\}$, and, for all $y \to y'$ in $\mathcal{R}^\dagger$, we can let $\alpha^\dagger_{y \to y'} := \alpha_{y \to y'}$. In this case, (13) will again obtain.

In either case, it is easy to see that $\alpha^\dagger$, taken with the original σ, suffices to establish the discordance of the subnetwork associated with the reaction set $\mathcal{R}^\dagger$.

Remark 4.3

In effect, Lemma 4.2 tells us that every fully open discordant network with reversible non-degradation reactions has a discordant fully open subnetwork in which no non-degradation reaction is reversible. Note that, apart from minor changes in labels within the reaction nodes (i.e., replacement of $y \rightleftarrows y'$ by $y \to y'$) the Species-Reaction Graph derived from the indicated discordant subnetwork is a subgraph of the Species-Reaction Graph drawn for the original network. As we explain at the beginning of Section 5, that subgraph will satisfy the conditions of Theorem 2.1 and its corollary if the parent Species-Reaction Graph does. These observations will help us simplify certain arguments that are otherwise complicated by the presence of reversible reactions.

PROOF OF THEOREM 2.1

The proof will be by contradiction. With this in mind, we assume hereafter the true-chemistry Species-Reaction Graph (SR Graph) for the fully open network under consideration has both attributes of the theorem statement and that, contrary to the assertion of the theorem, the fully open network is discordant.

In this case, Lemma 4.2 tells us that, when the true chemistry has reversible reaction pairs, then each such pair can be replaced by a certain irreversible reaction of the pair (or else removed completely) such that the resulting fully open network is again discordant. The SR Graph for the altered (totally irreversible) true chemistry might have fewer cycles than in the original SR Graph (but never more) because of removal of reversible reaction pairs. Similarly, there might be fewer *orientable* cycles (but never more) than in the original SR Graph as a result of replacement of reversible reaction pairs by single irreversible reactions. Moreover, there might be fewer critical subgraphs (but never more) than in the original SR Graph because of a loss of orientable cycles or because more cycles have acquired compulsory orientations. (Cycles that were even in the original SR Graph and persist in the new SR Graph remain even.) For all of these reasons, *the SR Graph for the totally irreversible subnetwork of the original "true" chemical reaction network will, like the SR*

Graph for the original network, satisfy the requirements of Theorem 2.1.

This is to say that there is no loss of generality in assuming, for the purposes of contradiction, that there is a "true" chemical reaction network, containing no reversible reactions, for which the SR Graph satisfies both requirements of Theorem 2.1 but for which the fully open extension is discordant.

Hereafter in the proof of Theorem 2.1, then, *we assume that all reactions in the "true" chemistry are irreversible, that the corresponding SR Graph satisfies both conditions of Theorem 2.1, and that, contrary to what the theorem asserts, the fully open extension of the true chemistry is discordant.*

Thus there exist fixed $\alpha \in \ker L$ and nonzero $\sigma \in \mathbb{R}^S$ satisfying the requirements of Definition 4.1. (Recall that the stoichiometric subspace for a fully open network is $\mathbb{R}^S$.) *It will be understood throughout the proof that all references to α and σ are relative to this fixed pair, so chosen.*

It will be helpful to divide the proof into subsections:

Preliminaries

We associate a sign with each species in the following way: *A species $s \in S$ is positive, negative, or zero* according to whether σ_s is positive, negative, or zero. By a *signed species* we mean one that is either positive or negative. Similarly, *a reaction $y \to y' \in \mathcal{R}$ is positive, negative, or zero* according to whether $\alpha_{y \to y'}$ is positive, negative, or zero. A *signed reaction* is one that is either positive or negative.

Remark 5.1

Note in particular, that for any "degradation reaction" of the kind $s \to 0$, Definition 4.1 requires that sgn $\alpha_{s \to 0}$ = sgn σ_s, so sgn $s \to 0$ = sgn s for every $s \in S$. (A similar situation obtains for any reaction of the form $ns \to y$, where n is a positive number.) For any "synthesis reaction" $0 \to s$, Definition 4.1 requires that $\alpha_{0 \to s} = 0$, so such reactions are unsigned.

The sign-causality graph and causal units

Hereafter we denote by $\mathcal{R}^*$ the set of all reactions not of the form $s \to 0$, $s \in S$. That is, $\mathcal{R}^*$ is the set of reactions that are not degradation reactions. Because α is a member of $\ker L$ we can write

$$\sum_{y \to y' \in \mathcal{R}^*} \alpha_{y \to y'} (y' - y) - \sum_{s \in \mathcal{S}} \alpha_{s \to 0} s = 0. \tag{14}$$

Thus, for any particular choice of species $s \in S$, we have

$$\sum_{y \to y' \in \mathscr{R}^*} \alpha_{y \to y'} (y'_s - y_s) = \alpha_{s \to 0}.$$

(15)

Now *suppose that in* (15), *the species s is positive*. From (15) and Remark 5.1 we have

$$\sum_{y \to y' \in \mathscr{R}^*} \alpha_{y \to y'} (y'_s - y_s) > 0,$$

(16)

in which case at least one term on the left must be positive. (At least one term on the left is "causal" for the inequality.[8]) Terms of this kind can arise in precisely two ways:

- There is a *positive* reaction $y \to y'$ (i.e., $\alpha_{y \to y'}$ is positive) with $s \in \text{supp } y'$ (so that y's>0, $y_s = 0$). Recall, however, that for $\alpha_{y \to y'}$ to be positive, the conditions of Definition 4.1 require that there be a *positive* species s' in supp y.

  In this case, reaction $y \to y'$ is "causal" for the sign of species s, while species s' is "causal" for the sign of reaction $y \to y'$. With this in mind, we write

$$\overset{+}{s'} \overset{y}{\rightsquigarrow} \overset{+}{y \to y'} \overset{y'}{\rightsquigarrow} \overset{+}{s} .$$

(17)

The signs above the species and the reactions indicate their respective signs. The complex labels above the "causal" arrows ($\rightsquigarrow$) indicate the complex in whose support the adjacent species resides.

- There is a *negative* reaction $y \to y'$ (i.e., $_{y \to y'}$ is negative) with $s \in$ supp y (so that $y_s > 0$, y's=0). But for $_{y \to y'}$ to be negative, the conditions of Definition 4.1 require that there be a *negative* species s' in supp y. As in case (i), reaction $y \to y'$ is causal for the sign of species s, while species s' is causal for the sign of reaction $y \to y'$. This time we write

$$\overset{-}{s'} \overset{y}{\rightsquigarrow} \overset{-}{y \to y'} \overset{y}{\rightsquigarrow} \overset{+}{s} .$$

(18)

On the other hand, *suppose that in* (15), *the species s is negative*. From (15) and Remark 5.1 we have

$$\sum_{y \to y' \in \mathscr{R}^*} \alpha_{y \to y'} (y'_s - y_s) < 0,$$

(19)

in which case at least one term on the left must be negative. (At least one term on the left is causal for the inequality.) Here again, terms of this kind can arise in precisely two ways:

- There is a *negative* reaction $y \to y'$ (i.e., $\alpha_{y \to y'}$ is negative) with $s \in$ supp y' (so that $y'_s > 0$, $y_s = 0$). For $\alpha_{y \to y'}$ to be negative, the conditions of Definition 4.1 require that there be a *negative* species s' in supp y. Here we write

$$\overset{-}{s'} \overset{y}{\leadsto} y \to y' \overset{y'}{\leadsto} \overset{-}{s} . \qquad (20)$$

- There is a *positive* reaction $y \to y'$ (i.e., $\alpha_{y \to y'}$ is positive) with $s \in$ supp y (so that $y_s > 0$, $y'_s = 0$). For $\alpha_{y \to y'}$ to be positive, the conditions of Definition 4.1 require that there be a *positive* species s' in supp y. We write

$$\overset{+}{s'} \overset{y}{\leadsto} y \to y' \overset{y}{\leadsto} \overset{-}{s} . \qquad (21)$$

The diagrams drawn in (17), (18), (20) and (21) can be viewed as edge-pairs in a bipartite directed graph:

The *sign-causality graph*, drawn for the network (relative to the α, σ pair under consideration) is constructed according to the following prescription: The vertices are the signed species and signed (non-degradation) reactions. An edge $\leadsto$ is drawn from a signed species s' to a signed reaction $y \to y'$ whenever s' is contained in supp y and the two signs agree; the edge is then labeled with the complex y. An edge $\leadsto$ is drawn from a signed reaction $y \to y'$ to a signed species s in either of the following situations: (i) s is contained in supp y' and the sign of s agrees with the sign of the reaction; in this case the edge carries the label y' or (ii) s is contained in supp y and the sign of s disagrees with the sign of the reaction; in this case the edge carries the label y. It is understood that the signed species and the signed reactions are labeled by their corresponding signs.

By a *causal unit* we mean a directed two-edge subgraph of the sign-causality graph of the kind $s' \leadsto R \leadsto s$, where R denotes a reaction. (We will often designate a reaction by the symbol R when there is no need to indicate the reactant and product complexes.) The species s' is the *initiator* of the causal unit $s' \leadsto R \leadsto s$, while s is its *terminator*. It is not difficult to see that the initiator and terminator of a causal unit must be *distinct* species.[9]

Causal units are of the four varieties shown in (17), (18), (20) and (21). Of these, (17) and (20) carry distinct complex labels on the two edges. On the other hand, (18) and (21) carry identical labels on the edges. By a *c-pair causal unit* we mean a causal unit of the second kind. (As with the Species-Reaction Graph, the term is meant to be mnemonic for *complex pair*.)

Remark 5.2

It is important to note that a c-pair causal unit results in a *change* of sign as the edges are traversed from the initiator species to the terminator species. Otherwise, a causal unit is characterized by *retention* of the sign.

Remark 5.3

It should be clear that, apart from the direction $\rightsquigarrow$ imparted to its edges, the sign-causality graph corresponding to α and σ can be identified with a subgraph of the undirected Species-Reaction Graph (which we will sometimes refer to as the sign-causality graph's "counterpart" in the Species-Reaction Graph). Moreover, every fixed-direction edge-pair in the SR Graph has direction consistent with that imparted by the $\rightsquigarrow$-relation (because each proceeds from a reactant species to an irreversible reaction to a product species).

The Sign-Causality Graph Must Contain a Directed Cycle, and All of Its Directed Cycles Are Even

By supposition σ is not zero, so there is at least one signed species, say s_1. From the discussion in Section 5.2 it is clear that s_1 is the terminator of a causal unit $s_2 \rightsquigarrow R_2 \rightsquigarrow s_1$, where s_2 is distinct from the initiator s_1. Because s_2 is also signed, it too must be the terminator of a causal unit $s_3 \rightsquigarrow R_3 \rightsquigarrow s_2$, where the signed species s_3 is distinct from s_2. Continuing in this way, we can see that there is a directed sequence of the form

$$\cdots \rightsquigarrow s_n \rightsquigarrow R_n \rightsquigarrow s_{n-1} \rightsquigarrow R_{n-1} \cdots \rightsquigarrow s_3 \rightsquigarrow R_3 \rightsquigarrow s_2 \rightsquigarrow R_2 \rightsquigarrow s_1, \tag{22}$$

with $s_i \neq s_{i+1}$.

Because the number of species is finite, two non-consecutive species in the sequence must in fact coincide, which is to say that *the sign-causality graph must contain a directed cycle.* Moreover, it is easy to see that each vertex in the sign-causality graph resides in a cycle or else there is a cycle $\rightsquigarrow$-upstream from it.

With this in mind, we record here some vocabulary that will be useful in the next section: A *source* is a strong component of the sign-causality graph whose vertices have no incoming edges originating at vertices outside that strong component. Clearly, every component of the sign-causality graph contains a source, and every vertex in a source resides in a directed cycle.

As with the SR Graph, we say that a (not necessarily directed) cycle in the sign-causality graph is *even* if it contains an even number of c-pairs.

Lemma 5.4

A (not necessarily directed) cycle in the sign-causality graph that is the union of causal units is even. In particular, every directed cycle in the sign-causality graph is even.

Proof

If we traverse the cycle beginning at a species s^0 and count the number of species-sign changes when we have returned to s^0, that number clearly must be even. But, if the cycle is the union of causal units, the number of sign changes is identical to the number of c-pair causal units the cycle contains (Remark 5.2). Clearly, every directed cycle in the sign-causality graph is the union of causal units.

Because a directed cycle in the sign-causality graph is even, its (orienttable) cycle counterpart in the undirected Species-Reaction Graph (Remark 5.3) will also have an even number of c-pairs. Because the sign-causality graph must contain a source, which in turn must contain a directed cycle, we now know that *a reaction network is concordant if its Species-Reaction Graph contains no orientable even cycles*.

Remark 5.5

In fact, *a source in the sign-causality graph, when viewed in the SR Graph, must be a critical subgraph*. That this is so follows from the fact that a source is a strong component of the sign-causality graph and therefore is the union of ∿-directed cycles. These even cycles, viewed in the SR Graph, have a consistent orientation, the orientation conferred by the directed-cycle ∿-orientation in the sign-causality graph, which in turn is consistent with any fixed-direction edge-pairs the SR Graph might contain.

In the next section we begin to consider what happens when the Species-Reaction Graph does contain at least one orientable even cycle (and therefore at least one critical subgraph). We will want to show that if the fully open network under consideration satisfies the hypothesis of Theorem 2.1 then the very existence of a source in the putative sign-causality graph becomes impossible.[10]

Inequalities Associated With a Source; Stoichiometric Coefficients Associated With Edges in The Sign-Causality Graph

Consider a source in the sign-causality graph having $\mathcal{S}_0 \subset \mathcal{S}$ as its set of species nodes and $\mathcal{R}_0 \subset \mathcal{R}^*$ as its set of reaction nodes. If s is a *positive* species in the source, then (16) can be written

$$\sum_{y \to y' \in \mathscr{R}_0} \alpha_{y \to y'}(y_s' - y_s) + \sum_{y \to y' \in \mathscr{R}^* \backslash \mathscr{R}_0} \alpha_{y \to y'}(y_s' - y_s) > 0.$$

$$(23)$$

Now suppose that a term in the second sum on the left, corresponding to reaction $\mathscr{I} \to \mathscr{I}$, is not zero. Because $\mathscr{I} \to \mathscr{I}$ is not a vertex of the source, any edge of the sign-causality graph that connects species s to reaction $\mathscr{I} \to \mathscr{I}$ must point *away* from s. Thus, the reaction $\mathscr{I} \to \mathscr{I}$ cannot be causal for the positive sign of species s, so the term

$$\alpha_{\bar{y} \to \bar{y}'}(\bar{y}_s' - \bar{y}_s)$$

must be negative. This implies that (23) can obtain only if we have

$$\sum_{y \to y' \in \mathscr{R}_0} \alpha_{y \to y'}(y_s' - y_s) > 0.$$

$$(24)$$

When s is a *negative* species in the source, we can reason similarly to write

$$\sum_{y \to y' \in \mathscr{R}_0} \alpha_{y \to y'}(y_s' - y_s) < 0.$$

$$(25)$$

Remark 5.6

Note that, for a particular $s \in \mathcal{S}_0$ a nonzero term in (24) or (25) might not correspond to any edge of the sign-causality graph (as when, for a particular reaction $y \to y'$, s is a member of supp y' and disagrees in sign with $\alpha_{y \to y'}$). If s is a positive species, then the term in question must be negative, and hence it can be removed from (24) without changing the sense of that inequality. Similarly, if s is a negative species, the term in question is positive and can be removed from (25) without changing the inequality's sense. *In what follows below, we shall assume that such removals have been made, so that every term in* (24) *or* (25) *corresponds to an edge in the sign-causality graph.*

Recall that a directed edge of the sign-causality graph from a species s to a reaction $y \to y'$ is always of the form

$$s \xrightarrow{y} y \to y'.$$

That is, the edge-label of such a species-to-reaction edge is invariably the reactant complex, y. Note that species s has a positive stoichiometric coefficient, y_s, in that complex. On the other hand, reaction-to-species edges of the sign-causality graph are of two kinds:

$$y \to y' \xrightarrow{y} s \quad \text{or} \quad y \to y' \xrightarrow{y'} s.$$

In either case, the species s has a positive stoichiometric coefficient (either y_s or y's) in the edge-labeling complex.

Hereafter, for a species-to-reaction edge $s \leadsto R$ of the sign-causality graph we denote by $e_{s \leadsto R}$ the (positive) stoichiometric coefficient of species s in the corresponding edge-labeling complex. For a reaction-to-species edge $R \leadsto s$ we denote by $f_{R \leadsto s}$ the (positive) stoichiometric coefficient of species s in its edge-labeling complex.

For the sign-causality graph source under consideration, we will in fact need a small amount of additional notation: For each species $s \in S_0$ we denote by $\mathcal{R}_0 \leadsto s$ the set of all edges of the source that are incoming to s and by $s \leadsto \mathcal{R}_0$ the set of all edges of the source that are outgoing from s. In light of Remark 5.6 and in view of notation we now have available, some analysis will indicate that the inequalities given by (24) and (25) can be written as a single system:

$$\sum_{\mathcal{R}_0 \leadsto s} f_{R \leadsto s} |\alpha_R| - \sum_{s \leadsto \mathcal{R}_0} e_{s \leadsto R} |\alpha_R| > 0, \quad \forall s \in \mathcal{S}_0. \tag{26}$$

Our aim will be to show that when the conditions of Theorem 2.1 obtain, the inequality system (26) cannot be satisfied. The following remark will play an important role.

Remark 5.7

Consider a source in the sign-causality graph, and suppose that

$$s_1 \leadsto R_1 \leadsto s_2 \leadsto R_2 \cdots \leadsto s_n \leadsto R_n \leadsto s_1 \tag{27}$$

is one of its $\leadsto$-directed (and therefore even) cycles. Note that the $\leadsto$-orientation of the cycle also provides an orientation of the cycle's counterpart in the Species-Reaction Graph, for it is consistent with directions of the fixed-direction edge-pairs in the SR Graph. (Remark 5.3)

Thus, when condition (i) of Theorem 2.1 is satisfied, the stoichiometric co-efficients of the directed cycle (27) *must satisfy the condition*

$$\frac{f_{R_1 \leadsto s_2} f_{R_2 \leadsto R_3} \cdots f_{R_n \leadsto s_1}}{e_{s_1 \leadsto R_1} e_{s_2 \leadsto R_2} \cdots e_{s_n \leadsto R_n}} \leq 1. \tag{28}$$

The Decomposition of A Source Into Its Blocks; The Block-Tree Of A Source

Here we draw on graph theory vocabulary that is more-or-less standard [18].[11] A *separation* of a connected graph is a decomposition of the graph into two edge-disjoint connected subgraphs, each having at least one edge, such that the two subgraphs have just one vertex in common. If a connected graph admits a separation (in which case it is *separable*), then the common vertex of the separation is called a *separating vertex* of the graph. A graph is *nonseparable* if it is connected and has no separating vertices. A maximal nonseparable subgraph of a graph is called a *block* of the graph. In rough terms, a connected graph is made up of its blocks, pinned together at the graph's separating vertices.

A source in the sign-causality graph can be decomposed into its blocks, which we call *source-blocks*. Because a source is strongly connected, each of its blocks is strongly connected (and non-separable). To illustrate some of these ideas we show in Figure 4 a hypothetical source with five blocks. (Although they are inconsequential to the block decomposition, the arrows in the figure are meant to connote the $\rightsquigarrow$-relation.) The separating vertices in the figure are R_2, S_3, S_4, and S_7. An example of a block is the subgraph having vertices $\{S_1, S_2, R_1, R_2, R_3\}$ and edges $\{S_1 \rightsquigarrow R_2, R_2 \rightsquigarrow S_2, S_2 \rightsquigarrow R_1, R_1 \rightsquigarrow S_1, S_1 \rightsquigarrow R_3, R_3 \rightsquigarrow S_2\}$.

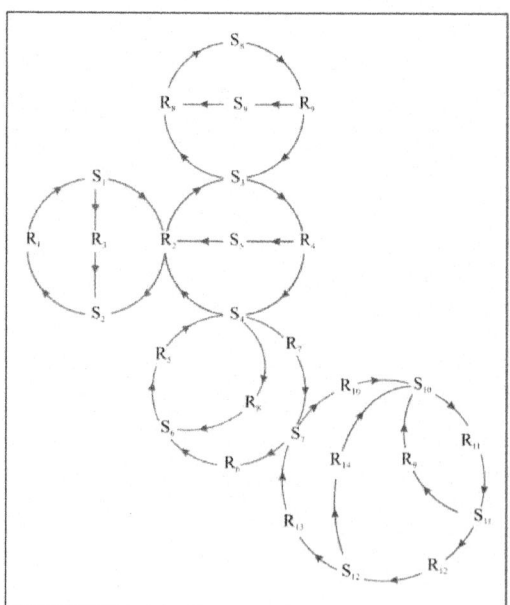

Figure 4: A hypothetical source with five blocks.

The *block-tree* of a connected graph depicts the way in which the various blocks of the graph are joined at its separating vertices. More precisely, the block-tree [18] of a connected graph is a bipartite graph whose vertices are of two kinds: symbols for the graph's blocks and symbols for the graph's separating vertices. If a block contains a particular separating vertex, then an edge is drawn from the block's symbol to the symbol for that separating vertex. In Figure 5 we show the block-tree for the hypothetical source depicted in Figure 4. For reasons that will be made clear later on, we have denoted the five blocks in the block-tree by the symbols $RB1$, $SB1$, $SB2$, $RB2$, and $RB3$. (We note that Figures 4 and and55 are somewhat unrepresentative, for it might happen that three or more source-blocks are adjacent to the same separating vertex.)

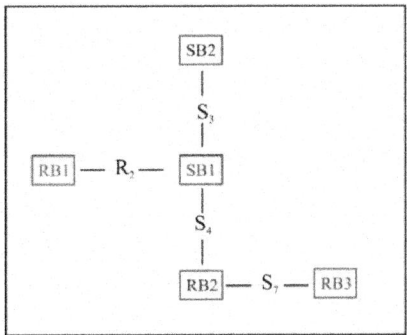

Figure 5: The block-tree for Figure 4.

A block is an *end block* if it contains no more than one separating vertex of the original graph. In this case, the block's symbol is a *leaf* of the block-tree graph. The end-blocks (leaves) in our example are those represented by $RB1$, $SB2$, and $RB3$.

Properties of Source-Blocks

Our aim in this subsection is to show that when condition (ii) of Theorem 2.1 is satisfied, the internal structure of source-blocks must have a certain degree of simplicity. Condition (ii) will exert itself primarily through the following proposition, which is proved in Appendix A.

Proposition 5.8

Suppose that, for the reaction network under consideration, the Species-Reaction Graph satisfies condition (ii) of Theorem 2.1. Then, in any source-block of the sign-causality graph, at most one of the following can obtain:

- There is a reaction vertex having more than two adjacent species vertices.
- There is a species vertex having more than two adjacent reaction vertices.

The proposition provides motivation for the following definition:

Definition 5.9

A source-block in the sign-causality graph is a *species block (S-block)* if each species node is adjacent to precisely two reaction nodes. A source-block in the sign-causality graph is a *reaction block (R-block)* if each reaction node is adjacent to precisely two species nodes.

Proposition 5.8 tells us that when condition (ii) of Theorem 2.6 is satisfied, every block within the sign-causality graph source is either an S-block or an R-block (or both in the case that the source-block is simply a single cycle). Note that in Figure 5 we have labeled the hypothetical source-blocks as *RBn* or *SBn* according to whether the corresponding source-block in Figure 4 is an R-block or an S-block.

We conclude this section with two more propositions. Neither is essential to the proofs of the main theorems of this paper, but they provide some additional and not-so-obvious properties of a sign-causality graph source.

The following proposition, proved in Appendix A, does not presuppose that condition (ii) of Theorem 2.1 is satisfied. Rather, it tells us about properties of R-blocks or S-blocks that might exist within a sign-causality-graph source. We already know that every directed cycle is even, but the proposition tells us that *all* cycles within a source's R-blocks and S-blocks are even.

Proposition 5.10

Every (not necessarily directed) cycle that lies within an S-block or an R-block of a sign-causality graph source is even.

The following proposition is a direct consequence of the two preceding ones:

Proposition 5.11

Suppose that, for the reaction network under consideration, the Species-Reaction Graph satisfies condition (ii) of Theorem 2.1. Then in any source of the sign-causality graph every cycle is even.

Properties of an end S-block

Recall that an *end* S-block in a source is an S-block that contains at most one separating vertex of the source. We consider properties of such an end S-block, designated ESB. There are three mutually exclusive possibilities: ESB contains no separating vertex at all; it contains just one separating vertex, and it is a reaction vertex; or it contains just one separating vertex, and it is a species vertex. For our purposes the first two possibilities can be treated together, while the third requires other considerations.

In fact, we show that when condition (i) of Theorem 2.1 holds, the first two possibilities cannot obtain; moreover, if the third obtains, we get sharpened information about the inequality in (26) corresponding to species at the block's separating vertex.

Possibilities 1 And 2: ESB Contains No Separating Vertex of The Source or It Contains A Separating Reaction Vertex Of The Source

Because it is strongly connected, ESB must contain a directed (and even) cycle, which we take to be

$$s_1 \rightsquigarrow R_1 \rightsquigarrow s_2 \rightsquigarrow R_2 \cdots \rightsquigarrow s_n \rightsquigarrow R_n \rightsquigarrow s_1. \tag{29}$$

Because ESB is a species-block and because the block contains no separating species-vertex of the source, each of the species in the block (and in the chosen cycle) is adjacent to at most two reactions of the block. Thus, the inequalities (26) corresponding to $s_1, s_2, \ldots, s_n$ reduce to

$$
\begin{aligned}
f_{R_n \rightsquigarrow s_1} |\alpha_{R_n}| - e_{s_1 \rightsquigarrow R_1} |\alpha_{R_1}| &> 0, \\
f_{R_1 \rightsquigarrow s_2} |\alpha_{R_1}| - e_{s_2 \rightsquigarrow R_2} |\alpha_{R_2}| &> 0, \\
&\vdots \\
f_{R_{n-1} \rightsquigarrow s_n} |\alpha_{R_{n-1}}| - e_{s_n \rightsquigarrow R_n} |\alpha_{R_n}| &> 0.
\end{aligned}
\tag{30}
$$

(Recall that, for a species-to-reaction edge $s \rightsquigarrow R$ of the sign-causality graph we denote by $e_{s \rightsquigarrow R}$ the stoichiometric coefficient of species s in the corresponding edge-labeling complex. For a reaction-to-species edge $R \rightsquigarrow s$ we denote by $f_{R \rightsquigarrow s}$ the stoichiometric coefficient of species s in its edge-labeling complex.) By sequentially invoking these inequalities from top to bottom, we can deduce from the last of them that

$$\left(\frac{f_{R_1 \rightsquigarrow s_2} f_{R_2 \rightsquigarrow R_3} \cdots f_{R_n \rightsquigarrow s_1}}{e_{s_1 \rightsquigarrow R_1} e_{s_2 \rightsquigarrow R_2} \cdots e_{s_n \rightsquigarrow R_n}} - 1 \right) |\alpha_{R_n}| > 0. \tag{31}$$

However, when condition (i) of Theorem 2.1 holds, (31) cannot obtain: By supposition $|\alpha_{R_n}|$ is positive. Given the $\rightsquigarrow$-orientation in the SR Graph, the even cycle (29) (viewed in the SR Graph) cannot be stoichiometrically expansive, so the first factor on the left of (31) is either zero or negative. (Recall (28).) Thus, we have a contradiction of (31).

We conclude, then, that when condition (i) of Theorem 2.1 holds, an end S-block must contain a separating*species* vertex of the source. We investigate next what happens in that case.

Possibility 3: ESB Contains a Separating Species Vertex Of The Source

Here again we consider a directed cycle in ESB, labeled as in (29). If ESB's (unique) separating species vertex is not in the cycle, then we would again obtain a contradiction, just as in § 5.7.1. We suppose, then, that species s_n is the separating vertex. Because ESB is a species-block, all other species vertices of the cycle are adjacent to precisely two reaction vertices, and those are in the cycle. Thus, all inequalities but the last in (30) remain unchanged. On the other hand, s_n is adjacent not only to R_{n-1} and R_n but also to certain reactions from nearby blocks sharing s_n as a species. Recall that $\mathcal{R}_0$ is the set of all reactions in the source under study. We denote by $\mathcal{R}_{ESB}$ the set of reactions in ESB and by $\mathcal{R}_0 \setminus \mathcal{R}_{ESB}$ the set of all source reactions not in ESB. Moreover, we let $\mathcal{R}_0 \setminus \mathcal{R}_{ESB} \rightsquigarrow s_n$ and $s_n \rightsquigarrow \mathcal{R}_0 \setminus \mathcal{R}_{ESB}$ be the sets of source edges residing outside of ESB that are, respectively, directed toward and away from species s_n.

In this case, the last inequality in (30) must be replaced by

$$f_{R_{n-1} \rightsquigarrow s_n} |\alpha_{R_{n-1}}| - e_{s_n \rightsquigarrow R_n} |\alpha_{R_n}| + \sum_{\mathcal{R}_0 \setminus \mathcal{R}_{ESB} \rightsquigarrow s_n} f_{R \rightsquigarrow s_n} |\alpha_R| - \sum_{s_n \rightsquigarrow \mathcal{R}_0 \setminus \mathcal{R}_{ESB}} e_{s_n \rightsquigarrow R} |\alpha_R| > 0.$$

(32)

Instead of (31), this time we deduce the inequality

$$\left(\frac{f_{R_1 \rightsquigarrow s_2} f_{R_2 \rightsquigarrow R_3} \cdots f_{R_n \rightsquigarrow s_1}}{e_{s_1 \rightsquigarrow R_1} e_{s_2 \rightsquigarrow R_2} \cdots e_{s_n \rightsquigarrow R_n}} - 1 \right) |\alpha_{R_n}| + \sum_{\mathcal{R}_0 \setminus \mathcal{R}_{ESB} \rightsquigarrow s_n} f_{R \rightsquigarrow s_n} |\alpha_R| - \sum_{s_n \rightsquigarrow \mathcal{R}_0 \setminus \mathcal{R}_{ESB}} e_{s_n \rightsquigarrow R} |\alpha_R| > 0.$$

(33)

When condition (i) of Theorem 2.1 obtains, however, the first term on the left cannot be positive for reasons given in §5.7.1. Thus, we arrive at the following inequality, *which relates entirely to source edges disjoint from ESB*:

$$\sum_{\mathcal{R}_0 \setminus \mathcal{R}_{ESB} \rightsquigarrow s_n} f_{R \rightsquigarrow s_n} |\alpha_R| - \sum_{s_n \rightsquigarrow \mathcal{R}_0 \setminus \mathcal{R}_{ESB}} e_{s_n \rightsquigarrow R} |\alpha_R| > 0.$$

(34)

For species s_n this amounts to a sharpened form of its counterpart in (26), a form we will draw upon later on.

Properties of an end R-block

We begin this subsection with an important proposition about R-blocks. A proof is provided.

Proposition 5.12

Suppose that, in a sign-causality graph source, an R-block has species set $\mathcal{S}^*$, and suppose that no directed cycle in the block is stoichiometrically expansive relative to the $\rightsquigarrow$ orientation. Then there is a set of positive numbers $\{M_s\}_{s \in \mathcal{S}^*}$ such that, for each causal unit $s \rightsquigarrow R \rightsquigarrow s'$ in the block, the following relation is satisfied:

$$f_{R \rightsquigarrow s'} M_{s'} - e_{s \rightsquigarrow R} M_s \leq 0.$$

(35)

Now we consider an end R-block in the source under consideration, designated ERB. We denote by $\mathcal{S}^*$ the set of species in ERB. In consideration of the source inequality system (26), we restrict our attention to just those inequalities corresponding to species in $\mathcal{S}^*$:

$$\sum_{\mathscr{R}_0 \rightsquigarrow s} f_{R \rightsquigarrow s} |\alpha_R| - \sum_{s \rightsquigarrow \mathscr{R}_0} e_{s \rightsquigarrow R} |\alpha_R| > 0, \quad \forall s \in \mathcal{S}^*.$$

(36)

We suppose that condition (i) of Theorem 2.1 is satisfied, so, by virtue of Remark 5.7, we can choose $\{M_s\}_{s \in \mathcal{S}}^*$ as in Proposition 5.12. If we multiply each inequality in (36) by the corresponding M_s and sum, we get the single inequality shown in (37).

$$\sum_{s \in \mathscr{S}^*} \left(\sum_{\mathscr{R}_0 \rightsquigarrow s} f_{R \rightsquigarrow s} M_s |\alpha_R| - \sum_{s \rightsquigarrow \mathscr{R}} M_s e_{s \rightsquigarrow \mathscr{R}} |\alpha_R| \right) > 0$$

(37)

As in §5.7 there are three possibilities: ERB contains no separating vertex; it contains just one separating vertex, and it is a reaction vertex; or it contains just one separating vertex, and it is a species vertex. We will show that neither of the first two possibilities can obtain. Then, as in §5.7.2, we will show that, if the third possibility is realized, the inequality in (26) corresponding to the species at the separating vertex can be sharpened to considerable advantage.

Possibilities 1 And 2: ERB Contains No Separating Vertex of The Source Or It Contains a Separating Reaction Vertex Of The Source

Because ERB is an R-block, each reaction is adjacent to precisely two species in the set S^*, which is to say that each reaction in ERB sits at the center of precisely one causal unit in ERB. (Recall that ERB is strongly connected.) Moreover, if either of the first two possibilities should obtain, no species is adjacent to a reaction not in ERB. In these cases, the inequality (37) can be rewritten. Let $\mathcal{U}$ be the set of causal units within ERB. Then (37) can be made to take the form shown in (38).

$$\sum_{s \rightsquigarrow R \rightsquigarrow s' \in \mathcal{U}} (f_{R \rightsquigarrow s'} M_{s'} - e_{s \rightsquigarrow R} M_s)|\alpha_R| > 0$$

(38)

However, (38) is contradicted by the attributes given to the set $\{M_s\}_{s \in S}^*$ in Proposition 5.12.

Thus, neither of the first two possibilities can obtain. ERB must contain a separating species vertex of the source. We examine next what can be said in that case.

Possibility 3: ERB Contains A Separating Species Vertex Of The Source

Suppose that ERB contains a species vertex s^* that is a separating vertex of the source under consideration. When condition (i) of Theorem 2.1 obtains, we can again choose positive numbers $\{M_s\}_{s \in S}^*$ to satisfy the requirements of Proposition 5.12, and the inequality (37) remains in force. On the other hand the passage from (37) to (38) becomes confounded by the fact that species vertex s^* is now adjacent to source edges not residing in ERB. We denote by $\mathcal{R}_{ERB}$ the set of reactions in ERB and by $\mathcal{R}_0 \setminus \mathcal{R}_{ERB}$ the set of all source reactions not in ERB. Moreover, we let $\mathcal{R}_0 \setminus \mathcal{R}_{ERB} \rightsquigarrow s^*$ and $s^* \rightsquigarrow \mathcal{R}_0 \setminus \mathcal{R}_{ERB}$ be the sets of source edges residing outside of ERB that are, respectively, directed toward and away from species s^*. As before, we let $\mathcal{U}$ be the set of causal units in ERB. In this case, (37) can be recast as (39).

$$\sum_{s \rightsquigarrow R \rightsquigarrow s' \in \mathcal{U}} (f_{R \rightsquigarrow s'} M_{s'} - e_{s \rightsquigarrow R} M_s)|\alpha_R| + M_{s^*}\left(\sum_{\mathcal{R}_0 \setminus \mathcal{R}_{ERB} \rightsquigarrow s^*} f_{R \rightsquigarrow s^*}|\alpha_R| - \sum_{s^* \rightsquigarrow \mathcal{R}_0 \setminus \mathcal{R}_{ERB}} e_{s^* \rightsquigarrow R}|\alpha_R| \right) > 0$$

(39)

Recall that the set $\{M_s\}_{s \in S}^*$ was chosen to satisfy the requirements of Proposition 5.12, so the first sum in (39) cannot be positive. Then, because M_s^* is positive, we must have

$$\sum_{\mathcal{R}_0 \backslash \mathcal{R}_{ERB} \leadsto s^*} f_{R \leadsto s^*} |\alpha_R| - \sum_{s^* \leadsto \mathcal{R}_0 \backslash \mathcal{R}_{ERB}} e_{s^* \leadsto R} |\alpha_R| > 0.$$

(40)

Note that this is a strengthened counterpart of the inequality in (26) corresponding to species s^*, *a counterpart that makes reference only to reactions external to the end reaction block ERB.*

The Concluding Argument: Leaf Removal

We begin this subsection with a review of what was established in §5.7 and §5.8: An end block in a sign-causality graph source, be it an S-block or an R-block, must contain a *species* separating vertex of the source. This implies that the hypothetical source depicted in Figure 4 (and having block-tree depicted in Figure 5) cannot in fact be a source, for it has an end block, corresponding to RB1 in Figure 5, that does not contain a separating *species* vertex of the putative source. (Each of the remaining end blocks does contain a separating species vertex.)

Moreover, we have established that the inequality in (26) corresponding to a separating species vertex s^* in an end block EB, be it an S-block or an R-block, can be strengthened to a form that makes no mention of reaction inside EB:

$$\sum_{\mathcal{R}_0 \backslash \mathcal{R}_{EB} \leadsto s^*} f_{R \leadsto s^*} |\alpha_R| - \sum_{s^* \leadsto \mathcal{R}_0 \backslash \mathcal{R}_{EB}} e_{s^* \leadsto R} |\alpha_R| > 0.$$

(41)

Here $\mathcal{R}_{EB}$ is the set of reactions in EB, and $\mathcal{R}_0 \backslash \mathcal{R}_{EB}$ the set of source reactions not in EB. Moreover, $\mathcal{R}_0 \backslash \mathcal{R}_{EB} \leadsto s^*$ and $s^* \leadsto \mathcal{R}_0 \backslash \mathcal{R}_{EB}$ are the sets of source edges residing outside of EB that are, respectively, directed toward and away from species s^*.

Now if EB is an end block in the source under consideration, we can replace the inequality in (26) corresponding to the unique separating species vertex s^* in EB with its strengthened form shown in (41). Thereafter, we can restrict the now-modified inequality system (26) to just those inequalities corresponding to s^* and to species residing outside EB. In effect, the resulting reduced system of inequalities corresponds to a subgraph of the source with the end block EB removed, but with s^* retained. Viewed in the source's block tree, this corresponds to removal of a leaf along with that leaf's adjacent species separating vertex (when that separating vertex is not adjacent to a different leaf).

It is not difficult to see that the arguments in §5.7 or §5.8 can then be applied to any end block EB' of the resulting source subgraph to produce a still smaller but strengthened inequality system corresponding to a still smaller source subgraph, a subgraph resulting from removal of EB'.

The process can be continued, amounting to a sequential pruning of the source's block-tree, with each stage corresponding to removal of a (perhaps new) leaf and perhaps its adjacent species separating vertex. The process will terminate when only one block remains, with the final source subgraph having no separating vertex at all. In this case, by arguments of §5.7.1 or §5.8.1 the corresponding reduced inequality system cannot be satisfied, and we have a contradiction.

This completes the proof of Theorem 2.1.

PROOF OF THEOREM 2.6

Here we prove Theorem 2.6, which is repeated below:

Theorem 2.6

A fully open reaction network is strongly concordant if its true-chemistry Species-Reaction Graph has the following properties:

- Every even cycle is an s-cycle.
- No two even cycles have a species-to-reaction intersection.

We begin by recalling the definition of strong concordance for an arbitrary reaction network $\{S, C, R\}$, not necessarily fully open, with stoichiometric subspace $S \subset \mathbb{R}^S$. Again we let $L : \mathbb{R}^R \to S$ be the linear map defined by

$$L\alpha = \sum_{y \to y' \in R} \alpha_{y \to y'} (y' - y).$$

$$(42)$$

Definition 6.1

A reaction network $\{S, C, R\}$ with stoichiometric subspace S is *strongly concordant* if there do not exist $\alpha \in \ker L$ and a non-zero $\sigma \in S$ having the following properties:

- For each $y \to y'$ such that $\alpha_{y \to y'} > 0$ there exists a species $s \in \mathrm{supp}\,(y - y')$ for which $\mathrm{sgn}\,\sigma_s = \mathrm{sgn}\,(y - y')_s$.
- For each $y \to y'$ such that $\alpha_{y \to y'} < 0$ there exists a species $s \in \mathrm{supp}\,(y - y')$ for which $\mathrm{sgn}\,\sigma_s = -\mathrm{sgn}\,(y - y')_s$.
- For each $y \to y'$ such that $\alpha_{y \to y'} = 0$, either (a) $\sigma_s = 0$ for all $s \in \mathrm{supp}\,y$, or

(b) there exist species s, $s' \in$ supp $(y - y')$ for which sgn σ_s = sgn $(y - y')_s$ and sgn $\sigma_{s'} = -$sgn $(y - y')_{s'}$.

The following lemma will take us just short of a proof of Theorem 2.6. Proof of the lemma is given in Appendix C.

Lemma 6.2

Suppose that a fully open reaction network with true chemistry $\{S, C, R\}$ is not strongly concordant. Then there is another true chemistry $\{S, C, R\}$ whose fully open extension is discordant and whose SR Graph is identical to a subgraph of the SR Graph for $\{S, C, R\}$, apart perhaps from changes in certain arrow directions within the reaction vertices.

Proof of Theorem 2.6 proceeds from Lemma 6.2 in the following way:

Proof of Theorem 2.6

Suppose that the SR Graph for a true chemistry $\{S, C, R\}$ satisfies conditions (i) and (ii) of the theorem but that, contrary to what is to be proved, the fully open extension of $\{S, C, R\}$ is not strongly concordant. Note that when the SR Graph of a true chemistry $\{S, C, R\}$ satisfies conditions (i) and (ii) of Theorem 2.6, so will any subgraph of that SR Graph. On the other hand, neither of those conditions depends upon the direction of arrows in the reaction nodes. Thus, the SR Graph of the true chemistry $\{S, C, R\}$ given by the lemma will also satisfy conditions (i) and (ii). But then, by Corollary 2.4 of Theorem 2.1, the fully open extension of $\{S, C, R\}$ could not be discordant, and we have a contradiction.

EXTENSIONS OF THE MAIN THEOREMS TO NETWORKS THAT ARE NOT FULLY OPEN

It is the purpose of this Section to elaborate on remarks made in §1.2.2.

When the SR Graph drawn for a true chemical reaction network satisfies the hypotheses of Theorems 2.1 or 2.6, these theorems tell us that the network's fully open extension is concordant (or strongly concordant). In this case, the fully open network inherits the many attributes described in [1] that accrue to all concordant (or strongly concordant) networks. We would like to know circumstances under which these theorems can be extended in their range to give concordance information about networks that are not fully open.

More generally, we would like to know conditions under which, for a given network, concordance or strong concordance of the network's fully open extension implies concordance or strong concordance of the network itself.[12] This is a question quite separate from SR Graph considerations. However, when the network satisfies such conditions and its underlying true chemistry SR Graph satisfies the hypothesis of Theorems 2.1 or 2.6, then the concordance properties ensured by those theorems for the fully open extension will be inherited by the original network. In [1] we showed that a *normal* network is concordant (strongly concordant) if its fully open extension is concordant (strongly concordant). Normality is a mild structural condition given in Definition 7.1 below. In [10] it was shown that *every weakly reversible network is normal.*

Theorems 7.10 and 7.11 below tell us that these same results also obtain for the still larger class of *weakly normal* networks. (See Definition 7.3.) In particular, a weakly normal network is concordant (strongly concordant) if its fully open extension is concordant (strongly concordant).

This improvement on results in [1] is helpful in itself, but it also has significance in another direction: We show in §7.1 that the class of weakly normal networks is synonymous with the very broad class of *nondegenerate* networks (Definition 7.6), which was described in the Introduction. Thus, *any nondegenerate reaction network with a concordant (strongly concordant) fully open extension is itself concordant (strongly concordant).* Moreover, we also show that, with respect to the possibility of concordance, *degenerate* reaction networks are not worth considering, for they are never concordant. In §7.3 we provide computational tests that serve to affirm network normality and weak normality (or, equivalently nondegeneracy).

NETWORK NORMALITY, WEAK NORMALITY, AND NONDEGENERACY

Definition 7.1

Consider a reaction network $\{S, C, R\}$ with stoichiometric subspace S. The network is *normal* if there are $q \in \mathbb{R}_+^S$ and $\eta \in \mathbb{R}_+^R$ such that the linear transformation $T : S \to S$ defined by

$$T\sigma := \sum_{y \to y' \in R} \eta_{y \to y'} (y * \sigma)(y' - y)$$

$$(43)$$

is nonsingular, where "$*$" is the scalar product in $\mathbb{R}^S$ defined by

$$x * x' := \sum_{s \in S} q_s x_s x'_s.$$

$$(44)$$

Remark 7.2

As indicated earlier, it was shown in [10] that every weakly reversible network is normal. Reference [10] also contains structural conditions that ensure normality for certain "partially open" networks that are not weakly reversible.

In preparation for the next definition we note that (43) can be written as

$$T\sigma := \sum_{y \to y' \in \mathscr{R}} (\eta_{y \to y'} y \circ q) \cdot \sigma (y' - y),$$

$$(45)$$

where "·" indicates the standard scalar product in $\mathbb{R}^{\mathcal{S}}$.

Definition 7.3

Consider a reaction network $\{\mathcal{S}, \mathcal{C}, \mathcal{R}\}$ with stoichiometric subspace S. The network is *weakly normal* if, for each reaction $y \to y'$, there is a vector $p_{y \to y'} \in \overline{\mathbb{R}}_+^{\mathcal{S}}$ with supp $p_{y \to y'}$ = supp y such that the linear transformation $T : S \to S$ defined by

$$\overline{T}\sigma := \sum_{y \to y' \in \mathscr{R}} P_{y \to y'} \cdot \sigma(y' - y)$$

$$(46)$$

is nonsingular. Here "·" is the standard scalar product in $\mathbb{R}^{\mathcal{S}}$.

Remark 7.4

A reaction network that is normal is also weakly normal. In fact, if $q \in \mathbb{R}_+^{\mathcal{R}}$ and $\eta \in \mathbb{R}_+^{\mathcal{R}}$ satisfy the requirements of Definition 7.1, then the choice $p_{y \to y'} := \eta_{y \to y'} y \circ q$, $\forall y \to y' \in \mathcal{R}$ will satisfy the requirements of Definition 7.3. On the other hand, a weakly normal network need not be normal. An example is given by

$$C \leftarrow A + B \to D \to 2A, \qquad (47)$$

which is weakly normal but not normal. Because every weakly reversible network is normal, it follows that every weakly reversible network is also weakly normal.

Remark 7.5

For readers familiar with standard language of chemical reaction network theory [16], a network cannot be normal if

$$t - \ell - \delta > 0, \qquad (48)$$

where t is the number of terminal strong linkage classes, ℓ is the number of linkage classes, and δ is the deficiency. This follows without much difficulty from [19]; see also [16]. For network (47), $t = 2$, $\ell = 1$, and $\delta = 0$, so normality is precluded by condition (48). Network (47) illustrates, however, that the same condition does not also preclude weak normality.

Definition 7.6

A reaction network is *nondegenerate* if there exists for it differentiably monotonic kinetics (§3.2) such that at some positive composition c^* the derivative of the species-formation-rate function $df(c^*) : S \to S$ is nonsingular. Otherwise, the network is *degenerate*.

Proposition 7.7

A reaction network is nondegenerate if and only if it is weakly normal.

Proof

Suppose that a network $\{S, C, \mathcal{R}\}$ is nondegenerate. Then there is for the network a kinetics $\mathcal{K}$ such that, at some composition $c* \in \mathbb{R}_+^{\mathcal{S}}$, the kinetics is differentiably monotonic and, moreover, the derivative of the species-formation-rate function, $df(c^*) : S \to S$, is nonsingular. In this case, for each $\sigma \in S$

$$df(c^*)\sigma = \sum_{y \to y' \in \mathscr{R}} \nabla \mathscr{K}_{y \to y'}(c^*) \cdot \sigma(y' - y),$$

(49)

where the components of $\nabla \mathscr{K}_{y \to y'}(c^*)$ have the non-negativity properties that follow from differentiable monotonicity (§3.2). By taking

$$p_{y \to y'} = \nabla \mathscr{K}_{y \to y'}(c^*), \quad \forall y \to y' \in \mathscr{R}$$

we can establish that the network is weakly normal.

On the other hand, suppose that the network $\{S, C, \mathcal{R}\}$ is weakly normal and, in particular, that the set $\{p_{y \to y'}\}_{y \to y' \in \mathcal{R}}$ satisfies the requirements of Definition 7.3. Let $\mathcal{K}$ be the (differentiably monotonic) kinetics defined by

$$p_{y \to y'} = \nabla \mathscr{K}_{y \to y'}(c^*), \quad \forall y \to y' \in \mathscr{R}$$

and let $c* \in \mathbb{R}_+^{\mathcal{S}}$ be such that $c*s = 1$ for each $s \in S$. Note that

$$\nabla \mathscr{K}_{y \to y'}(c^*) = p_{y \to y'}, \quad \forall y \to y' \in \mathscr{R}.$$

From this, (49), and the properties of the set $\{p_{y \to y'}\}_{y \to y' \in \mathcal{R}}$ given by Definition 7.3 it follows that $df(c^*) : S \to S$, is nonsingular, whereupon the network is nondegenerate.

Remark 7.8

Note that in the proof that nondegeneracy implies weak normality we did not actually require that the network be nondegenerate. In particular, we did not require that the kinetics $\mathcal{K}$ be differentiably monotonic at *all* positive compositions, only that it be differentiably monotonic at *one* composition, c^* (and, of course, that $df(c^*)$ be nonsingular). However, when these apparently milder conditions are invoked, the network *must* be nondegenerate nevertheless: The seemingly milder conditions result in weak normality, and, as the second part of the proof indicates, weak normality implies nondegeneracy. The following proposition indicates that a network that is not weakly normal (or, equivalently, is degenerate) has no chance of being concordant.

Proposition 7.9

A reaction network that is not weakly normal is discordant. Equivalently, every degenerate network is discordant.

Proof

Suppose that a reaction network $\{S, \mathcal{C}, \mathcal{R}\}$ is not weakly normal (and, in particular, is not normal). From Definition 7.3 it follows that, for the special choice $p_{y \rightarrow y'} = y$, $\forall y \rightarrow y' \in \mathcal{R}$, the corresponding map $T : S \rightarrow S$ given by (46) must be singular. This is to say that there is a nonzero $\sigma^* \in S$ such that

$$\sum_{y \rightarrow y' \in \mathcal{R}} y \cdot \sigma^* (y' - y) = 0.$$

Now let $\alpha \in \mathbb{R}^{\mathcal{R}}$ be defined by $\alpha_{y \rightarrow y'} = y \cdot \sigma^*$, $\forall y \rightarrow y' \in \mathcal{R}$. Then, in view of Definition 4.1, the pair consisting of α and σ^* serve to establish discordance of the network under consideration.

Concordance of a Network and Of Its Fully Open Extension

The following theorems about weakly normal networks amount to straight-forward extensions of theorems in [1] about normal networks; the proofs are almost identical. Although these theorems give the former ones a somewhat greater range, their main interest lies in the fact that they can be stated in terms of the more tangible, but equivalent, notion of network nondegeneracy.

Theorem 7.10

A weakly normal (or, equivalently, nondegenerate) network is concordant if its fully open extension is concordant. In particular, a weakly reversible network is concordant if its fully open extension is concordant.

Theorem 7.11

A weakly normal (or, equivalently, nondegenerate) network is strongly concordant if its fully open extension is strongly concordant. In particular, a weakly reversible network is strongly concordant if its fully open extension is strongly concordant. Taken together, Proposition 7.7, Proposition 7.9, Theorem 7.10, and Theorem 7.11 tell us that, in the class of networks that have a concordant (strongly concordant) fully open extension, the concordant (strongly concordant) ones are *precisely* the nondegenerate ones:

Corollary 7.12

Consider a reaction network that has a concordant (strongly concordant) fully open extension. Then the original network is concordant (strongly concordant) if and only if it is nondegenerate.

Remark 7.13

It is a consequence of Theorem 7.10 and results in [1] that the dynamical statements (i) – (iii) of Theorem 2.8 hold true for any nondegenerate reaction network whose fully open extension is concordant, not merely those that satisfy the SR Graph conditions of Theorem 2.1. In particular, the "all or nothing" observation of Remark 2.9 still obtains. The SR Graph conditions of Theorem 2.1 merely suffice to ensure concordance of the fully open extension.

Tests for Network Normality, Weak Normality, And Nondegeneracy

Here we provide some computational means to affirm normality and weak normality (or, equivalently, nondegeneracy) of a reaction network. Recall that the rank of a network is the rank of its set of reaction vectors.

Proposition 7.14

A reaction network $\{S, C, R\}$ of rank r is weakly normal (or, equivalently, nondegenerate) if there is a set of r reactions $\{yi \rightarrow y'i\}i=1...r$ and a set of vectors $\{pi\}i=1...r \subset \overline{R}\;S+$ with supp p_i = supp y_i, $i = 1 \ldots r$, such that the matrix

$$\left[p_i \cdot (y'_j - y_j) \right]_{i,j=1...r}$$

$$(50)$$

has nonzero determinant.

Proof of the proposition is provided at the end of this subsection. For a particular choice of r reactions one can readily construct the matrix (50) in terms of symbols for the species-wise components of the vectors $\{p_i\}_{i=1...r}$, and one can then calculate the determinant of the matrix as a polynomial in those same symbols. If, for *even one* choice of r reactions, the resulting determinant is not *identically* zero, then the network is weakly normal.

We remark in passing that a nonzero determinant will require that the r reaction vectors $\{y'_i - y_i\}$ i=1...r be linearly independent. Of course, such independent reaction vector sets will invariably exist for a network of rank r.

A special choice of $\{p_i\}_{i=1...r}$, one that obviates the need for symbolic computation, is invoked in the following corollary. This choice will often suffice to establish weak normality. In fact, when the condition in Corollary 7.15 below is satisfied, the network will not only be weakly normal but also normal. (See Remark 7.18 following the proof of Proposition 7.14.)

Corollary 7.15

A reaction network $\{S, C, R\}$ of rank r is weakly normal (and, in fact, normal) if there is a set of r reactions $\{y_i \rightarrow y'_i\}_{i=1...r}$ such that the matrix

$$\left[y_i \cdot (y'_j - y_j)\right]_{i,j=1...r}$$

$$(51)$$

has nonzero determinant.

Example 7.16

Here we apply Corollary 7.15 to network (1), the rank of which is 4. For the four-reaction set

$$\{P \rightarrow A+B, Q \rightarrow B+C, 2A \rightarrow C, C+D \rightarrow Q+E\}, \qquad (52)$$

it is easy to calculate that the matrix in the corollary has a determinant of 4, so the network is weakly normal. Recall that Theorem 2.1 established the concordance of the fully open extension of network (1). Because network (1) is weakly normal, concordance of its fully open extension extends to network (1) itself.

Remark 7.17

Corollary 7.15 is only one consequence of Proposition 7.14. There are other, more interesting ones (some with a graphical flavor) that we intend to take up in another article. For example, a network $\{S, C, R\}$ of rank r is weakly normal if there exist r reactions $\{y_i \rightarrow y_i'\}i=1\ldots r$ with the following property: There is a set of distinct species $S_* = \{s_1, \ldots, s_r\}$ with $s_i \varepsilon$ supp y_i, $i = 1 \ldots r$, such that

$$\det \left[s_i \cdot (y_j' - y_j) \right]_{i,j=1\ldots r} \neq 0.$$

(53)

It is not difficult to see that (53) has the following interpretation: Let $\{\bar{y}_i \rightarrow \bar{y}_i'\}_{i=1\ldots r}$ denote the set of "reactions" obtained from $\{y_i \rightarrow y_{r_i}\}i=1\ldots r$ by stripping away all species not in S_*. Then (53) is satisfied (whereupon the original network is weakly normal) precisely when the resulting set of "reaction vectors" $\{\bar{y}_i' - \bar{y}_i\}_{i=1\ldots r}$ is linearly independent.

Proof of Proposition 7.14

Suppose that the reaction vector set $\{\bar{y}_i \rightarrow \bar{y}_i'\}_{i=1\ldots r}$ and the set $\{p_i\}_{i=1\ldots r} \subset \bar{\mathbb{R}}_+'$ satisfy the conditions of the Proposition 7.14. In this case, it is not difficult to see that each of these sets must be linearly independent. In particular, the set $\{y_i' - y_i\}i=1\ldots r$ is a basis for S, the stoichiometric subspace for the network under consideration.

We begin by constructing the linear transformation $T_0 : S \rightarrow S$ defined by

$$T_0 \sigma = \sum_{i=1}^{r} p_i \cdot \sigma (y_i' - y_i), \quad \forall \sigma \in S,$$

(54)

which can be seen to be nonsingular in the following way: Suppose on the contrary that there is a nonzero $\sigma^* \in S$ such that $T_0 \sigma^* = 0$. Because the set $\{y_i' - y_i\}i=1\ldots r$ is linearly independent, we must have

$p_i \cdot \sigma^* = 0, i=1\ldots r.$

(55)

Because $\{y'j - yj\}j=1\ldots r$ is a basis for S and σ^* is a nonzero member of S, there must be $\xi_j, j = 1 \ldots r$, not all zero, such that

$$\sigma^* = \sum_{j=1}^{r} \xi_j (y_j' - y_j).$$

(56)

Insertion of this into (55) results in the system of r homogeneous equations

$$\sum_{j=1}^{r} p_i \cdot (y_j' - y_j) \xi_j = 0, \quad i=1\ldots r$$

(57)

that must be satisfied by the set $\{\xi_j\}_{j=1\ldots r}$. Since the determinant of the matrix (50) is nonzero, the only solution is $\xi_j = 0$, $j = 1 \ldots r$, which amounts to a contradiction. Thus, T_0 is nonsingular, and, as a result, $\det T_0 \neq 0$.

It remains to be shown that the requirements of weak normality are met by the network $\{S, C, R\}$. For this purpose, let R_o be the set of aforementioned reactions $\{y_i \to y'_i\} i=1 \ldots r$. Moreover, let $p_{y \to y'} \in \overline{\mathbb{R}}_+^{\mathcal{S}}$ be chosen to satisfy

$$p_{y_i \to y'_i} := p_i, \quad \forall y_i \to y'_i \in R_0$$
$$p_{y \to y'} := \varepsilon y, \quad \forall y \to y' \in R \backslash R_0,$$

where ε is a small positive number. Now let $T_\varepsilon : S \to S$ be defined by

$$\overline{T}_\varepsilon \sigma := \sum_{y \to y' \in R} p_{y \to y'} \cdot \sigma (y' - y)$$
$$= \overline{T}_0 \sigma + \varepsilon \left[\sum_{y \to y' \in R \backslash R_0} y \cdot \sigma (y' - y) \right]$$

Note that $T_\varepsilon|_{\varepsilon=0} = T_0|_0$. Because det $T_0 \neq 0$ and because det $T_\varepsilon|$ is continuous in ε, it follows that det $T_\varepsilon \neq 0$ for ε sufficiently small. Thus, for sufficiently small ε, T_ε is nonsingular, whereupon the network $\{S, C, R\}$ is weakly normal.

Remark 7.18

Corollary 7.15 derives from Proposition 7.14 by invoking the special choice $p_i = y_i$, $i = 1 \ldots r$. When the resulting condition in Corollary 7.15 is satisfied, the network at hand is not only weakly normal but also normal. To see this, it is enough to replace p_i by y_i everywhere in the proof of Proposition 7.14 and then invoke Definition 7.1 with $q_s = 1, \forall s \in S$.

A CONCLUDING REMARK

When their hypotheses are satisfied, the central theorems of this article permit one to affirm concordance of a particular reaction network from inspection of its Species Reaction Graph and, then, to invoke all of the powerful dynamical consequences that concordance implies [1]. At the same time, these theorems provide delicately nuanced insight into network attributes that give rise to concordance. And, because their hypotheses are fairly easy to satisfy, the theorems tell us more — that concordance in realistic chemical reaction networks is likely to be common. Moreover, the theorems also serve to make connections between concordance results in [1] and earlier, somewhat different SR-Graph-related results contained in [9, 4, 5, 6,7]. It should be remembered, however, that computational means for direct determination of whether a particular network — *fully open or otherwise* —is concordant *or discordant*, strongly concordant *or not strongly concordant*, are already available in user-friendly, freely-provided software [2] prepared in connection with [1]. In most instances, this software or a variant of it, will be the tool of choice.

ACKNOWLEDGMENTS

We are grateful to Daniel Knight for helpful discussions and to Uri Alon for his support and encouragement of this work.

REFERENCES

1. Shinar G, Feinberg M. Concordant chemical reaction networks. 2012 arXiv:1109.2923 (2011). Under revision for Mathematical Biosciences at the request of the editor.

2. Ji H, Ellison P, Knight D, Feinberg M. The chemical reaction network toolbox, version 2.1. 2011 Available at http://www.chbmeng.ohio-state.edu/~feinberg/crntwin/

3. Ji H. PhD thesis. Department of Mathematics, The Ohio State University; 2011. Uniqueness of Equilibria for Complex Chemical Reaction Networks.

4. Craciun G, Feinberg M. Multiple equilibria in complex chemical reaction networks. II. the species-reaction graph. SIAM Journal on Applied Mathematics. 2006;66:1321–1338.

5. Craciun G, Tang Y, Feinberg M. Understanding bistability in complex enzyme-driven reaction networks. Proceedings of the National Academy of Sciences. 2006;103:8697–8702. [PMC free article] [PubMed]

6. Banaji M, Craciun G. Graph-theoretic approaches to injectivity and multiple equilibria in systems of interacting elements. Communications in Mathematical Sciences. 2009;7:867–900.

7. Banaji M, Craciun G. Graph-theoretic criteria for injectivity and unique equilibria in general chemical reaction systems. Advances in Applied Mathematics. 2010;44:168–184. [PMC free article] [PubMed]

8. Schlosser PM. PhD thesis. University of Rochester; 1988. A Graphical Determination of the Possibility of Multiple Steady States in Complex Isothermal CFSTRs.

9. Schlosser PM, Feinberg M. A theory of multiple steady states in isothermal homogeneous CFSTRs with many reactions. Chemical Engineering Science. 1994;49:1749–1767.

10. Craciun G, Feinberg M. Multiple equilibria in complex chemical reaction networks: Semiopen mass action systems. SIAM Journal on Applied Mathematics. 2010;70:1859–1877.

11. Craciun G, Feinberg M. Multiple equilibria in complex chemical reaction networks. I. the injectivity property. SIAM Journal on Applied Mathematics. 2005;65:1526–1546.

12. Craciun G, Feinberg M. Multiple equilibria in complex chemical reaction networks: extensions to entrapped species models. IEE Proc Syst Biol. 2006;153:179–186. [PubMed]

13. Shinar G, Rabinowitz JD, Alon U. Robustness in glyoxylate bypass regulation. PLoS Comput Biol.2009;5:e1000297. [PMC free article] [PubMed]

14. Shinar G, Feinberg M. Structural sources of robustness in biochemical reaction networks. Science (New York, NY) 2010;327:1389–1391. [PubMed]

15. Shinar G, Feinberg M. Design principles for robust biochemical reaction networks: What works, what cannot work, and what might almost work. Mathematical Biosciences. 2011;231:39–48. [PMC free article] [PubMed]

16. Feinberg M. Lectures on chemical reaction networks. Written version of lectures given at the Mathematical Research Center, University of Wisconsin, Madison, WI. 1979 Available atwww.chbmeng.ohio-state.edu/~feinberg/LecturesOnReactionNetworks.

17. Biggs N. Algebraic Graph Theory. 2. Cambridge University Press; 1993.

18. Bondy A, Murty U. Graph Theory. Springer; 2010.

19. Feinberg M, Horn FJM. Chemical mechanism structure and the coincidence of the stoichiometric and kinetic subspaces. Archive for Rational Mechanics and Analysis. 1977;66:83–97.

20. Feinberg M. Necessary and sufficient conditions for detailed balancing in mass action systems of arbitrary complexity. Chemical Engineering Science. 1989;44:1819–1827.

21. Gale D. The Theory of Linear Economic Models. University of Chicago Press; 1960.

Chapter 9

REQUIRED LEVELS OF CATALYSIS FOR EMERGENCE OF AUTOCATALYTIC SETS IN MODELS OF CHEMICAL REACTION SYSTEMS

Wim Hordijk [1], Stuart A. Kauffman [2,3,4] and Mike Steel [5]

[1] Department of Ecology and Evolution, University of Lausanne, 1015 Lausanne, Switzerland

[2] Tampere University of Technology, Finland

[3] University of Vermont, 85 South Prospect Street Burlington, VT 05405, USA

[4] Santa Fe Institute, 1399 Hyde Park Road Santa Fe, NM 87501, USA

[5] Department of Mathematics and Statistics, University of Canterbury, Private Bag 4800, Christchurch, New Zealand

ABSTRACT

The formation of a self-sustaining autocatalytic chemical network is a necessary but not sufficient condition for the origin of life. The question of whether such a network could form "by chance" within a sufficiently complex suite of molecules and reactions is one that we have investigated for a simple chemical reaction model based on polymer ligation and cleavage. In this paper, we extend this work in several further directions. In particular, we investigate in more detail the levels of catalysis required for a self-sustaining autocatalytic network to form. We study the size of chemical networks within which we might expect to find such an autocatalytic subset, and we extend the theoretical and computational analyses to models in which catalysis requires template matching.

INTRODUCTION

In previous work we introduced and investigated a mathematical model of catalytic reaction systems and autocatalytic sets [1,2]. It was shown, both theoretically and computationally, that a linear growth rate in the level of catalysis (with increasing length n of the largest molecules in the system) is sufficient for autocatalytic sets to arise spontaneously [2,3] in a well-known binary polymer model of catalytic reaction systems [4,5].

In this paper we take a closer and more detailed look at our model and its results. First, we introduce a small modification to our mathematical definition of autocatalytic sets and the corresponding algorithm for finding them in general catalytic reaction systems (Section 3). This modification makes both the definition and the algorithm slightly simpler, and includes some specific (although probably rare) cases of autocatalytic sets which were previously left out. However, we show (formally, and in simulations) that this modified algorithm does not invalidate any previous results or conclusions.

Second, we show that there is a discrepancy between the theoretical and simulation results (Section 4). Both results show that a linear growth rate in level of catalysis is sufficient for the emergence of autocatalytic sets. However, there is a difference in the parameter values of these linear relations. Here, we recalculate and compare the required levels of catalysis in more detail and under different scenarios.

Third, we show how our model and algorithm can be used to answer other interesting questions relating to the emergence of autocatalytic sets (Section 5). In particular: What is the minimum required size of the molecule set for autocatalytic sets to emerge given a fixed (known) probability of a molecule catalyzing an arbitrary reaction?

Fourth, we show how more chemical realism can be included in our model, for example by considering template-based catalysis (Section 6). Even though this makes the model harder to analyze, it still generates interesting and useful results.

The next section briefly reviews our previously introduced model and definitions. The four sections following it will present the model modifications, extensions, and additional results mentioned above. The final section summarizes the main conclusions and discusses future directions. Mathematical proofs are provided in an .

Our study fits within a large and growing body of work that aims to formally model how self-sustaining biochemical systems necessary for life might have emerged. This is an area that has been investigated from many angles over the last three decades. Some approaches that are similar in scope but different in their specific details from the one we study here include models based on Petri-nets [6], algebraic approaches based on metabolic closure (such as Rosen's "(M,R) systems", [7–9]), computer simulations of autocatalytic networks involving artificial chemistry [10] or metabolic networks [11], differential equation modeling [12–15], and Erdös-Renyi style random graph theory [16]. The idea of autocatalytic sets as a precursor to life has certainly not been without criticism [17–19], but recent (exciting) experimental evidence shows that they are a real possibility [20–24]. Thus, we believe that our theoretical and computational studies and results are of direct relevance in the larger context of the origin of life [5,25,26].

RAF SETS

Autocatalytic sets may have played an important role in the origin of life [5,15,26,27], and are a necessary, although not sufficient, condition for life. Here, *autocatalytic sets* are defined more formally as *RAF sets* as follows (see [1,2] for the full mathematical definition and notations). Given a *catalytic reaction system* (CRS), *i.e.*, a network of molecule types and catalyzed chemical reactions, a (sub)set R of such reactions (plus the molecules involved in the reactions in R) is called:

1. Reflexively autocatalytic (RA) if every reaction in R is catalyzed by at least one molecule involved in any of the reactions in R;

2. F-generated (F) if every reactant in R can be constructed from a small "food set" F by successive applications of reactions from R;

3. Reflexively autocatalytic and F-generated (RAF) if it is both RA and F.

The food set *F* contains molecules that are assumed to be freely available in the environment. Thus, an RAF set formally captures the notion of "catalytic closure", *i.e.*, a self-sustaining set supported by a steady supply of (simple) molecules from some food set. Figure 1 shows a simple example. In [2], we also introduced a polynomial-time algorithm to find RAF sets in general catalytic reaction systems.

Note (as already stated earlier [26]) that this notion of an autocatalytic *set* is somewhat different from the (chemical) term autocatalytic *reaction* in which a molecule directly catalyzes its own production. With an autocatalytic set we do *not* mean a set of autocatalytic reactions, but rather a set of molecules and reactions which is collectively autocatalytic in the sense that all molecules help in producing each other (through mutual catalysis, and supported by a food set). Because of this confusion in terminology, we prefer to use the term "RAF set".

A Model of Catalytic Reaction Systems

In [4,5], a random CRS model was introduced using binary polymers (bit strings) of length at most n as molecule types, ligation and cleavage reactions, and a probability p of any molecule catalyzing any reaction (n and p are parameters of the model). It was argued that in a "sufficiently complex" system (*i.e.*, large enough diversity of molecule types), autocatalytic sets would appear "spontaneously". This was later criticized for requiring an exponential increase (with *n*) in the (average) number of reactions catalyzed by any one molecule [17], which is chemically unrealistic. However, in [2] it was shown computationally (by applying the RAF algorithm to instances of the random CRS model), and then confirmed theoretically in [3], that only a *linear* growth rate (with *n*) in the level of catalysis is sufficient for RAF sets to appear with high probability. Furthermore, with these results it is possible to quantify "sufficiently complex" in terms of the size of the molecule set (or maximum molecule length *n*) and the level of catalysis (average number of reactions catalyzed by any molecule). Here, we continue our investigations of this random CRS model, which we will refer to throughout as the *binary polymer model*.

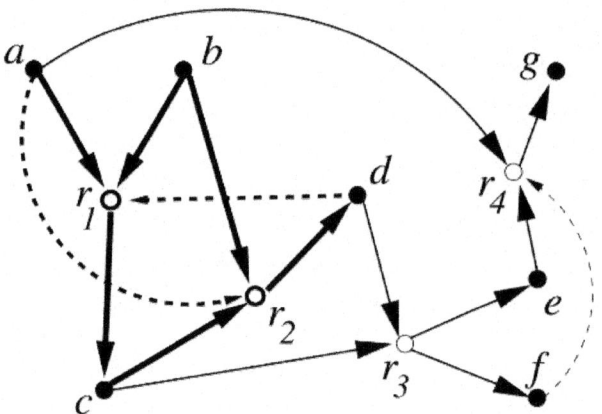

Figure 1: A simple example of a *catalytic reaction system* (CRS) with seven molecule types {*a, b, c, d, e, f, g*} (solid nodes) and four reactions {r_1, r_2, r_3, r_4} (open nodes). The food set is $F = \{a, b\}$. Solid arrows indicate reactants going into and products coming out of a reaction, dashed arrows indicate catalysis. The subset R = {r_1, r_2} (shown with bold arrows) is RAF.

MODIFICATION OF THE ORIGINAL DEFINITION AND ALGORITHM

In the original (mathematical) definition of RAF sets [2], the "RA" part is stated in terms of the *support* supp(R) of a set of reactions R, which is defined as all the molecules (reactants and products) that are involved in at least one of the reactions in R. For a set of reactions R to be reflexively autocatalytic (RA), it is required that every reaction in R is catalyzed by at least one molecule in supp(R).

Similarly, the "F" part of the definition is stated in terms of the *closure* $cl_R(F)$ of the food set *F* relative to the set of reactions R, which is defined as *F* together with all molecules that can be constructed from *F* by repeated applications of reactions in R. For a set of reactions R to be *F*-generated (*F*), it is required that the reactants of every reaction in R are in $cl_R(F)$.

Now imagine a situation where all reactions in R conform to the above two requirements, except for one reaction $r \in R$ which is catalyzed by a molecule $x \in F$ which is not involved in any other way in any reaction in R. So, $x \in cl_R(F)$, but $x \notin$ supp(R). In words, the molecule x is in the closure of the food set (by default, as it is part of the food set), but it is not in the support of the reaction set. So, according to the original definition, this set R is not RAF (it does not conform to the RA part of the definition), even though logically one would consider this case to be a proper autocatalytic set.

To remedy this, and make sure these (probably rare but relevant) cases are also included, we propose a slight modification to our original definition of RAF sets as follows:

RAF Definition

Given a catalytic reaction system Q = (*X*, R, *C*), with molecule set *X*, reaction set R, catalyzation *C* (a set of molecule-reaction pairs indicating which molecules catalyze which reactions), a nonempty subset R' of R is said to be:

- Reflexively autocatalytic (RA) if for all reactions r $\in$ R' there exists a molecule x $\in$ clR' (F) such that (x, r) $\in$ C;
- F-generated if ρ(R') $\subseteq$ clR' (F), where ρ(R') is the set of all reactants in R';
- Reflexively autocatalytic and F -generated (RAF) if R' is both RA and F.

In other words, a set of reactions R' is RAF if each reaction in R' is catalyzed by at least one molecule that can be produced by the set itself (starting from the food molecules), and all reactants (of all reactions in R') can also be produced by the set itself. So, instead of using supp(R) in the RA part of the definition and $cl_R(F)$ in the F part, as in the original definition, the modified definition uses $cl_R(F)$ in both the RA and the F part, thus simplifying it slightly. Note that an RAF in the original setting of [2] is still an RAF under this new definition by virtue of the following result, a proof of which is provided in the .

Lemma 3.1 Any set of reactions that forms an RAF under the earlier definition of [2] is an RAF in the modified definition above.

This may seem like a minor point, but it could be an important one. Consider, for example, the (reverse) citric acid cycle, which has been argued to have (possibly) been a major step in the origin of life by synthesizing the basic building blocks of organic molecules [28,29]. Even though at present the catalysts that drive the reactions in this cycle are enzymes (proteins), at the early stages these reactions could very well have been catalyzed by much simpler, naturally occurring elements, or ones that are easily synthesized (by basic chemistry) from freely available inorganic molecules [30]. Thus, these original catalysts can be considered food molecules, but they are not involved (as reactants or products) in any of the reactions in the (reverse) citric acid cycle (*i.e.*, they are not in the support). As a consequence, this important metabolic network of reactions would not be classified as an RAF with our original definition, but it is indeed a true RAF according to our modified definition.

Lemma 3.1 has several other desirable properties. First, it ensures that the recent result that Rosen's (M,R) systems can be viewed and studied within the RAF framework [9], still holds. Second, it tells us that if a system has no RAF under the new definition, then it clearly has none under the original one, nor, indeed, under more embellished definitions of an RAF that impose further conditions so as to avoid "trivialities". One such condition (from [1]) would be to require that not all reactions in an RAF are catalyzed by molecules in F (or in some larger subset S of molecules that can be generated from F by catalysis from F and resulting molecules). However, we can accommodate this additional condition within the new RAF framework as follows. The algorithm we will describes below constructs a unique maximal RAF (under the new definition), and from this one can easily check whether an RAF exists that satisfies the additional non-triviality condition (simply check whether this maximal RAF contains a reaction that does not have all its catalysts in F, or the larger set S). Third, spontaneous reactions (those that can proceed without any catalyst) can be allowed (and form part of an RAF) if we formally extend F by an extra "token" element with directed (catalysis) arrows from that token to all such spontaneous reactions. And there are various other extensions, restrictions, or variations that can be imposed on our more general framework.

Of course there are many more properties relevant to the origin of self-sustaining biochemistry beyond such ad-hoc conditions to exclude trivialities—for example, the dynamics of the reactions (the quantity of reagents and products (stoichiometry) along with thermodynamic considerations) and the effect of inhibition and degrading side reactions [18,19]. However, our view here is that RAF sets should be regarded very much as minimal *necessary* conditions for such systems, rather than in any sense *sufficient*. Identifying RAF sets, and establishing conditions for their existence is thus a natural question on the road to finding viable candidates for the origin of early biochemistry.

An additional benefit of our modified RAF definition is that it significantly simplifies the corresponding algorithm for finding RAF sets in general catalytic reaction systems, and its correctness proof. The original algorithm is based on repeatedly (and alternately) applying two reduction steps (starting from the full reaction set) [2]:

1. Remove all reactions that do not conform to the RA requirement;
2. Remove all reactions that do not conform to the F requirement.

However, since in our modified definition both the RA part and the F part are stated in terms of the closure, these two reduction steps can now be merged into one. As a consequence, the algorithm can be simplified to the following:

RAF Algorithm

1. Start with the complete set of reactions R and the food set F;
2. Compute the closure of the food set $cl_R(F)$ relative to the current set of reactions R;
3. For each reaction $r \in R$ for which (1) all catalysts, or (2) one or more reactants, are not in $cl_R(F)$, remove r from R;
4. Repeat steps 2 and 3 until no more reactions can be removed.

The resulting (reduced) reaction set R is either the (maximal) RAF set contained in the given catalytic reaction system, or it is empty, in which case there is no RAF. This was already proved for the original algorithm in [2], and a (simpler) proof for the modified version is provided. Note, however, that the overall running time of this modified algorithm is still the same as in the original case ($O(|R|^2 \log |R|)$ worst-case, but shown to be sub-quadratic on average in practice [2]), as it is dominated by step 2 (computing the closure of the food set). All results presented in this paper are generated with this new version of the RAF algorithm.

REQUIRED LEVELS OF CATALYSIS

In [2], we showed through computer simulations that a linear growth rate in the level of catalysis (with n, the length of the largest molecules in the system) appears to be sufficient for RAF sets to occur with high probability in the binary polymer model with ligation and cleavage reactions and random catalysis. This was subsequently confirmed theoretically in [3]. However, even though both the computational and theoretical results give a linear relation for the required level of catalysis, there appears to be a significant difference in the values of the parameters of these linear functions.

This discrepancy can partly be explained by the fact that the theoretical analysis in [3] actually assumes RAF sets that involve *all* molecule types in the system, *i.e.*, an RAF that contains the entire molecule set X (but not necessarily all reactions). This, of course, is a much stronger assumption than used in the simulation studies in [2], where the RAF algorithm was used to find *any* RAF set, regardless of how many molecules or reactions it contains. But the question remains whether the discrepancy can be explained entirely by this difference.

To answer this, we repeated the original simulations, using the RAF algorithm to find *any* RAF set, but this time with the modified RAF algorithm. Then we also applied the RAF algorithm to look for RAF sets that involve *all* molecule types in X. In both cases, we collected statistics for the average number $f(n)$ of reactions catalyzed by any molecule (*i.e.*, the level of catalysis) for which there is a probability $P_n = 0.50$ (or close to 0.50) of finding an RAF set in a number of instances of the random catalytic reaction model [31]. From these statistics, we then estimated a linear function $f(n) = a + bn$ using an ordinary least squares regression. We compare these results with the theoretical linear relation which can be calculated from Theorem 4.1 (ii) in [3] (using $P_n = 0.50$ and $k = t = 2$).

So, in short, we compare the required levels of catalysis for RAF sets to occur with high probability for three cases:

- Computational case for *any* RAF;
- Computational case for *all-molecule* RAFs;
- Theoretical case for *all-molecule* RAFs.

The computational values were calculated over 100 to 1000 (depending on the value of n) instances of the random catalytic reaction model, for $n = 7, \ldots, 20$ (because of an exponential increase in the number of molecules $|X|$ and reactions $|R|$ with n, we are computationally limited to about $n = 20$ in these simulations).

Table 1 presents the linear relations estimated from the simulation data or calculated from the theoretical analysis. There is a difference of almost 2 orders of magnitude between the observed (simulation) slope of the required growth rate (case A) and the theoretical one (case C). However, as case B shows, this difference cannot be fully explained by the fact that the theoretical analysis assumes RAFs involving *all* molecules. Even with this stronger assumption, the theoretical slope (case C) is still more than twice as large as the one from the simulations (case B).

Table 1: The empirical (cases A and B) and theoretical (case C) linear relations

A	$f_A(n) =$	$1.0970 + 0.0189n$
B	$f_B(n) =$	$-0.4736 + 0.7012n$
C	$f_C(n) =$	$1.6339n$

Figure 2 shows the complete data (dots for simulation values, lines for estimated or calculated linear relations). Clearly, even though the theoretically predicted value for $f(n)$ grows quite fast with n (slope = 1.6339), the actual value grows only very slowly (slope = 0.0189). So, for example, for $n = 20$ the theoretically expected value for the average number of reactions that need to be catalyzed by any molecule to have RAF sets occurring with probability at least $P_n = 0.50$ is $f_C(20) = 32.678$, which seems unrealistically high. However, the actual level of catalysis required is only $f_A(20) = 1.475$, which is chemically much more plausible.

REQUIRED SIZE OF THE MOLECULE SET

In addition to the level of catalysis required for RAF sets to emerge, we can use the RAF algorithm to answer other interesting, and related, questions. For example, one could ask what the minimum required size of the molecule set is (or, in the binary polymer model, the minimum size n of the largest molecules in the system) to get RAF sets with high probability, given a fixed (known) probability p of a molecule catalyzing an (arbitrary) reaction.

Suppose we fix the probability of catalysis at $p = 0.00001$, or perhaps even $p = 0.000001$ (one in a million), which is, for example, roughly the probability in phage display of a random peptide binding an arbitrary ligand [5,32], and "the ease of polypeptide evolution with a small number of arbitrary sequences indicates that a significant fraction of all possible sequences may have functions, at least binding activity in correlation with catalytic activity" [33]. In the binary polymer model, with these probabilities of catalysis, what value of n is required to get a probability of, say, $P_n \geq 0.50$ of RAF sets occurring?

Figure 3 shows the results of applying the (modified) RAF algorithm to instances of the random binary polymer model (Section 2) for the two given values of p and for $n = 5, \ldots, 20$. With $p = 0.00001$ (left curve in Figure 3), $P_n = 0$ for $n \leq 12$, but $P_{13} = 0.982$, and $P_n = 1$ for $n \geq 14$. So, in this case a value of at least $n = 13$ is required to get RAF sets with high probability. For only a slightly higher probability of catalysis ($p = 0.00002$), a value of $n = 12$ would be sufficient (results not shown).

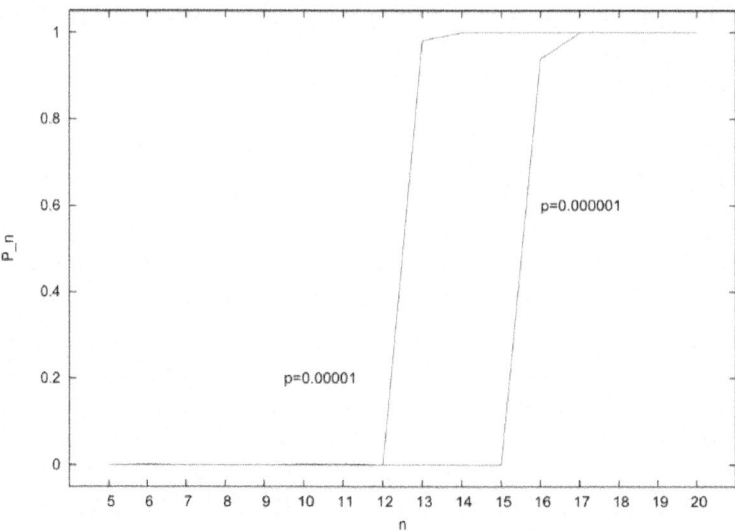

Figure 3: The probability P_n of finding RAF sets for different (fixed) catalysis probabilities p and values of n.

Similarly, with $p = 0.000001$ (right curve in Figure 3), $P_n = 0$ for $n \leq 15$, $P_{16} = 0.939$, and $P_n = 1$ for $n \geq 17$. So, in this case a value of at least $n = 16$ is required to get RAF sets with high probability. Again, for only a slightly higher probability of catalysis ($p = 0.000002$), a value of $n = 15$ would be sufficient (results not shown). The size of the molecule set in this case would be $|X| = 65534$.

TEMPLATE-BASED CATALYSIS

One could argue that the binary polymer model used in our studies so far is perhaps somewhat oversimplified to be biologically or chemically realistic. However, the model serves as a useful starting point with which precise mathematical statements can be formulated and proved, or at least tested computationally. Furthermore, our RAF definition and algorithm are independent of the particular model that is used, and can in principle also be applied to real catalytic reaction systems (for example metabolic networks, of which the already mentioned citric acid cycle is a core element). And, equally importantly, it is actually not very difficult to add more chemical realism into our mathematical models.

As one particular example, we have considered *template-based catalysis* [5]. The idea here is that, in order to act as a catalyst, a molecule must match at least a certain area around the reaction site according to some template-based matching rule. Similar to, for example, base-pair complementarity in RNA and

DNA, we could require a catalyst to match the complement of at least four positions around the reaction site of a ligation or cleavage (two on either side). Consider the ligation reaction

$00101 + 0011 \rightarrow 001010011$

The reaction site template in this case is 0100 (the last two bits of the first binary polymer, plus the first two bits of the second polymer). A given molecule can only act as a catalyst for this reaction if, somewhere along its binary string representation, it contains the complement of this template, *i.e.*, 1011.

In line with the original random CRS model, and some initial simulations with such template-based catalysis [4,5], we have included this idea as follows. For each combination of a molecule $x \in X$ and a reaction $r \in R$, if x matches (anywhere along its length) the complement of the reaction site template of length four of r, then with probability p the pair (x, r) is included in the set of catalyzation C. We present analytical results for this model in Section 6.1 below. However, in our simulations (Section 6.2 below), we used a slightly less constrained version as follows: If a molecule happens to be shorter than length four, then it only has to match part of the complement of the reaction site template, but we require catalysts to be of at least length two. So, the molecule 01 could also catalyze the reaction in the above example. This is mainly done to give some of the food molecules (which are all bit strings up to length two in our simulations) also a chance to act as catalysts.

Note that this template matching requirement is almost the same as in the original simulations [5,34–36], except that we do not allow partial matches here. So, we are considering a slightly "stronger", or more constrained case. Some initial results on similar simulations were reported in [37].

Theoretical Results

Theorem 4.1 (ii) of Mossel and Steel (2005) shows that for polymers of length up to n over an alphabet of size $\kappa \geq 2$, and with a food set of polymers of length up to t, the probability P (n) that there exists an all-molecule RAF is at least:

$$P(n) \geq 1 - \frac{\kappa(\kappa e^{-\lambda})^t}{1 - \kappa e^{-\lambda}}$$

when each molecule catalyzes on average (at least) λn reactions (provided also that $\lambda > \log_e(\kappa)$). Notice that P (n) can be chosen as close to 1 as we wish by selecting λ large enough (and independently of n!). Thus, this result justifies the statement that the average number of reactions each molecule catalyzes needs to grow only linearly with n in order for there to be a given (high) probability of generating an RAF.

We now describe how this result modifies if catalysis is required to be template-based, as described above. Suppose that a polymer x can catalyze a given reaction only if x contains a substring of length s = s1 + s2 that is complementary to the end-segment (of length s1) and the initial segment (of length s2) of the two molecules involved in the cleavage or ligation. Thus, for the above set-up we have: s1 = s2 = 2 and so s = 4. We also assume that the probability that a molecule xcatalyzes a reaction r with a complementary template just depends on x and not on r (this is the analogue of the template-free model assumption (R_2) in [3]).

The following result shows that RAFs will still arise with high probability under linear growth in the average number of reactions each molecule catalyzes, provided the constant involved is increased by a factor of κs (this factor would be 16 for the binary model with template size four). More precisely we have the following result, whose proof is provided in the.

Theorem 6.1 Let Ps(n) be the probability that there exists an all-molecule RAF under this template matching model. Suppose that each molecule catalyzes (on average) at least λsn reactions, where $\lambda_s = \kappa^s \lambda$ and $\lambda > \log_e(\kappa)$. Then Ps(n) satisfies the same inequality as P (n), namely

$$P_s(n) \geq 1 - \frac{\kappa (\kappa e^{-\lambda})^t}{1 - \kappa e^{-\lambda}}$$

COMPUTATIONAL RESULTS

As with the original (random) model (see Section 4), we expect there to be a difference between the theoretically predicted required level of catalysis and the observed level (from simulations) in case of template-based catalysis. Unfortunately, we are computationally even more restricted with this template-based catalysis case as with the original model. The running time of our RAF algorithm is polynomial in the size of the reaction set |R|, as was shown in [2]. However, since |R| is *exponential* in the size of the largest molecules *n*, *i.e.*, $|R| \propto 2^n$, we can only go up to about $n = 20$ to get computational results in a reasonable amount of time (even on a parallel cluster) for the original model (Section 4). But for the template-based catalysis as described above, it is even worse. We now have to check for each pair of molecule $x \in X$ and reaction $r \in$ R whether there is a template match between any part of the molecule and the complement of the (4-site) reaction template. Since both $|X|$ and |R| are $\propto 2^n$, this means that $|X \times R| \propto 2^{2n}$, and in practice this means that we can only go up to about $n = 16$ with our template-based catalysis simulations.

Figure 4 shows the results of these simulations, using our modified algorithm to find RAF sets (of any size) in the template-based catalysis case, compared to the original model (case A in Section 4). As can be expected, for smaller values of n, a higher level of catalysis is needed to find RAF sets with high probability (again, $P_n = 0.50$ is taken as the transition point) in the case of template-based catalysis compared to the purely random model. Since each molecule type $x \in X$ is now restricted, to some extent, in terms of which reactions it can catalyze, the system as a whole is more constrained, and it will be harder to get RAF sets.

However, for longer and longer molecule types, this restriction becomes less of a problem, as a longer molecule has an increasingly higher probability of matching a given 4-site template somewhere along its length. Indeed, the required level of catalysis for the template-based case tapers off as n increases, and converges to that of the original (purely random) model, reaching the same level at $n = 16$. Given that the purely random model is the "limiting" case for the template-based model, we expect that the level of catalysis for this template-based model will follow the same pattern as the base model for $n > 16$, and follow a linear relation.

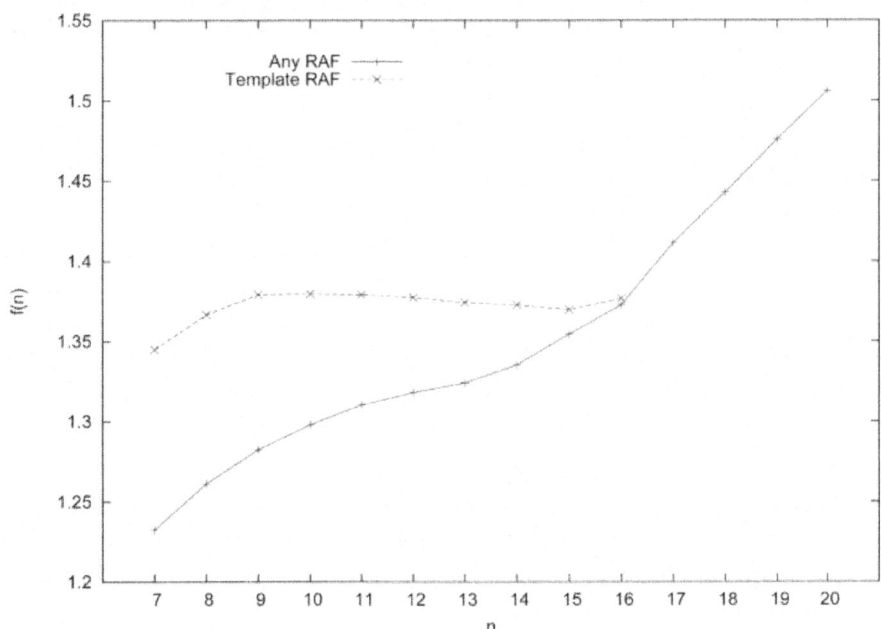

Figure 4: The level of catalysis $f(n)$ required for the template-based catalysis case compared to the original (purely random) case, for different values of n.

CONCLUSIONS

Building on our previous work, we have investigated in more detail the levels of catalysis required for the emergence of autocatalytic sets in models of chemical reaction systems. First, we have shown that there is a discrepancy between the theoretically predicted levels and the computationally observed ones. Although both results yield a linear relation between the required level of catalysis and the size n of the largest molecules in the system, in practice this required level is almost two orders of magnitude smaller than the predicted one. However, this discrepancy cannot be fully explained by the fact that the theoretical result is based on a much stronger assumption (requiring the RAF sets to contain *all* molecules in the system). Even for large systems ($n = 20$, containing several millions of molecule types), each molecule only needs to catalyze (on average) between one and two reactions to have RAF sets appear with high probability, which is chemically highly plausible.

Next, we looked at the minimum size of the molecule set (*i.e.*, number of different types of molecules) necessary to get RAF sets given a fixed probability of catalysis. With a (realistic) probability on the order of one in a million of any molecule catalyzing any given reaction, we only need molecules up to length $n = 15$ or $n = 16$, or about 65,000 different molecule types. Again, this number is well within a plausible (experimental) range.

Finally, we studied an extension of the original model, including template-based catalysis. We established formally that in this case a linear growth rate in the level of catalysis also suffices for RAF sets to appear with arbitrary high probability. However, the simulations show that for smaller values of n (the length of the largest molecules in the system), this linear relation is not exact, and that a higher level of catalysis is necessary compared to the original (purely random) model to get RAF sets with a similar probability. But, as n increases, the template matching constraint becomes less of an issue, and the required levels of catalysis converge to those of the original model. This example shows how more chemical realism can be included in our model.

We intend to continue studying the emergence of autocatalytic sets in chemical reaction systems under various scenarios, models, and extensions. However, so far we have mainly studied the (static) underlying graph structures of such systems. One particularly important issue we hope to address in the future is the actual molecular dynamics in a given (catalytic) reaction system. Also, next to studying models of chemical reaction systems, we would like to apply our RAF framework to real (bio)chemical systems such as metabolic networks, or the collection of all known (organic) substrates and reactions. It is our hope that this line of work will help provide more insight into the (possible) origin of life in general.

ACKNOWLEDGMENTS

The computations were performed at the Vital-IT (http://www.vital-it.ch) center for high-performance computing of the Swiss Institute of Bioinformatics. We thank Jotun Hein for stimulating discussions. MS also thanks the Royal Society of New Zealand (James Cook fellowship).

REFERENCES

1. Steel, M. The emergence of a self-catalysing structure in abstract origin-of-life models. *Appl. Math. Lett* 2000, *3*, 91–95.

2. Hordijk, W; Steel, M. Detecting autocatalytic, self-sustaining sets in chemical reaction systems. *J. Theor. Biol* 2004, *227*, 451–461.

3. Mossel, E; Steel, M. Random biochemical networks: The probability of self-sustaining autocatalysis. *J. Theor. Biol* 2005, *233*, 327–336.

4. Kauffman, SA. Autocatalytic sets of proteins. *J. Theor. Biol* 1986, *119*, 1–24.

5. Kauffman, SA. *The Origins of Order*; Oxford University Press: New York, NY, USA, 1993.

6. Sharov, A. Self-reproducing systems: Structure, niche relations and evolution. *BioSystems*1991, *25*, 237–249.

7. Letelier, JC; Soto-Andrade, J; Abarzúa, FG; Cornish-Bowden, A; Cárdenas, ML. Organizational invariance and metabolic closure: Analysis in terms of (M,R) systems. *J. Theor. Biol* 2006, *238*, 949–961.

8. Cornish-Bowden, A; Cárdenas, ML; Letelier, JC; Soto-Andrade, J. Beyond reductionism: Metabolic circularity as a guiding vision for a real biology of systems. *Proteomics* 2007, *7*, 839–845.

9. Jaramillo, S; Honorato-Zimmer, R; Pereira, U; Contreras, D; Reynaert, B; Hernández, V; Soto-Andrade, J; Cárdenas, M; Cornish-Bowden, A; Letelier, J. (M,R) Systems and RAF Sets: Common Ideas, Tools and Projections. Proceedings of the Alife XII Conference, Odense, Denmark, 19–23 August, 2010, 94–100.

10. Flamm, C; Ullrich, A; Ekker, H; Mann, M; Hoegerl, D; Rohrschneider, M; Sauer, S; Scheuermann, G; Klemm, K; Hofacker, IL; *et al.* Evolution of metabolic networks: A computational framework. *J Syst Chem* 2010, *1*.

11. Kun, A; Papp, B; Szathmáry, E. Computational identification of obligatorily autocatalytic replicators embedded in metabolic networks. *Genome Biol* 2008, *9*. []

12. Awazu, A; Kaneko, K. Discretness-induced transition in catalytic reaction networks. *Phys Rev E* 2007, *76*, 041915:1–041915:8.

13. Brogioli, D. Marginally stable chemical systems as precursors of life. *Phys Rev Lett* 2010, *105*, 058102:1–058102:4.

14. Bartsev, SI; Mezhevikin, VV. On initial steps of chemical prebiotic evolution: Triggering autocatalytic reaction of oligomerization. *Adv. Space Res* 2008, *42*, 2008–2013.

15. Dyson, FJ. A model for the origin of life. *J. Mol. Evol* 1982, *18*, 344–350.

16. Bollobas, B; Rasmussen, S. First cycles in random directed graph processes. *Discret. Math*1989, *75*, 55–68.

17. Lifson, S. On the crucial stages in the origin of animate matter. *J. Mol. Evol* 1997, *44*, 1–8.

18. Szathmary, E. The evolution of replicators. *Philos. Trans. R. Soc. Lond. B* 2000, *355*, 1669–1676.

19. Orgel, LE. The implausibility of metabolic cycles on the prebiotic earth. *PLoS Biol* 2008, *6*, 5–13.

20. Sievers, D; von Kiedrowski, G. Self-replication of complementary nucleotide-based oligomers. *Nature* 1994, *369*, 221–224.

21. Lee, DH; Severin, K; Ghadiri, MR. Autocatalytic networks: The transition from molecular self-replication to molecular ecosystems. *Curr. Opin. Chem. Biol* 1997, *1*, 491–496.

22. Ashkenasy, G; Jegasia, R; Yadav, M; Ghadiri, MR. Design of a directed molecular network.*Proc. Nat. Acad. Sci. USA* 2004, *101*, 10872–10877.

23. Hayden, EJ; von Kieddrowski, G; Lehman, N. Systems chemistry on ribozyme self-construction: Evidence for anabolic autocatalysis in a recombination network. *Angew. Chem. Int. Ed* 2008, *120*, 8552–8556.

24. Lincoln, TA; Joyce, GF. Self-Sustained Replication of an RNA Enzyme. *Science* 2009, *323*, 1229–1232.

25. Penny, D. An interpretive review of the origin of life research. *Biol. Philos* 2005, *20*, 633–671.

26. Hordijk, W; Hein, J; Steel, M. Autocatalytic sets and the origin of life. *Entropy* 2010, *12*, 1733–1742.

27. Dyson, FJ. *Origins of Life*; Cambridge University Press: Cambridge, UK, 1985.

28. Morowitz, HJ; Kostelnik, JD; Yang, J; Cody, GD. The origin of intermediary metabolism. *Proc. Nat. Acad. Sci. USA* 2000, *97*, 7704–7708.

29. Smith, E; Morowitz, HJ. Universality in intermediary metabolism. *Proc. Nat. Acad. Sci. USA* 2004, *101*, 13168–13173.

30. Morowitz, HJ; Srinivasan, V; Smith, E. Ligand field theory and the origin of life as an emergent feature of the periodic table of elements. *Biol. Bull* 2010, *219*, 1–6.

31. It was shown in [2] that $P_n = 0.50$ can be taken as the "transition point" between RAFs not existing at all and RAFs occurring with high probability.

32. Scott, JK; Smith, GP. Searching for peptide ligands with an epitope library. *Science* 1990, *249*, 286–390.

33. Yamauchi, A; Nakashima, T; Tokuriki, N; Hosokawa, M; Nogami, H; Arioka, S; Urabe, I; Yomo, T. Evolvability of random polypeptides through functional selection within a small library. *Protein Eng* 2002, *15*, 619–626.

34. Bagley, RJ. *A Model of Functional Self Organization*; PhD Thesis; University of California, San Diego: CA, USA, 1991.

35. Bagley, RJ; Farmer, JD. Spontaneous Emergence of a Metabolism. In *Artificial Life II*; Langton, CG, Taylor, C, Farmer, JD, Rasmussen, S, Eds.; Addison-Wesley: Upper Saddle River, NJ, USA, 1991; pp. 93–140.

36. Bagley, RJ; Farmer, JD; Fontana, W. Evolution of a metabolism. In *Artificial Life II*; Langton, CG, Taylor, C, Farmer, JD, Rasmussen, S, Eds.; Addison-Wesley: Upper Saddle River, NJ, USA, 1991; pp. 141–158.

37. Andersen, IT; Nan, L; Kjaersgaard, MIS. Search for life in catalytic reaction systems. Available online: http://www.stats.ox.ac.uk/research/genome/projects/pastprojects (accessed on 5 May 2011).

Chapter 10

ENGINEERING MODEL REDUCTION AND ENTROPY-BASED LYAPUNOV FUNCTIONS IN CHEMICAL REACTION KINETICS

Katalin M. Hangos

Process Control Research Group, Computer and Automation Research Institute, Kende u. 13-17, 1111 Budapest, Hungary; Phone: +36 1 279 6101

ABSTRACT

In this paper, the structural properties of chemical reaction systems obeying the mass action law are investigated and related to the physical and chemical properties of the system. An entropy-based Lyapunov function candidate serves as a tool for proving structural stability, the existence of which is guaranteed by the second law of thermodynamics. The commonly used engineering model reduction methods, the so-called quasi equilibrium and quasi steady state assumption based reductions, together with the variable lumping are formally defined as model transformations acting on the reaction graph. These model reduction transformations are analysed to find conditions when (a) the reduced model remains in the same reaction kinetic system class, (b) the reduced model retains the most important properties of the original one including structural stability. It is shown that both variable lumping and quasi equilibrium based reduction preserve both the reaction kinetic form and the structural stability of reaction kinetic models of closed systems with mass action law kinetics, but this is not always the case for the reduction based on quasi steady state assumption.

INTRODUCTION

Chemical Reaction Networks (CRNs) form a wide class of positive (or non-negative) systems attracting significant attention not only among chemists but in numerous other fields such as physics, or even pure and applied mathematics where nonlinear dynamical systems are considered [1]. Beside pure chemical reactions, CRNs are often used to model the dynamics of intracellular processes, metabolic or cell signalling pathways [2]. The increasing interest towards reaction networks as a well-defined special class of positive nonlinear systems is clearly shown by recent tutorial and survey papers [3,4,5] in the systems and control community.

The above approach of general systems and control theory can be usefully complemented with a thermodynamic approach taking into account that reaction kinetic systems are special types of process systems [6], where most often stability is well-established in the thermodynamic approach to analyse their dynamical properties (see e.g., [7,8,9]). The thermodynamic approach recognizes, that the reaction kinetic equations originate from dynamic conservation balances constructed for component masses, and utilizes the second law of thermodynamics [10] to associate an entropy-based Lyapunov function candidate to investigate the stability of the system.

The traditional way of describing the structure of a reaction kinetic system for the purpose of analysing its dynamic properties originate from the classical work of Feinberg (see e.g., [11,12]), who introduced the notion of deficiency that is based upon the structure of the so-called reaction graph.

Based on the above interdisciplinary foundations, the *aim of this work is to precisely define and analyse model reduction schemes based on the thermodynamic and system theoretical understanding, combined with engineering intuition.*

Mathematical models of reaction kinetic systems based on detailed kinetic studies are most often too large for dynamic analysis, model-based control, diagnosis or parameter estimation purposes. Therefore, the need arises to derive more simple versions from these detailed dynamic models that can be handled by the tools and techniques of nonlinear systems and control theory, and possess the same dynamic properties. These detailed dynamic models are simplified by reducing the number of their state (concentration) variables using engineering judgement and operating experience on the relative time constants of the different reaction steps present in the system. The most common model reduction or model simplification steps applied in practice include *the removal of variables being in quasi steady state and lumping together variables with similar dynamics.*

The model reduction based on quasi steady state approximation is well-studied in biochemical kinetics, because this is the dominant way of obtaining simpler equations. The applicability of quasi steady state approximation was investigated by Tzafiri and Edelman [13] for enzyme kinetics, but they disregard the conservation of reaction kinetic forms of the equations. This type of model reduction was also analysed in signal transduction pathways [14]. It is clear from the studies of quasi steady state approximation that this simplification method changes not only the reaction kinetic form of the equations resulting in reaction rate expressions in rational function form, but also the dynamic properties of the model as well. One of the motivations of the present work is to theoretically analyse the conditions that lead to the distortion of dynamical properties during model reduction.

The model reduction using variable lumping is less popular in practice, but relatively well-defined mathematically. Linear kinetic lumping schemes were defined and analysed by Farkas [15], while general nonlinear lumping in chemical kinetics was investigated by Lee *et al.* [16]. Although both of the above model reduction methods are quite widespread, none of them has been analysed from a thermodynamic or from a system theoretical viewpoint yet.

The outline of the paper is as follows. First the considered reaction kinetic model classes obeying the mass action law are described, and then their structural properties are briefly discussed. The model reduction using variable lumping, quasi equilibrium and quasi steady state assumptions are presented in the following sections. Finally conclusions are given.

REACTION KINETIC MODEL CLASSES

In this section, the characteristic structural properties of various reaction kinetic model classes are identified. The structural dynamic properties of reaction kinetic models are determined by two sets of *descriptors*, the so called General (G) and Reaction (R) descriptors, introduced in this section and used for the analysis of the model reduction schemes in the second part of the paper.

Basic assumptions The original physical and chemical picture underlying the reaction kinetic system class is a *closed system under isothermal and isobaric conditions*, where chemical components or species $X_i, i=1,...,n$ take part in r'chemical reactions. The system is assumed to be perfectly stirred, i.e. concentrated parameter in the simplest case. The concentrations $x_i, i=1,...,n$ of the chemical components form the state vector, the elements of which are positive by nature.

For the sake of simplicity, physico-chemical properties are assumed to be constant. Because the system is assumed to be thermodynamically closed, its total mass (and its volume because of the constant density) is also constant. In addition, the*closed system assumption* enables to use the principles of the classical and irreversible thermodynamics [10] when analysing the equilibrium state(s) and their stability (see later in Section 3).

Chemical reactions A complex chemical reaction mechanisms is composed of *elementary reaction steps* in the following form:

$$\sum_{i=1}^{n} \alpha_{ij} X_i \rightarrow \sum_{i=1}^{n} \beta_{ij} X_i \quad j = 1, ..., r'$$

$$(1)$$

where α_{ij} is the so-called *stoichiometric coefficient* of component X_i in the *j*th reaction, i.e. the number of colliding Ximolecules, and β_{it} is the stoichiometric coefficient of the product X_t. Note that *the stoichiometric coefficients are always non-negative integers in classical reaction kinetic systems.*

The set of components with non-zero stoichiometric coefficients α_{ij} or β_{ij} on a side of a reaction form a so-called complex $C_v, k=1,...,m$ with *m* being the number of complexes, that is, $C_k = \sum_{i=1}^{n} \beta_{ij} X_i$ for some *j*. The vector $v(k)$ consisting of the stoichiometric coefficients α_{ij} or β_{ij} (i=1,...,n) characterizes the complex C_k. In the general case one may have less complexes than reactions when some of the reactions have the same reactant complex.

The zero complex is a special complex with a zero stoichiometric vector $v(k)=0$. This appears in the reaction schemes where one considers either an external source or sink of some components, i.e. elementary reaction steps of the form

$$\emptyset \rightarrow \sum_{i=1}^{n} \beta_{ij} \mathbf{X}_i \quad \text{or} \quad \sum_{i=1}^{n} \alpha_{ij} \mathbf{X}_i \rightarrow \emptyset$$

Note that the zero complex is not component specific: there is only a single sink or source term for all components that represent component mass inflows and outflows. In these cases *the systems is thermodynamically **open**,* therefore the conservation of the total mass in the describing reaction kinetic system model does not hold.

Descriptors (G) The above number of components, reactions and complexes (*n*, *r'* and *m*) form the first group of descriptors of reaction kinetic systems. The stoichiometric coefficient vectors $v(k), k=1,...,m$ of the complexes belong also to this group.

General Reaction Kinetic Differential Equations

In the most general case the dynamics of a reaction kinetic system in the concentration (state) space can be described with the following set of ordinary differential equations (ODEs) that *originate from the dynamic conservation balance equations [6] constructed for component masses*:

$$\frac{dx_i}{dt} = F_i(x) = -x_i f_i(x) + g_i(x) \; , \quad i = 1, ..., n$$

(2)

where fi and gi are non-negative polynomials, i.e., polynomials with positive coefficients, and Fi is a polynomial right-hand side (RHS) function [1,17].

Although this is the most general class of reaction kinetic systems, it can be shown that *a model (2) can always be realized with a reaction kinetic system obeying the mass action law* [17]. Note however, that *the zero complex may appear in the realization, so the closed system assumption does not always hold in this general case.*

Mass Action Law (MAL) Kinetics

The mass action law is based on the extended collision picture of the elementary reaction steps, where a reaction occurs if the reactants collide. Thus the rate of the above elementary reaction steps can be described as

$$\rho_j = k_j \prod_{i=1}^{n} [X_i]^{\alpha_{ij}} = k_j \prod_{i=1}^{n} x_i^{\alpha_{ij}} \; , \quad j = 1, ..., r'$$

(3)

where $[X_i] = x_i$ is the concentration of the component X_i, and $k_j > 0$ is the *reaction rate constant* of the jth reaction, that is always positive (see [18]).

Algebraic characterization The reaction rates are described using the so-called *reaction monomials* associated to the complexes in the form

$$\varphi_j(x) = \prod_{i=1}^{n} x_i^{y_{ij}}$$

(4)

where the elements of the matrix Y are the stoichiometric coefficients of the components $i, i = 1, ..., n$ in the complexes $j, j = 1, ..., m$, such that the stoichiometric vector of the jth complex forms the jth column of matrix Y, i.e. $[Y]._j = \nu^{(j)}$.

Note that the stoichiometric coefficients α_{ij} of the reactants in the irreversible reaction steps (1) appear in matrix Y, while the reaction monomials are the principal factors in the MAL reaction rate expression (3).

The reaction graph The vertices V of the reaction graph G=(V,E) correspond to the complexes, and the edges E to the reactions. Two complexes C_k and C_ℓ are connected by a directed edge, if a reaction in the form of

$C_k \rightarrow C_\ell$ (5)

exists. Edge weights can be associated to the edges that are the reaction rate constants $kk_\ell > 0$, thus the reaction graph is a weighted directed graph.

The *Kirchhoff matrix of the reaction graph* $A_k \in \mathbb{R}^{m \times m}$ uniquely describes the reaction graph with

$$[A_k]_{ij} = \begin{cases} -\sum_{l=1}^{m} k_{il} & \text{if} & i = j \\ k_{ji} & \text{if} & i \neq j \text{ and } \exists \ C_j \rightarrow C_i \\ 0 & \text{otherwise} \end{cases}$$

(6)

The Kirchhoff matrix A_k is a column conservation matrix with non-positive diagonal and non-negative off-diagonal, where the sum of the elements in a column is equal to zero.

The reaction kinetic equations In order to construct the dynamic state equations of a reaction kinetic system, the information on the composition of the complexes that are coded in the stoichiometric matrix Y is needed together with the Kirchhoff matrix Ak. The dynamic model that describes the evolution of the reaction kinetic system in its *component or state space* is given by

$$\frac{dx}{dt} = Y \cdot A_k \cdot \varphi(x) = N \cdot \varphi(x)$$

(7)

It is important to note that the matrices Y and Ak uniquely determine the reaction kinetic system, because the stoichiometric coefficients in Y determine the reaction monomials in $\varphi(x)$.

It should be remarked that the solution of the inverse *realization problem*, i.e., to find matrices Y and Ak to a given state matrix N in (7) such that they describe a reaction kinetic system, is not at all unique for systems with higher than zero deficiency, see e.g., in [19].

Descriptors (R) The general descriptors (G) determine the matrix Y, therefore only the Kirchhoff matrix A_k constitutes the group of the reaction network descriptors.

One can form a complex combinatorial structure that uniquely describes a reaction kinetic system with MAL kinetics if vertex weights are also associated to the vertices of the reaction graph in such a way that the column of the matrix Y that belongs to the complex C_j is associated to the vertex as its weight. Then *any formal transformation applied to the reaction kinetic system can be described as a graph transformation applied to the vertex weighted reaction graph.*

MAL Kinetics With Reversible Reactions

A special class of reaction kinetic systems is the case of reversible reactions, the rate equations of which obey the mass action law (MAL). This case was first investigated by Bykov *et al.* [20].

The reaction scheme consists of r reversible reactions of the form

$$\sum_{i=1}^{n} a_{ij} X_i \rightleftharpoons \sum_{i=1}^{n} \beta_{ij} X_i \quad j = 1, ..., r \tag{8}$$

Here we have 2r complexes from which there can be identical ones, i.e., k=1,...,m,m≤2r. Note that the reversible (8) can be realized as r'=2r irreversible reaction steps in the form of (1).

The mass action law type reaction rates can also be applied to this reversible case by considering the rate of the jth reversible step in the form:

$$W_j(x) = W_j^+(x) - W_j^-(x) = k_j^+ \prod_{i=1}^{n} x_i^{\alpha_{ij}} - k_j^- \prod_{r=1}^{n} x_i^{\beta_{ij}} \tag{9}$$

Here both *reaction rate constants* $k_j^+ > 0 \text{ and } k_j^- > 0$ are strictly positive. The terms $W_j^+(x) \text{ and } W_j^-(x)$ are the reaction rates of the forward and backward directed reaction steps, respectively.

The reaction kinetic equations of a reaction kinetic system with reversible MAL kinetics are in the form

$$\frac{dx}{dt} = \mathcal{N}W(x) \tag{10}$$

where $N \in R^{n \times r}$ is the stoichiometric matrix, and $W \in R^r$ is the reaction rate vector described in Equation (9).

The *stoichiometric matrix* N is constructed form the stoichiometric coefficients α_{ij} and β_{ij} in the following way. To each complex C_k a column vector $v^{(k)} \in R^n$ is associated in the usual way, as in the stoichiometric matrix Y. Note that precisely two complexes take part in a reaction (see Equation (8)), thus one can form two matrices N(α) from the complexes of the left hand sides of the r reactions, and N(β) from that of the right hand sides by collecting the column vectors v(k) of the corresponding complexes. Thus N is simply the difference of the two, i.e., N=N(β)−N(α) where the jth column vector of N, $\mu(j) \in R^n$ contains the difference of the stoichiometric coefficients of the jth reaction. Note that the parameters that describe a reaction kinetic system with reversible MAL kinetics are given by the stoichiometric matrix N and the reaction rate coefficients $k^+ = [k_1^+ ... k_r^+]^T \text{ and } k^- = [k_1^- ... k_r^-]^T$.

Examples

Three simple yet characteristic examples are introduced here for illustrating the concepts and algorithms in the later sections.

Linear kinetics Let us consider a simple reaction kinetic system consisting of two reversible first order steps and three components

$$X_1 \rightleftarrows X_2 \rightleftarrows X_3$$

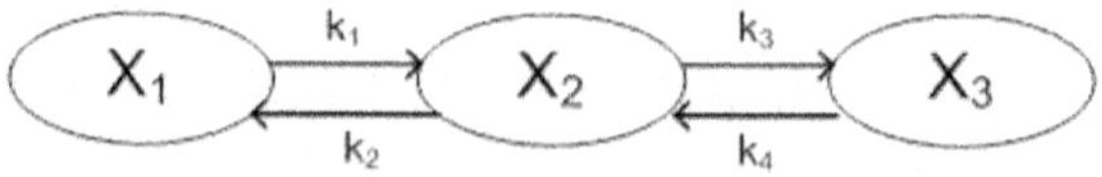

Figure 1: Reaction graph of the linear kinetics system.

The reaction graph consisting of a single connected component is seen in Figure 1.

The dynamic state equations are as follows.

$$
\begin{aligned}
\frac{dx_1}{dt} &= -k_1 x_1 + k_2 x_2 \\
\frac{dx_2}{dt} &= k_1 x_1 - k_2 x_2 - k_3 x_2 + k_4 x_3 \\
\frac{dx_3}{dt} &= k_3 x_2 - k_4 x_3
\end{aligned}
\tag{11}
$$

From this the representation matrices and vectors are easy to derive

$$
\varphi(x) = \begin{bmatrix} x_1 \\ x_2 \\ x_3 \end{bmatrix}, \quad
Y = \begin{bmatrix} 1 & 0 & 0 \\ 0 & 1 & 0 \\ 0 & 0 & 1 \end{bmatrix}, \quad
N = A_k = \begin{bmatrix} -k_1 & k_2 & 0 \\ k_1 & -(k_2 + k_3) & k_4 \\ 0 & k_3 & -k_4 \end{bmatrix}
\tag{12}
$$

The system consists of only reversible reactions, thus the reversible representation form also exists with r=2 and

$$
W(x) = \begin{bmatrix} -k_1 x_1 + k_2 x_2 \\ -k_3 x_2 + k_4 x_3 \end{bmatrix}, \quad
\mathcal{N} = \begin{bmatrix} -1 & 0 \\ 1 & -1 \\ 0 & 1 \end{bmatrix}
\tag{13}
$$

Descriptors The general descriptors of this example are

m=n=3,r'=4,r=2 $\qquad$ (14)

and the reaction descriptor is in Equation (12).

Michealis–Menten kinetics The Michaelis–Menten kinetics is the simplest example of an enzyme kinetic reaction that is widely used for kinetic and model reduction studies. The overall reaction equation is

E+S⇌E+P $\qquad$ (15)

where E is the enzyme, S is the substrate and P is the product of the reaction. The enzyme acts as a catalyst. The mechanism assumes that the substrate forms a complex ES with the enzyme in a reversible reaction step that can react further to form the product P and giving back the unchanged enzyme E. There is a substrate inhibition step in the scheme, too.

The kinetic scheme consists of three reversible reactions

$$E+S \rightleftarrows ES \quad ES \rightleftarrows E+P \quad ES+S \rightleftarrows ESS \tag{16}$$

The reaction kinetic model now consists of the component mass balances for the species S, E, ES, ESS and P. These can be written in the following ODE form with $x=[x1,x2,x3,x4,x5]^T=[[E],[S],[ES],[ESS],[P]]^T$.

$$\frac{dx_1}{dt} = -k_1^+ x_1 x_2 + k_1^- x_3 - k_2^- x_1 x_5 + k_2^+ x_3 \tag{17}$$

$$\frac{dx_2}{dt} = -k_1^+ x_1 x_2 + k_1^- x_3 - k_3^+ x_2 x_3 + k_3^- x_4 \tag{18}$$

$$\frac{dx_3}{dt} = +k_1^+ x_1 x_2 - k_1^- x_3 + k_2^- x_1 x_5 - k_2^+ x_3 - \\ -k_3^+ x_2 x_3 + k_3^- x_4 \tag{19}$$

$$\frac{dx_4}{dt} = +k_3^+ x_2 x_3 - k_3^- x_4 \tag{20}$$

$$\frac{dx_5}{dt} = -k_2^- x_1 x_5 + k_2^+ x_3 \tag{21}$$

The reaction graph consisting of two connected components is seen in Figure 2.

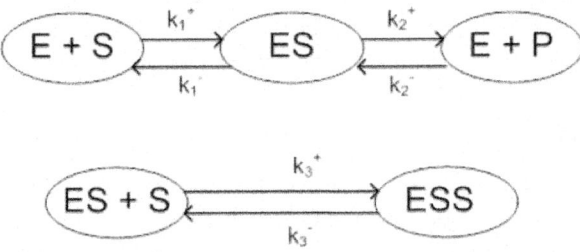

Figure 2: Reaction graph of the Michaelis-Menten kinetic scheme

Descriptors The general descriptors of this example are

$$m=n=5, r'=6, r=3 \tag{22}$$

The stoichiometric and Kirchhoff matrices are in the form

$$Y = \begin{bmatrix} 1 & 0 & 1 & 0 & 0 \\ 1 & 0 & 0 & 1 & 0 \\ 0 & 1 & 0 & 1 & 0 \\ 0 & 0 & 0 & 0 & 1 \\ 0 & 0 & 1 & 0 & 0 \end{bmatrix}, \quad A_k = \begin{bmatrix} -k_1^+ & k_1^- & 0 & 0 & 0 \\ k_1^+ & -(k_1^- + k_2^-) & k_2^- & 0 & 0 \\ 0 & k_2^- & -k_2^- & 0 & 0 \\ 0 & 0 & 0 & -k_3^+ & k_3^- \\ 0 & 0 & 0 & k_3^+ & -k_3^- \end{bmatrix} \tag{23}$$

Sequential Reactions A reaction network of three consecutive reaction steps are considered, the reaction graph of which is depicted in Figure 3.

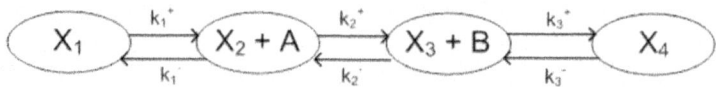

Figure 3: Reaction graph of the sequential kinetic scheme.

The kinetic equations are as follows using the concentration vector $x=[x1,x2,x3,x4,a,b]^T=[[X1],[X2],[X3],[X4],[A],[B]]^T$

$$\frac{dx_1}{dt} = -k_1^+ x_1 + k_1^- x_2 a$$

$$\frac{dx_2}{dt} = k_1^+ x_1 - k_1^- x_2 a - k_2^+ x_2 a + k_2^- x_3 b = \frac{da}{dt}$$

$$\frac{dx_3}{dt} = k_2^+ x_2 a - k_2^- x_3 b - k_3^+ x_3 b + k_3^- x_4 = \frac{db}{dt}$$

$$\frac{dx_4}{dt} = k_3^+ x_3 b - k_3^- x_4$$

$$(24)$$

PROPERTIES OF REACTION KINETIC MODELS

The most important properties and their physical and chemical background is presented in this section, which will be used later when analysing the effect of the engineering model reduction transformations. As is well-known, the reaction rate equations (2) originate from the component mass balances [6] of the underlying lumped process system, while there is no need for any energy balance because of the constant temperature and pressure assumption.

Basic Properties

It can be shown that all MAL reaction kinetic models are *non-negative*, i.e., they remain in the positive orthant or on its boundary if the initial conditions are non-negative [5]. A reaction network is said to be *persistent* if it starts in the positive orthant and does not approach the boundary of the orthant, in other words, if provided that every species is present at the start of the reaction, no species will tend to be eliminated in the course of the reaction. Conditions for persistence are given by Angeli *et al.* [21].

Originally, the theory of reaction network was developed for closed and isothermal systems where the conservation of overall mass holds. Open systems can be described by using the *zero complex* that describes an infinite source or sink representing the environment. If a zero complex is present in a reaction kinetic model, it implies that the systems is open, and thus the conservation of the overall mass does not hold.

Reversibility is a property of the reaction graph as a directed graph. A reaction network is reversible, if for every reaction $C_j \to C_k$ the system contains its reverse, $C_k \to C_j$. *Weak reversibility* is a weaker notion, which requires that each complex C_k lies on a directed circle in the reaction graph.

The stability of reaction networks (10) can be examined using the notion of deficiency (see [3,11,18]). It is an integer number that depends on the properties of matrix Y and the structure of the reaction graph. The *deficiency* δ is defined as:

$$\delta = m - L - s \tag{25}$$

where m is the number of complexes and L is the number of connected components (linkage classes) in the reaction graph, while s is the dimension of the stoichiometric sub-space, i.e., $s = \mathrm{rank}(R)$. The stoichiometric subspace R is spanned by the so-called reaction vectors ε_k, $k = 1, \ldots, r'$, that are associated to the reaction $C_i \to C_j$ such that

$$\varepsilon_k = [Y] \cdot, j - [Y] \cdot, i$$

where $[Y] \cdot, j$ is the jth column of matrix Y.

Note that precisely the reaction vectors, one from a pair of the reversible reaction pairs, appear in matrix N of Equation (10) in the case of reversible MAL kinetics.

Conservations and first integrals

Given the dynamic state equations of a reaction kinetic system obeying the mass action law (7), it is easy to show that the solution of it remains on a linear manifold, on the so-called *reaction simplex* determined by the initial conditions, assuming that all stoichiometric coefficients are non-negative. Each vector $e \in \ker(N^T)$ generates a **linear invariant** for the system (10) since

$$e^T N = 0 \;\Rightarrow\; e^T \frac{dx}{dt} = 0 \;\Rightarrow\; e^T x(t) = const = e^T x_0$$

The linear invariants correspond to conservation of the overall mass or that of another species (e.g., atoms of a given type). Therefore, *closed reaction kinetic systems have at least one linear invariant describing the conservation of the overall mass*. The form of this invariant is determined by the unit of the component concentrations x_i. If x_i is in mole per unit mass (or volume), then $\sum_{i=1}^{n} \gamma_i x_i = M = const$, where M is the overall mass of the components that take part in the chemical reactions (there may be inert components present) and γ_i is their molecular weight. If x_i is measured in mole fractions, then M is simply equal to 1, assuming that no inert component is present.

In addition to the linear invariants, there can be other nonlinear algebraic equations that determine the manifold on which the system dynamics evolves. These are nonlinear first integrals of the reaction equations, and can be consequences of model reduction transformations, as we shall see later.

Stability

With the thermodynamically closed system assumption, the most important property characterizing the qualitative behaviour of the time evolution of the system is its stability. Stability is related to the number of equilibrium (steady state) points, as only systems with a single unique equilibrium point can be globally asymptotically stable.

Unlike many other nonlinear systems that exhibit different stability properties depending on the value of their parameters, the global stability of a reaction kinetic system with zero deficiency does not depend on the value of its parameters, i.e., on the value of its reaction rate constants, but only on its structural descriptors. Therefore, its stability can be called *structural stability*.

The deficiency zero theorem [11] gives a condition for reaction kinetic systems with zero deficiency to be *structurally stable*, i.e., globally stable irrespectively of the actual values of its reaction rate constants. *If the reaction network with zero deficiency is (weakly) reversible then there exists within each reaction simplex precisely one equilibrium, and that equilibrium is asymptotically stable if the dynamics is restricted to the reaction simplex to which the equilibrium point belongs.* Consequently, the original system in the concentration space is globally stable in the positive orthant. Therefore, having zero deficiency is a very strong structural property.

Since reversible reaction kinetic systems with MAL kinetics and with independent reaction vectors fall into the deficiency zero class, they *are proved to be structurally stable* in agreement with [11].

The entropy-based Lyapunov function The structural stability of reversible reaction networks is proved by using a Lyapunov function of thermodynamic origin in [20]. This is applied to the dynamic state equation of a *closed reaction kinetic system* with reversible reactions given in the form of (10), where $\dim\ker(N)=0$. This means that the reaction vectors, which form the columns of the matrix N, one from each pair of a reversible reaction, are linearly independent, which—together with the closed nature of the system with reversible reactions—guarantees the existence of a unique positive equilibrium point x* in the concentration (state) space.

*The **structural stability theorem** states that the equilibrium point x* of such a system is globally stable with the Lyapunov function*

$$B_{x^*}(x) = \sum_{i=1}^{n} x_i \left(\ln \left(\frac{x_i}{x_i^*} \right) - 1 \right) + x_i^* = \sum_{i=1}^{n} x_i (\ln x_i - \ln x_i^*) - (x_i - x_i^*)$$

(26)

The basic idea behind the above function is to use the second law of thermodynamics [8] to derive a Lyapunov function of the form

$$\mathcal{L} = \overline{Z}^T w^* - \overline{S}$$

(27)

where $\overline{Z}$ is the vector of conserved extensive variables (internal energy, volume, and component mass of chemical constituents) expressed as a deviation from the reference state $\overline{Z} = Z - Z^*$, $w^* = \frac{\partial S}{\partial Z}|_{Z=Z^*}$ is the vector of intensive driving force variables (temperature, pressure and chemical potential) evaluated at the reference state Z^*, and $\overline{S}$ is the entropy deviation from the reference state. The reference state is usually chosen as a thermodynamic equilibrium state.

In the case of *isothermal, isobaric and closed reaction kinetic systems* only the masses of components and their chemical potential appear in Z and w, respectively, and the elements of these vectors can be expressed as functions of the component concentrations x_i as $Z_i = Mx_i$ with M being the total constant mass of the system, and $w_i = -C_w l_{nxi}$ with the constant $C_w = RT$ where T is the constant temperature and R is the universal gas constant. Note that one has to assume perfect mixing and ideal mixtures for such a simple expression for the chemical potentials [10].

The entropy of the system can then be given as

$$S = -C_w \sum_{i=1}^{n} x_i \ln x_i = -C_w \sum_{i=1}^{n} x_i \mu_i$$

(28)

where $\mu_i = \ln x_i$. With the above expression, the thermodynamically motivated Lyapunov function (27) becomes

$$\mathcal{L}(x) = -C_w \left(\sum_{i=1}^{n} (x_i - x_i^*)\mu_i^* - (x_i\mu_i - x_i^*\mu_i^*) \right) = C_w \left(\sum_{i=1}^{n} x_i(\ln x_i - \ln x_i^*) \right)$$

(29)

This shows that L contains the driving force of the reactions $(\ln x_i - \ln x_i^*)$ for each component, which also appears in the Lyapunov function (26) of the structural stability theorem.

Note that the above Lyapunov functions can be seen as thermodynamic analogues of relative entropies or divergences defined in the theory of stochastic processes (see [22]).

In order to show the identity of the two Lyapunov functions L(x) and Bx*(x), one can use the fact that *in closed systems* $\sum_{i=1}^{n} x_i = \sum_{i=1}^{n} x_i^* = 1 = \text{const}$ if x_i is measured in mole fractions. With this, Equation (26) can be transformed to the form

$$B_{x^*}(x) = \sum_{i=1}^{n} x_i \left(\ln x_i - \ln x_i^* \right) - \left(\sum_{i=1}^{n} x_i - \sum_{i=1}^{n} x_i^* \right) = \sum_{i=1}^{n} x_i \left(\ln x_i - \ln x_i^* \right)$$

The proof of the structural stability theorem [20] sheds light to the thermodynamic background of the Lyapunov function (26). Here we assume the so called detailed balance condition [23], i.e., reversible reactions with independent reaction vectors, and we first define the auxiliary vector μ as

$$\mu(x) = [\ln(x_1), ..., \ln(x_1)]^T$$

that is the chemical potential vector (up to a multiplicative constant), and denote the jth column of the stoichiometric matrices $\mathcal{N}^{(\alpha)}$ by $\alpha^{(j)}$ and $\mathcal{N}^{(\beta)}$ by $\beta^{(j)}$, respectively. The relations

$$\mu^T \beta^{(j)} = \beta_1^{(j)} \ln(x_1) + ... + \beta_n^{(j)} \ln(x_n) = \ln \left(x_1^{\beta_1^{(j)}} ... x_n^{\beta_n^{(j)}} \right) = \ln \left(\frac{1}{k_j^-} W_j^-(x) \right)$$

$$\mu^T \alpha^{(j)} = \ln \left(\frac{1}{k_j^+} W_j^+(x) \right)$$

(30)

are used to compute the time-derivative of B_{x^*} as

$$\frac{dB_{x^*}}{dt} = \frac{\partial B_{x^*}}{\partial x} \mathcal{N} W(x) = (\mu - \mu^*)^T \sum_{j=1}^{r} (\beta^{(j)} - \alpha^{(j)})(W_j^+(x) - W_j^-(x))$$

(31)

$$= \sum_{j=1}^{r} \left(\mu^T \beta^{(j)} - \mu^T \alpha^{(j)} - \mu^{*T} \beta^{(j)} + \mu^{*T} \alpha^{(j)} \right) (W_j^+(x) - W_j^-(x))$$

(32)

$$= \sum_{j=1}^{r} \ln \left(\frac{W_j^-}{W_j^+} \right) (W_j^+(x) - W_j^-(x)) \leq 0$$

(33)

Properties of The Example Kinetic Systems

Linear kinetics This simple reaction kinetic system is closed, it contains reversible reactions in a single linkage class. Its deficiency is zero, and it has a single linear invariant $M = x_1 + x_2 + x_3$ corresponding to the conservation of the overall mass in the system.

Therefore, the system is structurally stable because of the deficiency zero theorem.

Michealis–Menten kinetics This kinetic system is a closed, reversible system, and its deficiency is zero.

Linear invariants It is easy to see from the physical interpretation that there exists linear conservation invariants for the "E" and the "S" type groups as follows:

$$[E]^{tot}=[E]+[ES]+[ESS]=x_1+x_3+x_4 \tag{34}$$

$$[S]^{tot}=[S]+[ES]+2[ESS]+[P]=x_2+x_3+2x_4+x_5 \tag{35}$$

where [E]tot and [S]tot are constants, besides the linear invariant for the overall mass conservation.

Structural stability The system is structurally stable because of the deficiency zero theorem.

MODEL REDUCTION USING VARIABLE LUMPING

Model reduction aims at finding a "reduced" or "more simple" model to a dynamic system that describes its dynamical behaviour (relatively) well but with a smaller number of dynamic variables and/or equations. One possible and quite widespread way of reducing the number of variables is to *lump them together, i.e., to consider only an aggregate "averaged" variable instead of at least two original ones* in the reduced dynamic model.

Variable lumping in the above sense is widely used in reducing reaction kinetic and other process models. Farkas [15] has analysed kinetic lumping schemes consisting of linear projective transformation of the original concentrations, and gave conditions for the transformation and its pseudo-inverse that preserve the reaction kinetic form of the reduced model. The general nonlinear exact and approximate lumping of reaction kinetic models has been analysed by Li and co-workers [16], who also treated the model reduction as a nonlinear projective model transformation.

For cascade type lumped process models Leitold *et al.* [24] developed and analysed the variable lumping transformation. The transformation is represented as *context sensitive graph transformation* acting on the structure graphs of the dynamic process models. In the paper [24] it was shown that the variable lumping transformation preserves the structural controllability and observability of process models, but nothing was reported about the stability of the reduced model.

Following this line, the *variable lumping transformation is treated* in this paper *as a graph transformation acting on the vertex-weighted reaction graph* by identifying its effect on the describing matrices Y and A_k. This allows us to detect if this transformation preserves the structural stability of the reduced reaction kinetic system model.

Mathematical and Physical Characterization

The engineering two-variable version of variable lumping can be applied when the dynamic response of two concentration variables (x_i and x_j) are of the same character. Then both of them can be substituted by a single one $\overline{x}_i$ such that

$$x_i(t) = \chi_i \overline{x}_i(t) \ , \ \ x_j(t) = \chi_j \overline{x}_i(t)$$

$$(36)$$

where both χ_i and χ_j are constants. The above relations imply that $x_i(t) = const \cdot x_j(t)$.

A kinetic lumping scheme in the sense of [15] arises when $\chi_i = \chi_j = \frac{1}{2}$. . Assume that we perform a permutation of the concentration variables such that the two variables to be lumped are x_n and x_{n-1}, i.e., they are the last two ones. The so-called *lumping matrix* forms the reduced set of concentration variables $\overline{x} = \mathcal{M}x$, with $\dim \overline{x} = \dim x - 1$. Then the lumping matrix and its generalized inverse are of the form

$$\mathcal{M} = \begin{bmatrix} \emptyset^T & 1 & 1 \\ I & 0 & 0 \end{bmatrix} \ , \ \ \overline{\mathcal{M}} = \begin{bmatrix} \emptyset & I \\ \frac{1}{2} & 0 \\ \frac{1}{2} & 0 \end{bmatrix}$$

$$(37)$$

where I is the $(n-2) \times (n-2)$ unit matrix and $\emptyset$ is a $(n-2)$-dimensional zero column vector.

It is important to note that the *generalized inverse of a lumping matrix is not unique, the one we selected in (37) reflects the assumption* $x_i(t) = x_j(t)$ *that is behind the lumping of the two variables.*

The lumped new concentration variable $\overline{x}_{n-1}$ and the RHS of its kinetic equation can be computed as

$$\overline{x}_{n-1} = x_{n-1} + x_n \ , \ \ \frac{d\overline{x}_{n-1}}{dt} = \frac{dx_{n-1}}{dt} + \frac{dx_n}{dt} = F_{n-1}(x) + F_n(x)$$

$$(38)$$

while

$$x_n = \frac{1}{2}\overline{x}_{n-1} \ , \ \ x_{n-1} = \frac{1}{2}\overline{x}_{n-1}$$

$$(39)$$

With these equations the model reduction transformation consists of three consecutive steps.

L1 Substitute $\overline{x}_{n-1}$ for x_n and x_{n-1} into the RHS of the kinetic equations whenever they appear using Equation (39).

L2 Form the RHS of the kinetic equation of the lumped variable $\overline{x}_{n-1}$ as

$$\overline{F}_{n-1}(\overline{x}) = F_{n-1}(\overline{x}) + F_n(\overline{x})$$

from Equation (38).

Replace the last two equations, the nth and $(n-1)$th ones, of the original model with the newly formed equation

$$\frac{d\overline{x}_{n-1}}{dt} = \overline{F}_{n-1}(\overline{x})$$

The primary effect of variable lumping is that one component disappears from the equations, i.e., $n^- = n-1$.

Effect on The Model Properties

The effect of the three transformation steps L1, L2 and L3 is analysed both on the Y and A_k matrices.

L1 By substituting the lumped variable into the two original ones, each complex that contains them will change. Formally, the nth row $[Y]n,\cdot$ is added to the $(n-1)$th one, and the nth is deleted, while the number of columns in Y does not change.

L2,L3 These steps do not influence the structure of the matrices Y and A_k but implement the changes in the reaction equations caused by step L1. Note that the reaction rate constants in A_k will change.

In *degenerate special cases it may happen*, that some columns of Y become identical, i.e., *some complexes may become identical*. Then the number of complexes also decreases, and the dimension of the stoichiometric space s may decrease, too. This case requires that the *two vertices* corresponding to these identical complexes be *condensed* (united) in the reaction graph. As the new vertex inherits all directed edges from the ones it is condensed from, the reversibility of the reaction graph remains unchanged.

Effect on structural stability If no condensation of the reaction complexes occurs, then neither the structure of the reaction graph nor the dimension of the stoichiometric space change. These facts imply that the structural stability of the reaction kinetic system remains unchanged during this model reduction transformation.

The condensation of some of the vertices in the reaction graph does not change its reversibility, but may change its deficiency. Therefore, the conservation of structural stability can only be guaranteed in the reversible case.

Examples

Variable lumping is performed on both of the examples introduced in subSection 2.4, and the properties of the reduced model are investigated.

Linear kinetics Let us lump components X_2 and X_3. This implies

$$\bar{x}_2 = x_2 + x_3 \ , \quad x_2 = \frac{1}{2}\bar{x}_2 \ , \quad x_3 = \frac{1}{2}\bar{x}_2$$

Substituting this into the original model (11) we obtain

$$\frac{dx_1}{dt} = -k_1 x_1 + \frac{1}{2}k_2\bar{x}_2$$
$$\frac{d\bar{x}_2}{dt} = k_1 x_1 - \frac{1}{2}k_2\bar{x}_2$$

(40)

The reduced reaction graph is in the form

$$X_1 \leftrightarrows \bar{X}_2$$

The structural descriptors change to $\bar{n} = n - 1 = 2, \bar{m} = m - 1 = 2$ and $\bar{r}' = r' - 2 = 2$. The representation matrices and vectors are also easy to derive

$$\varphi(x) = \begin{bmatrix} x_2 \\ x_3 \end{bmatrix}, \quad Y = \begin{bmatrix} 1 & 0 \\ 0 & 1 \end{bmatrix}, \quad N = A_k = \begin{bmatrix} -k_1 & \frac{1}{2}k_2 \\ k_1 & -\frac{1}{2}k_2 \end{bmatrix}$$

(41)

The deficiency of the reduced model with reversible reactions is zero, thus it remains structurally stable.

Michaelis–Menten kinetics The two intermediate components ES and ESS with similar dynamics are lumped to form the lumped pseudo-component ES with concentration $\bar{x}_3 = [E_S]$. Then the equations needed for the transformation are

$$\bar{x}_3 = x_3 + x_4 \ , \quad x_3 = \frac{1}{2}\bar{x}_3 \ , \quad x_4 = \frac{1}{2}\bar{x}_3$$

and the concentration variable x_4 is left out from the model.

The reduced kinetic equations are

$$\frac{dx_1}{dt} = -k_1^+ x_1 x_2 + \frac{1}{2} k_1^- \overline{x}_3 + \frac{1}{2} k_2^+ \overline{x}_3 - k_2^- x_1 x_5 \tag{42}$$

$$\frac{dx_2}{dt} = -k_1^+ x_1 x_2 + \frac{1}{2} k_1^- \overline{x}_3 + \frac{1}{2} k_3^- \overline{x}_3 - \frac{1}{2} k_3^+ x_2 \overline{x}_3 \tag{43}$$

$$\frac{d\overline{x}_3}{dt} = k_1^+ x_1 x_2 - \frac{1}{2} k_1^- \overline{x}_3 - \frac{1}{2} k_2^+ \overline{x}_3 + k_2^- x_1 x_5 \tag{44}$$

$$\frac{dx_5}{dt} = \frac{1}{2} k_2^+ \overline{x}_3 - k_2^- x_1 x_5 \tag{45}$$

It is important to note that the number of complexes has also been reduced by one, as both ES and ESS formed complexes that were replaced by a new complex formed by ES. Thus the new stoichiometric matrix Y is in the form

$$Y = \begin{bmatrix} 1 & 0 & 0 & 0 \\ 1 & 0 & 0 & 1 \\ 0 & 1 & 0 & 1 \\ 0 & 0 & 1 & 0 \end{bmatrix}$$

and the reaction graph is depicted in Figure 4

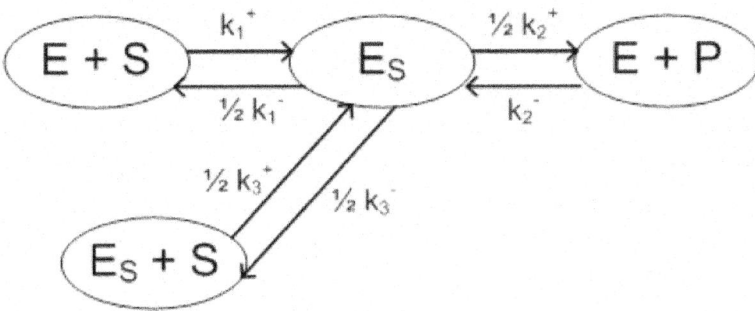

Figure 4: Reaction graph of the lumped Michaelis-Menten scheme

Deficiency and structural stability Because of the condensation of two complexes, the number of complexes is now $\overline{m}$ =4, and the number of linkage classes is also reduced by one, i.e., $\overline{L}$ =1. The dimension of the stoichiometric space remains s=3, therefore the reduced model is of zero deficiency, and therefore it is structurally stable.

MODEL REDUCTION USING QUASI EQUILIBRIUM AND QUASI STEADY STATE ASSUMPTIONS

For both the quasi equilibrium and quasi steady state assumptions, the underlying physical picture determines the conditions under which it can be applied. Similar to the case of variable lumping, this is translated into a mathematical description in the form of a model reduction transformation that is then formally applied to the reaction kinetic model.

Mathematical and Physical Characterization

The model reduction transformations using quasi equilibrium and quasi steady state assumptions are closely related and imply each other in certain cases. In both of the cases one adds polynomial type algebraic equation(s) to the original differential ones, and possibly omits a differential equation from the original model in case of the quasi steady state assumption. The additional algebraic equations can be considered as nonlinear first integrals that restrict the dynamic evolution of the system in the state (concentration) space to a lower dimensional manifold determined by these first integrals.

It is important to note, however, that the reaction kinetic form may be, and in most of the cases will be, destroyed when one simply eliminates one of the concentration variables by expressing it from the algebraic equations, and substitutes the resulting expression into the differential ones. Therefore, this substitution will be circumvented here by choosing carefully the model reduction transformation.

Quasi equilibrium Here one assumes that there exists a *fast* (compared to the others) *reversible reaction step* consisting of the jth and ℓth irreversible reaction step for which

$$k_j \prod_{i=1}^{n} x_j^{\alpha_{ij}} = k_j \varphi_j(x) = k_\ell \varphi_\ell(x) = k_\ell \prod_{i=1}^{n} x_\ell^{\alpha_{i\ell}}$$

$$(46)$$

The condition when this assumption can be applied is that both reaction rate coefficients k_j and k_ℓ are at least several orders of magnitude larger than all the other ones such that the reaction steps in both directions can take place much faster than the other reaction steps.

Formally one adds the defining Equation (46) to the kinetic equations that constrains its dynamics to a manifold determined by this equation. Equation (46) can be considered as a additional nonlinear first integral of the system. However, it is important to note that *the substitution of* $k_{\ell \varphi \ell}(x)$ *into the term* $k_{j \varphi j}(x)$ *may result in a negative cross-effect if the complex behind* $k_{j \varphi j}(x)$ *appears in other reaction(s) but the one in quasi equilibrium, too. This implies that the reduced model will loose its reaction kinetic form.*

In order to preserve the reaction kinetic form of the reduced model, the *reduction transformation* is performed in an unusual way by introducing a new variable that forms a complex itself instead of the two complexes that appear in the reaction in quasi equilibrium. Therefore, the following transformation sub-steps are to be carried out:

E_1

introducing a new *pseudo concentration variable* $x_{j\ell}$ in the form of

$$x_{j\ell} = k_\ell \varphi_\ell(x) = k_j \varphi_j(x)$$

(47)

E_2

substituting the new concentration variable into the terms $k_{\ell \varphi \ell}(x)$ and $k_{j\varphi j}(x)$ whenever they appear,

E_3

condensing (uniting) the complexes C_ℓ and C_j into a new complex $C_{j\ell}$ corresponding to the new variable $x_{j\ell}$ in the reaction graph, and *forming a new pseudo component mass balance*, i.e. a kinetic equation for $x_{j\ell}$ by adding together all the RHS functions $F_k(\overline{x})$ where any of the terms $k_{\ell \varphi \ell}(x)$ and $k_{j\varphi j}(x)$ appear, and taking its negative.

It is to be noted that the defining Equation (47) can be interpreted as introducing the new pseudo variable as

$$x_{j\ell} = (k_\ell \varphi_\ell(x) \cdot k_j \varphi_j(x))^{1/2}$$

This, together with the equality (46) leads to the defining equation, but shows a rationale for the complex vertex condensation in the reaction graph performed in step E3 above.

The *primary effect of this reduction transformation* is that two complexes and one reaction-pair disappear from the reaction kinetic system, and one complex and one component appear. The descriptors will change accordingly, thus

$$\overline{m} = m - 1 \ , \ \overline{n} = n + 1 \ , \ \overline{r}' = r' - 2$$

(48)

It will be seen later in the examples, however, that the number of components may even decrease if a component forms a complex itself and takes part only in the reaction that is in quasi equilibrium.

Quasi equilibrium example on the sequential kinetic scheme The above reduction transformation is applied to the sequential kinetic scheme with kinetic equations (24) and reaction graph in Figure 3 assuming quasi steady state for the reaction $X_2 + A \leftrightarrows X_3 + B.$. The equilibrium assumption implies to introduce the pseudo component concentration

$$x_{23} = k_2^+ x_2 a = k_2^- x_3 b$$

The reduced kinetic equations are

$$\frac{dx_1}{dt} = -k_1^+ x_1 + \frac{k_1^-}{k_2^+} x_{23}$$

$$\frac{dx_4}{dt} = \frac{k_3^+}{k_2^-} x_{23} - k_3^- x_4$$

$$\frac{dx_{23}}{dt} = k_1^+ x_1 - \frac{k_1^-}{k_2^+} x_{23} + k_3^- x_4 - \frac{k_3^+}{k_2^-} x_{23}$$

$$(49)$$

Here 4 original components, X_2, X_2, A and B disappear, and one new, X_{23} appears. Thus there are only 3 complexes and 3 components in the reduced model that are connected by 4 reactions. The resulting reaction graph is seen in Figure 5.

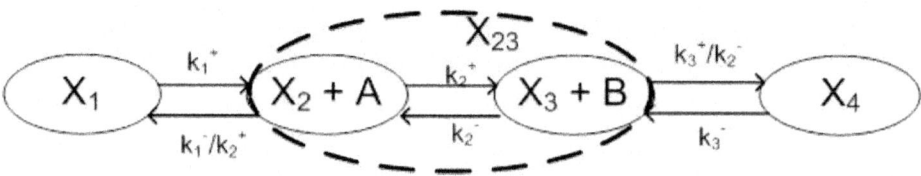

Figure 5: Reaction graph of the reduced sequential kinetic scheme

As *a side effect of a quasi equilibrium assumption, the component(s) that only appear in the complexes C_j or C_ℓ will be in quasi steady state*, because their component mass balance will have the form

$$\frac{dx_i}{dt} = k_j \varphi_j(x) - k_\ell \varphi_\ell(x) = 0$$

(50)

that implies x_i=const. In this case the number of components will decrease, too, giving rise to $\bar{n} \leq n$.

Quasi steady state This is the most "popular" engineering model reduction assumption [6], when the concentration of a certain component in the model, say the jth one is assumed to be constant, formally

$$\frac{dx_j}{dt} = 0 \ , \ x_j = x_j^* = \text{const}$$

(51)

The physical picture behind this assumption is that there exists a reservoir of infinite capacity attached to the system that can enforce the concentration to be constant; in other words, the component X_j is "in great excess". This normally *contradicts to the conservation of the overall mass in the system, thus to the closed system assumption.*

The *model reduction transformation* consists of two consecutive formal steps:

SS1

eliminating the component X_j by setting its concentration x_j to a constant x^*_j whenever it appears,

SS2

substituting the resulting algebraic equation from Equation (51), e.g., Fj(x)=0 into the differential ones.

Without the second step SS2, the resulting reduced system would have an additional nonlinear first integral described by the algebraic equation

$$F_j(x) = 0 = \sum_{j=1}^{m} [Y]_{ij} \sum_{\ell}^{m} [A_k]_{j\ell} \varphi_\ell(x)$$

(52)

and the system would evolve on a manifold determined by this first integral.

The usual way of substituting the above algebraic equation to the differential ones is to express one concentration from it, and substitute the resulting expression—a fraction of two polynomials in the general case—into the expressed concentration. This way, the number of components can be further reduced, but on the price of obtaining a differential equation with non-polynomial right-hand sides (e.g., see [13,14]). This means that the reduced model will no longer belong to the reaction kinetic class.

Another possible way proposed here is to express a whole reaction monomial $k\ell^*\varphi\ell^*(x)$ *from Equation (52) and substitute this expression, a* polynomial, whenever it appears in the equations. This way the number of components will not decrease further but the polynomial form of the right-hand sides remain. Unfortunately, it can happen that the reduced model will contain negative cross-effects as a results of this substitution, so it will not be reaction kinetic any more. On the price of an increased number of complexes, however, one can transform this polynomial equation into a reaction kinetic form [25] using a simple state-dependent time-rescaling [26].

The *primary effect of this model reduction transformation* is that one component disappears from the equations, i.e., $\bar{n} = n-1$.

Effect on The Model Properties

The effect of the two model reduction transformation is related, and they may imply each other in special cases.

Quasi equilibrium Both complexes C_j and $C\ell$ and one reaction pair $C_j \leftrightarrows C_\ell$ (say the κth reaction pair) disappear from the reaction network, and a new complex $C_{j\ell}$ appears, which is in itself a new component $X_{j\ell}$. Therefore

- the columns $[Y]\cdot,j$ and $[Y]\cdot,\ell$ are deleted from matrix Y, and a new column $[Y]\cdot,j\ell$ is inserted being a unit column vector,

- the vertices C_j and $C\ell$ in the reaction graph are condensed (united), such that the united remaining vertex $C_{j\ell}$ inherits every incoming and outgoing directed edges (all reactions) that were adjacent to C_j and C_ℓ. Formally this can be performed by adding the jth row to the ℓth one in the Kirchhoff matrix Ak, and the jth column to the ℓth one (except for the diagonal entries), and then delete the jth row and column (and recompute $[Ak]\ell\ell$).

A *special "degenerate" case* arises if any of the components, say the ith, appear only in the reaction pair in quasi equilibrium, because then this component will be in quasi-steady state, and its row $[Y]i,\cdot$ is deleted from matrix Y, too.

Because of the above vertex condensation in the reaction graph, *the reversibility of the reaction network does not change*, while the number of connected components L either remains unchanged or decreases by one. The dimension of the stoichiometric space s may also decrease by one in the non-degenerate case, because precisely one reversible reaction, a pair of reaction vectors $(\varepsilon_\kappa, -\varepsilon_\kappa)$ is deleted, where κ identifies the the reaction pair $C_j \leftrightarrows C_\ell$. As the number of complexes m is also decreased by one, the *deficiency of the reaction kinetic system may change*.

Quasi steady state The effect of the two transformation steps SS1 and SS2 is analysed on both the Y and Ak matrices.

SS1

The row $[Y]k,\cdot$ belonging to the component in quasi steady state is deleted from the matrix Y.

SS2

The substitution of one reaction monomial $k_{\ell*\varphi\ell*}(x)$ into the RHSs using Equation (52) decreases the number of complexes, which implies to delete the column $[Y]\cdot,\ell_*$ from the matrix Y, to omit the vertex belonging to C_ℓ^* from the reaction graph, and to introduce directed edges into the graph.

Step SS1 may result in identical transformed columns, $[Y]\cdot,j$ and $[Y]\cdot,i$, i.e., *identical transformed complexes* $\bar{C}_j$ and $\bar{C}_i$ in degenerate special cases. In addition, the *zero complex may also appear* in the transformed matrix $\bar{Y}$, when the deleted component formed a complex itself. In the *general non-degenerate cases, however, neither the number of complexes m nor the reaction vectors change, thus the dimension of the stoichiometric space s remains unchanged* as a result of this step.

Step SS2 results in a decrease of the number of complexes and in a drastic change of the reaction graph that *may destroy its reversibility*.

Effect on structural stability The effect of the *quasi equilibrium model reduction transformation* on the basic properties shows that it *preserves structural stability* in the general non-degenerate cases.

This is, unfortunately, not the case for the quasi steady state transformation when we cannot guarantee that structural stability remains unchanged.

Examples

Model reduction is performed on both of the examples introduced in subSection 2.4 using a quasi equilibrium and a quasi steady state assumption.

Linear kinetics Let us assume first *quasi equilibrium* for the reaction $X1 \leftrightarrows X2$. This implies

$$k_1 x_1 = k_2 x_2$$

Substituting this to the original model (11) we obtain

$$\frac{dx_1}{dt} = 0 \;, \quad x_1 = \text{const}$$
$$\frac{dx_2}{dt} = -k_3 x_2 + k_4 x_3$$
$$\frac{dx_3}{dt} = k_3 x_2 - k_4 x_3$$

(53)

It is important to observe that the quasi equilibrium assumption implies the *quasi steady state* for the component X_1, because it only participates in the reaction for which the equilibrium assumption holds.

The reduced reaction graph is in the form

$$X_2 \leftrightarrows X_3$$

The   structural   descriptors   change   to
$\bar{n} = n - 1 = 2$, $\bar{m} = m - 1 = 2$ and $\bar{r}' = r' - 2 = 2$ The
representation matrices and vectors are also easy to derive

$$\varphi(x) = \begin{bmatrix} x_2 \\ x_3 \end{bmatrix}, \quad Y = \begin{bmatrix} 1 & 0 \\ 0 & 1 \end{bmatrix}, \quad N = A_{\mathbf{k}} = \begin{bmatrix} -k_3 & k_4 \\ k_3 & -k_4 \end{bmatrix}$$

(54)

The resulting reduced model preserves its structural stability.

Michaelis–Menten kinetics First two *quasi equilibrium assumptions* are investigated.

1.   Quasi equilibrium for the reaction E+S⇆ES

The equation to be used is $k_1^+ x_1 x_2 = k_1^- x_3$, the complex (E+S) is eliminated. The resulting reduced model is in a reaction kinetic form

$$\frac{dx_1}{dt} = -k_2^- x_1 x_5 + k_2^+ x_3$$

(55)

$$\frac{dx_2}{dt} = -k_3^+ x_3 x_2 + k_3^- x_4 \qquad ,$$

(56)

$$\frac{dx_3}{dt} = k_2^- x_1 x_5 - k_2^+ x_3 - k_3^+ x_3 x_2 + k_3^- x_4$$

(57)

$$\frac{dx_4}{dt} = k_3^+ x_3 x_2 - k_3^- x_4$$

(58)

$$\frac{dx_5}{dt} = -k_2^- x_1 x_5 + k_2^+ x_3$$

(59)

that remains structurally stable.

•   Quasi equilibrium for the reaction ES+S⇆ESS

Here we use the equation $k_3^+ x_3 x_1 = k_3^- x_4$ for the reduction, that implies quasi steady state for the component ESS, that is, $x_4 = [ESS] = x_4^*$. In addition, as the reaction ES+S⇆ESS forms a connected component in the reaction graph in itself, this connected component disappears from the reaction graph completely. Performing the reduction formally, one obtains the following reduced model in reaction kinetic form

$$\frac{dx_1}{dt} = -k_1^+ x_1 x_2 + k_1^- x_3 - k_2^- x_1 x_5 + k_2^+ x_3$$

(60)

$$\frac{dx_2}{dt} = -k_1^+ x_1 x_2 + k_1^- x_3 \tag{61}$$

$$\frac{dx_3}{dt} = k_1^+ x_1 x_2 - k_1^- x_3 + k_2^- x_1 x_5 - k_2^+ x_3 \tag{62}$$

$$\frac{dx_5}{dt} = -k_2^- x_1 x_5 + k_2^+ x_3 \tag{63}$$

that preserves it structural stability.

The reduced reaction graphs are seen in Figure 6.

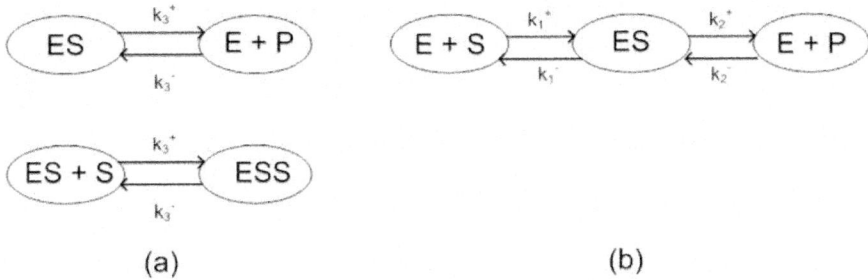

(a)   (b)

Figure 6: Reaction graph of the reduced Michelis–Menten kinetics, (a) quasi equilibrium for E+S⇋ES and (b) quasi equilibrium for ES+S⇋ESS

It is easy to check using the deficiency zero theorem that both of the above reduced models have structural stability.

Finally, *model reduction using two quasi steady state assumptions* is presented.

- Quasi steady state for the component ES

Let us denote the steady state value of component ES by x^*_3, and let us notice that ES forms a complex in itself. Therefore, the reduced stoichiometric matrix $\overline{Y}$ becomes

$$\overline{Y} = \begin{bmatrix} 1 & 0 & 1 & 0 & 0 \\ 1 & 0 & 0 & 1 & 0 \\ 0 & 0 & 0 & 0 & 1 \\ 0 & 0 & 1 & 0 & 0 \end{bmatrix}$$

that shows that the zero complex appears in the reaction kinetic system (see 2nd column of $\overline{Y}$). The reduced kinetic equations are

$$\frac{dx_1}{dt} = -k_1^+ x_1 x_2 + \overline{k}_1^- - k_2^- x_1 x_5 + \overline{k}_2^+ \tag{64}$$

$$\frac{dx_2}{dt} = -k_1^+ x_1 x_2 + \overline{k_1^-} - \overline{k_3^+} x_2 + k_3^- x_4 \tag{65}$$

$$\frac{dx_4}{dt} = \overline{k_3^+} x_2 - k_3^- x_4 \tag{66}$$

$$\frac{dx_5}{dt} = -k_2^- x_1 x_5 + \overline{k_2^+} \tag{67}$$

where the new reaction rate constants are formed as $\overline{k_i^\pm} = k_i^\pm x_3^*$.. For the zero complex that appears in the place of ES there is no component mass balance, but the presence of constant positive terms indicate that *the systems becomes open.* The reaction graph of the reduced system is seen in Figure 7. It is important to note that structural stability of the reduced model cannot be guaranteed.

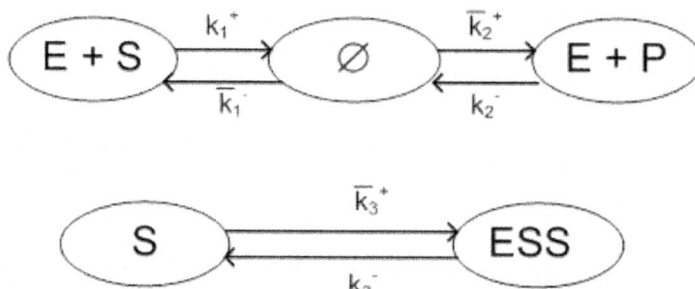

Figure 7. Reaction graph of the reduced Michelis–Menten kinetics with ES in quasi steady state 2.

Quasi steady state for the component E

With the steady state value x_i^* of component E the transformed reaction rate constants are formed as $\overline{k_i^\pm} = k_i^\pm x_1^*$, and the reduced kinetic equations are as follows.

$$0 = -\overline{k_1^+} x_2 + k_1^- x_3 + k_2^+ x_3 - \overline{k_2^-} x_5 \tag{68}$$

$$\frac{dx_2}{dt} = -\overline{k_1^+} x_2 + k_1^- x_3 - k_3^+ x_2 x_3 + k_3^- x_4 \tag{69}$$

$$\frac{dx_3}{dt} = \overline{k_1^+} x_2 - k_1^- x_3 - k_3^+ x_2 x_3 + k_3^- x_4 \tag{70}$$

$$\frac{dx_4}{dt} = k_3^+ x_2 x_3 - k_3^- x_4 \tag{71}$$

$$\frac{dx_5}{dt} = -\overline{k_2^-} x_5 + k_2^+ x_3$$

(72)

The reduced stoichiometric matrix is now

$$\overline{Y} = \begin{bmatrix} 1 & 0 & 0 & 1 & 0 \\ 0 & 1 & 0 & 1 & 0 \\ 0 & 0 & 0 & 0 & 1 \\ 0 & 0 & 1 & 0 & 0 \end{bmatrix}$$

It can be easily seen that the algebraic equation (68) restricts the system dynamics to a 3-dimensional manifold in the 4-dimensional component space. However, the reaction kinetic form would be lost if one expresses a variable from Equation (68) and substitutes it into the differential ones because of the appearance of negative cross-effects.

At the same time, structural stability of the model remains unchanged, because the reduced "unrestricted model" (Equations (69)–(72)) is in a reaction kinetic form with reversible reactions.

CONCLUSIONS

In this paper, the structural properties of chemical reaction systems obeying the mass action law were investigated and related to the physical and chemical properties of the system. An entropy-based Lyapunov function candidate serves as a tool for proving structural stability for closed reaction kinetic system obeying the MAL. The equivalence of the special form of this Lyapunov function used in reaction kinetics [20] was shown to be equivalent to the thermodynamically motivated Lyapunov function used in process systems engineering [7,8].

The commonly used engineering model reduction methods, the so-called quasi equilibrium and quasi steady state assumption based reductions, together with the variable lumping, were formally defined as model transformations acting on the reaction graph, and their effect on the structural descriptors and on structural stability were investigated.

It was shown that both variable lumping and quasi equilibrium based reduction preserves both the reaction kinetic form and the structural stability of reaction kinetic models of closed systems, but this is not always the case for the reduction based on quasi steady state assumption.

ACKNOWLEDGEMENTS

This research was partially supported by the Hungarian Scientific Research Fund and the National Office for Research and Technology through grant no. K67625 and by the Control Engineering Research Group of HAS at Budapest University of Technology and Economics.

The author is indebted to Dr. Gábor Szederkényi for his constructive comments and discussions.

REFERENCES

1. E´rdi, P.; To´th, J. *Mathematical Models of Chemical Reactions. Theory and Applications of Deterministic and Stochastic Models*; Manchester University Press, Princeton University Press: Manchester, UK and Princeton, NJ, USA, 1989.

2. Haag, J.; Wouver, A.; Bogaerts, P. Dynamic modeling of complex biological systems: a link between metabolic and macroscopic description. *Math. Biosci.* 2005, *193*, 2549.

3. Sontag, E. Structure and stability of certain chemical networks and applications to the kinetic proofreading model of T-cell receptor signal transduction. *IEEE Trans. Autom. Control* 2001, *46*, 1028–1047.

4. Angeli, D. A tutorial on chemical network dynamics. *Eur. J. Control.* 2009, *15*, 398–406.

5. Chellaboina, V.; Bhat, S.P.; Haddad, W.M.; Bernstein, D.S. Modeling and analysis of mass-action kinetics – nonnegativity, realizability, reducibility, and semistability. *IEEE Control. Syst. Mag.* 2009, *29*, 60–78.

6. Hangos, K.M.; Cameron, I.T. *Process Modelling and Model Analysis*; Academic Press: London, 2001.

7. Alonso, A.; Ydstie, B. Process systems, passivity and the second law of thermodynamics. *Comp. Chem. Eng* 1996, *20*, 1119–1124.

8. Hangos, K.M.; Alonso, A.A.; Perkins, J.D.; Ydstie, B.E. Thermodynamic approach to the structural stability of process plants. *AIChE Journal* 1999, *45*, 802–816.

9. Alonso, A.; Fernandez, C.; Banga, J. Dissipative systems: From physics to robust nonlinear control. *Int. J. of Robust Nonlinear Contr.* 2004, *14*, 157–159.

10. Callen, H.B. *Thermodynamics and an Introduction to Thermostatistics*; John Wiley and Sons: New York, 1980.

11. Feinberg, M. *Lectures on Chemical Reaction Networks*; Notes of lectures

given at the Mathematics Research Center, University of Wisconsin: Wiscosin, 1979.

12. Feinberg, M. Chemical reaction network structure and the stability of complex isothermal reactorsI. The deficiency zero and deficiency one theorems. *Chem. Eng. Sci.* 1987, *42*, 2229–2268.

13. Tzafiri, A.; Edelman, E. The total quasi-steady-state approximation is valid for reversible enzyme kinetics. *J. Theor. Biol.* 2004, *226*, 303–313.

14. Millat, T.; Bullinger, E.; Rohwer, J.; Wolkenhauer, O. Approximations and their consequences for dynamic modelling of signal transduction pathways. *Math. Biosci.* 2007, *47*, 40–57.

15. Farkas, G. Kinetic lumping schemes. *Chem. Eng. Sci.* 1999, *54*, 3909–3915.

16. Li, G.; Rabitz, H.; To´th, J. A general analysis of exact nonlinear lumping in chemical kinetics. *Chem. Eng. Sci.* 1994, *49*, 343–361.

17. Ha´rs, V.; To´th, J. Qualitative theory of differential equations. North Holland, 1981; chapter On the inverse problem of reaction kinetics; pp. 363–379.

18. Feinberg, M. On chemical kinetics of a certain class. *Arch. Rational Mech. Anal.* 1972, *46*, 1–41.

19. Szederke´nyi, G. Computing sparse and dense realizations of reaction kinetic systems. *J. Math. Chem.* 2009. *to appear*, available online.

20. Bykov, V.; Gorban, A.N.; Yablonskii, G.; Elokhin, V. *Kinetic Models of Catalytic Reactions (Comprehensive Chemical Kinetics, V.32, ed. by R.G. Compton)*; Elsevier: Amsterdam, 1991.

21. Angeli, D.; Leenheer, P.; Sontag, E. A Petri net approach to the study of persistence in chemical reaction networks.*Math. Biosci.* 2007, *210*, 598–618.

22. Gorban, S.; Gorban, P.; Judge, G. Entropy: the Markov ordering approach. *Entropy* 2010. *to appear*.

23. Feinberg, M. Necessary and sufficient conditions for detailed balancing in mass action systems of arbitrary complexity. *Chem. Eng. Sci.* 1989, *44*, 1819–1827.

24. Leitold, A.; Hangos, K.M.; Tuza, Z. Structure simplification of dynamic process models. *J. Process Control* 2002, *12*, 69–83.

25. Szederke´nyi, G. Computing reaction kinetic realizations of positive nonlinear systems using mixed integer programming. In IFAC Symposium on Nonlinear Control Systems (NOLCOS), 2010.

26. Szederke´nyi, G.; Hangos, K.; Magyar, A. On the time-reparametrization of quasi-polynomial systems. *Phys. Lett. A*2005, *334*, 288–294.

Chapter 11

CONCORDANT CHEMICAL REACTION NETWORKS

Guy Shinar[1] and Martin Feinberg[2]

[1]InBrain Therapeutics Ltd., 12 Metzada St., Ramat Gan 52235, Israel

[2]The William G. Lowrie Department of Chemical & Biomolecular Engineering and Department of

Mathematics, Ohio State University, 140 W. 19th Avenue, Columbus, OH, USA 43210

ABSTRACT

We describe a large class of chemical reaction networks, those endowed with a subtle structural property called concordance. We show that the class of concordant networks coincides precisely with the class of networks which, when taken with any weakly monotonic kinetics, invariably give rise to kinetic systems that are injective — a quality that, among other things, precludes the possibility of switch-like transitions between distinct positive steady states. We also provide persistence characteristics of concordant networks, instability implications of discordance, and consequences of stronger variants of concordance. Some of our results are in the spirit of recent ones by Banaji and Craciun, but here we do not require that every species suffer a degradation reaction. This is especially important in studying biochemical networks, for which it is rare to have all species degrade.

INTRODUCTION

Although chemical reaction networks can be highly intricate and their induced differential equations can be extraordinarily complex, it is typically the case that dynamical behavior resulting from them is tame.

Surprisingly often, for example, there is precisely one stationary state, even when kinetic parameters range over all feasible values. This is in contrast to the behavior of complex systems of differential equations in general, for which multiple stationary states are common and, in fact, are to be expected. Viewed against a background of differential equations generally, then, those that derive from chemical reaction networks appear to admit bistable behavior far less often than might normally be supposed. This dullness of dynamics seems extant across very large classes of reaction networks and persists robustly, even against large changes in parameter values within kinetic models for the rates of individual chemical reactions.

The implications for the design of biological circuitry are especially compelling. Evolution has certainly fashioned special reaction network modules with dynamical behavior sufficiently rich as to underlie, for example, the bistable switching devices or the oscillatory timekeeping mechanisms necessary for the machinery of life. But these special actors probably require a fairly stable and reliable, yet dynamic, chemical background against which they can play their roles. In this sense, biology-in-the-large would seem to require that only exceptional reaction networks have the capacity to exhibit very rich dynamics, while most others execute their functions in steadier, less surprising ways.

To understand, in a general way, the design principles by which biology operates, it therefore becomes important to understand the perhaps subtle distinctions between reaction network architectures that, on one hand, might underlie specialized dynamically-rich devices and, on the other hand, architectures that, by their very nature, enforce duller, more restricted behavior despite what might be great intricacy in the interplay of many species, even independently of values that kinetic parameters might take.

This article is about a large class of chemical reaction networks, called concordant networks, for which a certain dullness of behavior is enforced for quite general kinetics. (Concordance is a property of the network itself, divorced from any assignment of a kinetics.) Whether a particular network is concordant can be readily checked, most easily with help of software of the kind made available in [1][3], and, as we show in a separate article [3], it can sometimes be affirmed by inspection of what has come to be known as the network'sSpecies-Reaction Graph [4, 5, 6].

Taken with what we call a weakly monotonic kinetics — essentially a kinetics in which an increase in the rate of a particular reaction requires that there be an increase in the concentration of at least one of its reactant species — a concordant network has the following properties (among others):

1.   Not only can the resulting kinetic system not admit two distinct stoichiometrically-compatible positive[4]equilibria, it cannot even admit two distinct stoichiometrically-compatible equilibria, at least one of which is positive. This is to say that if the resulting kinetic system admits a positive equilibrium, it cannot admit any second equilibrium, positive or otherwise, that is stoichiometrically compatible with the first.

2.   If the network is weakly reversible and conservative, and if the kinetics is continuous, then among all compositions that are stoichiometrically compatible with some positive composition — including "boundary compositions" (i.e., those for which at least one species concentration is zero) — there willindeed be precisely one equilibrium, and it is positive. In fact, the boundary of the set of all such compositions is repulsive against approach from a positive composition in the following sense: At every boundary composition, the production rate of each species with concentration zero is non-negative, and there is at least one such species whose production rate is strictly positive. Moreover, in at least some instances — when the kinetics is smooth and the network's "fully open extension" is also concordant (Section 7) — we can assert that the unique equilibrium has a degree of stability: every real eigenvalue associated with it is negative.

In describing these features of concordant networks, we have only told part of the story that will unfold later on, but it is perhaps already clear that, for concordant networks, the underlying network structure enforces a very circumscribed kind of behavior. No matter how intricate a concordant network might be — and concordant networks can be highly intricate — it cannot, for example, provide the basis for a bistable switch involving two equilibria, at least one positive, so long as the kinetics is weakly monotonic.

This is not to say that concordance in a network precludes all kinds of rich behavior over all possible choices of weakly monotonic kinetics. It does not. (See Remarks 8.5 and 8.6.) What is surprising, however, is that concordance — or the lack of it — tells us so much.

The remainder of this article is organized as follows: In Section 2 we discuss some earlier, related results and, in particular, connections to the striking work of Banaji and Craciun [7, 5]. In Section 3 we provide the chemical reaction network theory ideas necessary for the rest of the article.

In Section 4 we provide the definitions of concordance, weakly monotonic kinetics, and injectivity of the species formation rate function. In the same section we show in a very simple way that for a concordant reaction network, taken with weakly monotonic kinetics, the species formation rate function is injective. A direct consequence of this is the impossibility of two distinct stoichiometrically-compatible equilibria, at least one of which is positive. In fact, we show that, for weakly monotonic kinetics, injectivity is, in a certain sense, synonymous with concordance: A reaction network has injectivity in all weakly monotonic kinetic systems derived from it if and only if the network is concordant. With the results of Section 4 behind us, we return briefly in Section 5 to a more fully articulated discussion of connections to other work.

In Section 6 we discuss persistence properties of kinetic systems derived from concordant networks, in particular boundary behavior of the species formation rate function. Most of the results obtained there rely only on very basic, natural features of kinetic rate functions and do not invoke further requirements such as weak monotonicity.

Section 7 is about connections between concordance of a network and concordance of a network's "fully open extension" — roughly, the network obtained by requiring that every species s of the original network suffers a degradation reaction s → 0. It is in this section that we begin to make statements about eigenvalues associated with equilibria. In particular, we show how network concordance enforces a certain degree of stability of positive equilibria for all (differentiably) monotonic kinetics. By way of contrast, in Section 8 we discuss consequences of network discordance beyond those stated in Section 4, in particular the guarantee ofunstable positive equilibria for at least certain instances of differentiably monotonic kinetics. Among other things we show that for every weakly reversible discordant network there is a differentiably monotonic kinetics such that the resulting kinetic system admits a positive unstable equilibrium.

In Section 9 we discuss some variations of the definition of concordance and their consequences. There we consider very broad kinetics, in which the rate of a particular reaction might even be influenced by the concentration of a species that, for the reaction, is neither a reactant nor a product. As an important special case, we introduce the notion of strong concordance and show that, when the kinetics is similar to the kind admitted by Banaji and Craciun [5], then strong concordance suffices for ensuring injectivity of the species-formation rate function (and the absence of multiple stoichiometrically-compatible equilibria, at least one of which is positive). This is also discussed in a preliminary way in Section 2. Section 10 contains a concluding discussion, with special emphasis on distinctions between fine-grained and approximate coarse-grained reaction network descriptions of enzyme-driven biochemistry.

SOME BACKGROUND AND A LOOK AHEAD

Chemical reaction network theory has already been able to delineate large classes of networks for which dull behavior is ensured, at least with respect to certain dynamical features (e.g., the absence of multiple positive stoichiometrically-compatible steady states), mostly when the individual reaction rates are presumed to be governed by mass action kinetics. In this section we discuss two quite separate strands of reaction network theory, and then, by way of a preview, we briefly discuss the standing of concordance theory in relation to these.

Deficiency-oriented Theory

The earliest results were centered about the classification of reaction networks by means of a non-negative integer index called the deficiency. Thus, there are networks of deficiency zero, of deficiency one, and so on. The deficiency is not a measure of a network's size. In fact, a very large complex network can have a deficiency of zero. Nevertheless, the Deficiency Zero Theorem indicates the sense in which all deficiency zero networks taken with mass action kinetics give rise to quite dull dynamics, including the absence of multiple stoichiometrically-compatible positive equilibria, the absence of periodic positive composition trajectories, and the absence of an unstable positive equilibrium. The Deficiency One Theorem describes a large class of mass action networks for which multiple stoichiometrically-compatible positive equilibria are precluded, regardless of rate constant values. Still other parts of deficiency-oriented theory translate questions about a mass action network's capacity for multiple stoichiometrically-compatible positive equilibria — an apparently nonlinear problem—into question about systems of linear inequalities. For a survey of some deficiency-oriented reaction network theory see [8, 9, 10, 11, 12, 13]. The deficiency-oriented theory continues to evolve [2, 1].

Theory Aimed at "fully-open" Reaction Networks

A rather different strain of reaction network theory began with the Ph.D. work of Paul Schlosser [14, 15] and the advent of what was called the Species-Complex-Linkage Graph (SCL Graph). When a reaction network's SCL Graph satisfies certain conditions, involving its cycles, then the hypothesis of a certain technical proposition is also satisfied, and this proposition, in turn, ensures that multiple positive steady states are impossible.

It is important to understand that this work was tailored very specifically to determining when a reaction network, taken with mass action kinetics, has the capacity to admit multiple positive steady states in the context of what chemical engineers call the (isothermal) continuous flow stirred tank reactor (CFSTR), in which all species are present in an effluent stream. In reaction network theory terms, one studies such reactors by adjoining to the network of "true" chemical reactions additional "degradation reactions" of the form s → 0, one for each species s, to account for the efflux of species s in the reactor's effluent [8]. (There are sometimes additional reactions of the form 0 → s to account for infusion of species s in the feed stream.) Thus, work originating in the Schlosser thesis was aimed specifically at understanding the capacity for multiple positive equilibria for fully-open reactors, characterized by "fully-open reaction networks," in which there is a degradation reaction corresponding to each and every species.

This work was later broadened in [16, 4, 6] by Craciun and Feinberg to give results which are, in many respects, more powerful, using different theoretical underpinnings, but again for mass action kinetics and again for reaction networks in which there is a degradation reaction corresponding to each species. If the kinetics is mass action, and, for the network at hand, it is the case that the Jacobian of the species-formation-rate function is nonsingular at every positive composition and for every assignment of rate constants, then it was shown that the species-formation-rate function is injective (relative to positive compositions) for every assignment of rate constants. In this case, multiple positive equilibria become impossible.

Two techniques were deployed to determine when a network satisfies "the Jacobian condition." One was analytical [16], relying on a symbolic determinant expansion of the Jacobian with the aim of establishing that, for the network under study, every term in the expansion is non-negative with at least one positive (whereupon the Jacobian must be nonsingular). The other [4, 6], building on the first, was graphical: When the network's Species-Reaction Graph (SR Graph) satisfies certain mild conditions, closely resembling the earlier ones for the SCL Graph, then the Jacobian condition is satisfied, and the species-formation-rate function is injective.

Remark 2.1

For the benefit of readers already familiar with results based on the Species-Reaction Graph, we reiterate those conditions here: Every even cycle of the network's Species-Reaction Graph is an s-cycle and no two even cycles have a species-to-reaction intersection In a separate article [3] we show that a "normal" reaction network (Section 7) is concordant — in fact, strongly concordant — if these same conditions are satisfied by the network's Species-Reaction Graph. (Moreover, in [3] we show that concordance also follows from substantially weaker conditions.)

In remarkable and highly surprising subsequent work, Banaji and Craciun [7, 5] were able to extend results in [4, 6] to a far broader class of kinetics, but again for networks in which there is a degradation reaction for every species. The kinetics studied by Banaji and Craciun — which they call nonautocatalytic (NAC) kinetics — is required to have the property that, for a particular reaction, an increase [decrease] in the concentration of one of its reactant [product] species — keeping all other concentrations fixed — cannot result in a decrease of the reaction rate. These conditions are expressed in an essential way in terms of derivatives, so, for them, the kinetics is required also to have a degree of smoothness.

The Banaji-Craciun work built upon earlier work by Banaji and co-workers [17], in which it was shown that if the so-called stoichiometric matrix of a reaction network has the "SSD property" — i.e, the property thatall of its square sub-matrices are either singular or else "sign-nonsingular" — and if there is a degradation reaction for every species, then, when the kinetics is NAC, the species formation rate function is injective (and multiple equilibria are excluded).

Banaji and Craciun [5] then established, surprisingly, that the Species-Reaction Graph properties shown to ensure injectivity for certain networks in the mass-action case also suffice to ensure that the stoichiometric matrix has the "SSD property," which implies in turn that, for fully open networks, one has injectivity of the species-formation rate function even for the very broad class of NAC kinetics.

The requirement that there be a degradation reaction for every species figures heavily in the Banaji-Craciun result and in the earlier work by Banaji and co-workers: It plays an essential role in connecting the "SSD property" of the stoichiometric matrix, via the Gale-Nikaido theorem [18], to injectivity of the species-formation rate function. In contrast to the deficiency-oriented results described in the preceding subsection, this strong reliance on the fully-open setting is a feature common to all of the research described in this subsection, beginning with the Schlosser work, reflecting its origins in the study of continuous-flow stirred-tank reactors.

A Look Ahead: Stoichiometry, Concordance, and Strong Concordance

In part, fully-open networks become easier to study because, in that setting, stoichiometric constraints become far less of a concern. Reaction networks typically studied in biology are not fully open — they rarely contain a "degradation reaction" for every species — and, for such networks, stoichiometric balances come into play more forcefully. Thus, for example, when asking about the possibility of multiple equilibria, it is usually sensible to ask only about the possibility of two distinct equilibria that are stoichiometrically compatible; see Section 3. For this reason, the fully-open results have limited range. (Nevertheless, they have been used to infer the absence of multiple stoichiometrically-compatible equilibria in more general settings [19, 20], but the most incisive of the resulting theorems [20] have heretofore been restricted to mass-action kinetics.)

Unlike the more narrowly focused fully-open theory described in §2.2, the deficiency-oriented theory of §2.1 was fashioned to deal with stoichiometric-compatibility issues from the outset. For example, when the kinetics is massaction, the Deficiency Zero and Deficiency One Theorems assert the impossibility of two distinct stoichiometrically-compatible positive equilibria, regardless of rate constant values, for any reaction network that satisfies their hypotheses. No degradation reactions are required to be present.

As we shall see in Section 4, the definition of network concordance has a built-in pre-disposition to stoichiometric compatibility; it is there from the very beginning. Only in Section 7 will we be concerned with the concordance of a reaction network's "fully open extension" — that is, the network formed by adjoining to the original network a full supply of degradation reactions — but this is because we can sometimes infer from concordance of the fully open extension properties of the original network. (This is similar in spirit to [20], but here there is no requirement that the kinetics be mass-action.)

In a sense, then, the theory of concordant networks shares with the deficiency-oriented theory (§2.1) its indifference to the "fully open requirement," but it also shares with the fully open theory (§2.2) an affinity for special inferences that can be made when there is a full supply of degradation reactions.

We close this section with an explanation of our choice to defer, completely, the notion of strong concordance to the article's end: To the extent that they can be compared, the very broad NAC kinetics admitted by Banaji and Craciun is significantly broader than the (also broad) weakly monotonic kinetics we study in most of this article. (Both subsume mass action kinetics.) We show among other things in Section 4 that, when the kinetics is weakly monotonic, concordance of a reaction network suffices for the impossibility of multiple stoichiometrically-compatible equilibria, at least one of which is positive. However, in Section 9 we show that the same can also be asserted for what we call two-way weakly monotonic kinetics, which is very similar to NAC kinetics, provided that the network is strongly concordant. (In both cases, there is no requirement that the network be fully open.)

We chose to center this article around the weaker and broader notion of concordance because it already gives rise to a variety of interesting and important consequences, some having nothing at all to do with either choice of kinetics. By deferring to Section 9 the discussion of strong concordance and its implications for kinetics of the kind studied by Banaji and Craciun, we tried to keep the already powerful consequences of simple concordance clearly at center stage. In Section 9 we also discuss even more general kinetics, in which the rate of particular reaction can be influenced in quite arbitrary ways even by the concentrations of species that do not appear in that reaction.

REACTION NETWORK THEORY PRELIMINARIES

Here we provide the reaction network theory concepts required for stating and proving the results in this article. Much of this section is also contained in a more extended form, with more discussion and examples, in [10]. We begin with notational conventions.

Notation

In the study of networks, one sometimes has a number associated with each member of a set I of objects in the network, where I might be the set of edges or the set of vertices. For example, one might have a current associated with each of the edges in the network. In this case, it might be advantageous to represent the state of the network by a "current vector" that carries information about the currents on the various edges. One could certainly work in the standard vector space $\mathbb{R}^N$, where N is the number of edges, with the ith component of the current vector corresponding to the current associated with the ith edge. However, this requires an ordering of the edges, so that one can speak of the 3rd or the 5th edge. Such an artificially imposed order is often superfluous to the physical problem at hand and unnecessarily intrusive to its analysis.

In the literature on networks and graphs — see, for example, [21] — there is a standard way to circumvent this problem, by working not in $\mathbb{R}^N$ but, rather, in $\mathbb{R}^I$, the vector space of real-valued functions with domain I. Suppose, for example, that I is the set of edges. If we consider the function $x : I \to \mathbb{R}$ that assigns the number (e.g., current) x_i to the element (e.g., edge) $i \in I$, then x, viewed as a vector in $\mathbb{R}^I$, represents that assignment without requiring that members of I be ordered. In effect, a vector in $\mathbb{R}^I$ is very much like a vector in $\mathbb{R}^N$, but its components are indexed by the members of I itself rather than by the first N integers.

Hereafter, when I is a finite set (e.g., species, reactions), we denote the vector space of real-valued functions with domain I by $\mathbb{R}^I$. The subset of $\mathbb{R}^I$ consisting of those functions that take only positive (nonnegative) values is denoted $\mathbb{R}^I_+ (\overline{\mathbb{R}}^I_+)$. For $x \in \mathbb{R}^I$ and $i \in I$, the symbol x_i denotes the value assigned to i by x.

For each $x \in \mathbb{R}^I$, the symbol exp(x) denotes the element of RI+ defined by

$$(\exp(x))_i := \exp x_i, \forall i \in I.$$

For each $x \in \mathbb{R}^I$ and for each $z \in \overline{\mathbb{R}}^I_+$, the symbol x^z denotes the real number defined by

$$x^z := \prod_{i \in I} (x_i)^{z_i},$$

with the understanding that $0^0 = 1$. For each x, x' $\in \mathbb{R}^I$, the symbol x∘x' denotes the element of $\mathbb{R}^I$ defined by

$$(x \circ x')_i := x_i x_i', \forall i \in I.$$

For each z$\in \mathbb{R}^I_{++}$, the symbol $\frac{1}{z}$ denotes the element of $\mathbb{R}^I_{++}$ defined by

$$\left(\frac{1}{z}\right)_i := \frac{1}{z_i}, \forall i \in I.$$

For each $i \in I$, we denote by ω_i the element of $\mathbb{R}^I$ such that $(\omega_i)_j = 1$ whenever $j = i$ and $(\omega_i)_j = 0$ whenever $j \neq i$.

The standard basis for $\mathbb{R}^I$ is the set $\{\omega_i \in \mathbb{R}^I : i \in I\}$. Thus, for each $x \in \mathbb{R}^I$, we have the representation $x = \Sigma_{i \in I} x_i \omega_i$. The standard scalar product in $\mathbb{R}^I$ is defined as follows: If x and x' are elements of $\mathbb{R}^I$, then

$$x \cdot x' = \sum_{i \in I} x_i x_i'.$$

Note that the standard basis of $\mathbb{R}^I$ is orthonormal with respect to the standard scalar product. It will be understood that $\mathbb{R}^I$ carries the standard scalar product and the norm derived from the standard scalar product. It will also be understood that $\mathbb{R}^I$ carries the corresponding norm topology.

Whenever U is a linear subspace of $\mathbb{R}^I$, we denote by $U^\perp$ the orthogonal complement of U in $\mathbb{R}^I$ with respect to the standard scalar product.

By the support of $x \in \mathbb{R}^I$, denoted supp x, we mean the set of indices $i \in I$ for which x_i is different from zero. When ξ is a real number, the symbol sgn (ξ) denotes the sign of ξ. When x is an element of $\mathbb{R}^I$, sgn (x) denotes the function with domain I defined by

$$(\text{sgn}(x))_i := \text{sgn} x_i, \forall i \in I.$$

Some Definitions

We will use the reaction network displayed in (1) to motivate some of our definitions. Following Horn and Jackson [22], we call the objects at the heads and tails of the reaction arrows—2A,B,C,C+D, and E in (1)—the complexes of the network. In this way, we can view the network as a directed graph, with complexes playing the role of the vertices and reaction arrows playing the role of the edges.

$$2A \;\rightleftarrows\; B$$

$$\nwarrow \quad \swarrow$$

$$C$$

$$C + D \rightleftarrows E \qquad\qquad (1)$$

Remark 3.1

We shall frequently be working in $\mathbb{R}^S$, where S is the set of species in a network. In this special case, it is advantageous to replace symbols for the standard basis of $\mathbb{R}^S$ with the names of the species themselves. Thus, if the species in the network are given by $S = \{A,B,C,D,E\}$ then a vector such as $\omega_C + \omega_D \in \mathbb{R}^S$ can instead be written as $C + D$, and $2\omega_A$ can be written as $2A$. In fact, $\mathbb{R}^S$ can then be identified with the vector space of formal linear combinations of the species. In this way, the complexes of a reaction network with species set S can be identified with vectors in $\mathbb{R}^S$.

Definition 3.2

A chemical reaction network consists of three finite sets:

1. a set S of distinct species of the network;
2. a set $\mathscr{C} \subset \mathbb{R}^S$ of distinct complexes of the network;
3. a set $\mathcal{R} \subset \mathcal{C} \times \mathcal{C}$ of distinct reactions, with the following properties:

    a. $(y, y) \notin \mathcal{R}$ for any $y \in \mathcal{C}$;

    b.   for each $y \in \mathcal{C}$ there exists $y' \in \mathcal{C}$ such that $(y, y') \in \mathcal{R}$ or such that $(y', y) \in \mathcal{R}$.

If (y, y') is a member of the reaction set $\mathcal{R}$, we say that y reacts to y', and, following the usual notation in chemistry, we write $y \rightarrow y'$ to indicate the reaction whereby complex y reacts to complex y'. We call the complex situated at the tail of a reaction arrow the reactant complex of the corresponding reaction, and the complex situated at the head of a reaction arrow the reaction's product complex.

The set of species of the network depicted in (1) is $S = \{A,B,C,D,E\}$. The set of complexes of the network is $\mathcal{C} = \{2A,B,C,C+D,E\}$. The set of reactions of the network is $\mathcal{R} = \{2A \rightarrow B, B \rightarrow 2A, B \rightarrow C, C \rightarrow 2A, C+D \rightarrow E, E \rightarrow C+D\}$.

The diagram in (1) is an example of a standard reaction diagram: each complex in the network is displayed precisely once, and each reaction in the network is indicated by an arrow in the obvious way. In very few places in this article we will refer to the linkage classes of a reaction network, which for our purposes can be identified with the connected components of the network's standard reaction diagram. Thus, in network (1)there are two linkage classes, containing, respectively, the complexes {2A,B,C} and {C+D,E}. For a formal definition of a linkage class see [10].

More important is the idea of weak reversibility. The following definition provides some preparation.

Definition 3.3

A complex $y \in \mathcal{C}$ ultimately reacts to a complex $y' \in \mathcal{C}$ if any of the following conditions is satisfied:

1. $y \rightarrow y' \in \mathcal{R}$;
2. There is a sequence of complexes y(1), y(2),..., y(k) such that
    $y \rightarrow y(1) \rightarrow y(2) \rightarrow ... \rightarrow y(k) \rightarrow y'$.

In our example, the complex 2A ultimately reacts to the complex C, but the complex C does not ultimately react to the complex C +D.

Definition 3.4

A reaction network $\{\mathcal{S}, \mathcal{C}, \mathcal{R}\}$ is called weakly reversible if for each y, y' $\in$ $\mathcal{C}$, y' ultimately reacts to y whenever y ultimately reacts to y'. A network is called reversible if y' $\rightarrow$ y $\in \mathcal{R}$ whenever y $\rightarrow$ y' $\in \mathcal{R}$.

Network (1) is an example of a weakly reversible reaction network that is not reversible. Note that any reversible network is also weakly reversible. Note also that whenever a weakly reversible reaction network is displayed as a standard reaction diagram, every arrow in the diagram belongs to a directed cycle of arrows.

Definition 3.5

The reaction vectors for a reaction network $\{\mathcal{S}, \mathcal{C}, \mathcal{R}\}$ are the members of the set

$$\{y' - y \in \mathbb{R}^{\mathcal{S}} : y \rightarrow y' \in \mathcal{R}\}.$$

In our example the reaction vector corresponding to the reaction 2A $\rightarrow$ B is B−2A, the reaction vector corresponding to the reaction C+D $\rightarrow$ E is E−C−D, and so on.

Definition 3.6

The stoichiometric subspace S of a reaction network $\{S, C, R\}$ is the linear subspace of $\mathbb{R}^S$ defined by

$$S := \mathrm{span}\{y' - y \in \mathbb{R}^S : y \rightarrow y' \in \mathscr{R}\}. \tag{2}$$

We note that, for a reaction network $\{S, C, R\}$, the stoichiometric subspace S will often be a proper subspace of $\mathbb{R}^S$. In other words, it will often be the case that the dimension of S will be smaller than the number of species in the network. For example, in network (1) we have dim S = 3 while dim $\mathbb{R}^S$ = #(S) = 5. In fact, the stoichiometric subspace will be a proper subspace of $\mathbb{R}^S$ whenever the network is conservative:

Definition 3.7

A reaction network $\{S, C, R\}$ is conservative whenever the orthogonal complement $S^\perp$ of the stoichiometric subspace S contains a strictly positive member of $\mathbb{R}^S$:

S⊥∩RS+≠∅.

Network (1), for example, is conservative: it is easy to verify that the strictly positive vector (A+2B+2C+D+3E)∈ $\mathbb{R}^S_+$ is orthogonal to each of the reaction vectors of (1).

If $\{S, C, R\}$ is a reaction network, then a mixture state will generally be represented by a composition $c \in \overline{\mathbb{R}}^S_+$, where, for each s ∈ S, we understand c_s to be the molar concentration of species s. By a positive composition we mean a strictly positive composition — that is, a composition in $\mathbb{R}^S_+$.

Definition 3.8

A kinetics $\mathcal{K}$ for a reaction network $\{S, C, R\}$ is an assignment to each reaction y $\rightarrow$ y' ∈ R of a rate function $\mathscr{K}_{y \rightarrow y'} : \overline{\mathbb{R}}^S_+ \rightarrow \overline{\mathbb{R}}_+$ such that

$\mathscr{K}_{y \rightarrow y'}(c) > 0$ if and only if supp$y \subset$ supp$c$.

Definition 3.9

A kinetic system $\{S, C, R, \mathcal{K}\}$ is a reaction network $\{S, C, R\}$ taken with a kinetics $\mathcal{K}$ for the network.

Remark 3.10

In chemical reaction network theory the rate functions are generally presumed at the outset to be continuous, but we have not insisted on that here. This is because some of the main results in this article (in particular those having to do with injectivity) do not require it. For certain other formally stated results, we will sometimes say specifically that we are assuming continuity or differentiability. On the other hand, when we discuss in an informal way the differential equations associated with a kinetic system, it will be understood implicitly that there is smoothness sufficient to ensure that the commonly presumed properties of differential equations are present.

In this article we shall often refer to mass action kinetics and to mass action kinetic systems. Both are formally defined below:

Definition 3.11

A kinetics $\mathcal{K}$ for a reaction network $\{\mathcal{S}, \mathcal{C}, \mathcal{R}\}$ is mass action if, for each reaction $y \rightarrow y' \in \mathcal{R}$, there is a positive number $k_{y \rightarrow y'}$ such that the rate function $\mathcal{K}_{y \rightarrow y'}$ takes the form

$$\mathcal{K}_{y \rightarrow y'}(c) = k_{y \rightarrow y'} c^y.$$

(3)

The positive number $k_{y \rightarrow y'}$ is the rate constant for reaction $y \rightarrow y'$.

Definition 3.12

A mass action kinetic system is a reaction network taken together with a mass action kinetics for the network.

Definition 3.13

The species formation rate function for a kinetic system $\{\mathcal{S}, \mathcal{C}, \mathcal{R}, \mathcal{K}\}$ with stoichiometric subspace S is the map $f: \mathbb{R}_+^{\mathcal{S}} \rightarrow S$ defined by

$$f(c) = \sum_{y \rightarrow y' \in \mathcal{R}} \mathcal{K}_{y \rightarrow y'}(c)(y' - y).$$

(4)

For a kinetic system $\{\mathcal{S}, \mathcal{C}, \mathcal{R}, \mathcal{K}\}$ whose underlying network is our example network (1), the species formation rate function has the following species-wise form:

$$f_A(c) = -2\mathcal{K}_{2A \rightarrow B}(c) + 2\mathcal{K}_{B \rightarrow 2A}(c) + 2\mathcal{K}_{C \rightarrow 2A}(c),$$

(5)

$$f_B(c) = -\mathscr{K}_{B \to C}(c) + \mathscr{K}_{2A \to B}(c) - \mathscr{K}_{B \to 2A}(c),$$

$$f_C(c) = -\mathscr{K}_{C \to 2A}(c) + \mathscr{K}_{B \to C}(c) - \mathscr{K}_{C+D \to E}(c) + \mathscr{K}_{E \to C+D}(c),$$

$$f_D(c) = -\mathscr{K}_{C+D \to E}(c) + \mathscr{K}_{E \to C+D}(c),$$

$$f_E(c) = -\mathscr{K}_{E \to C+D}(c) + \mathscr{K}_{C+D \to E}(c).$$

DEFINITION 3.14

The differential equation for a kinetic system with species formation rate function f($\cdot$) is given by

$$\dot{c} = f(c). \tag{6}$$

Let $\{S, C, R, K\}$ be a kinetic system. From equations (2), (4), and (6) we observe that the vector $\dot{c}$ will invariably lie in the stoichiometric subspace S of the network $\{S, C, R\}$. Thus, the difference of any two compositions $c \in \overline{\mathbb{R}}_+^{\mathscr{S}}$ and $c' \in \overline{\mathbb{R}}_+^{\mathscr{S}}$ that lie along the same solution of (6) will always reside in S. This motivates the following definition:

Definition 3.15

Let $\{S, C, R\}$ be a reaction network with stoichiometric subspace S. Two compositions c and c' in $\overline{\mathbb{R}}_+^{\mathscr{S}}$ are called stoichiometrically compatible if c' $- c \in S$.

We note that stoichiometric compatibility is an equivalence relation. As such, it partitions $\overline{\mathbb{R}}_+^{\mathscr{S}}+$ into equivalence classes that we call stoichiometric compatibility classes. Thus, the stoichiometric compatibility class containing an arbitrary composition c, denoted $(c+S) \cap \overline{\mathbb{R}}_+^{\mathscr{S}},+$, is given by

$$(c+S) \cap \overline{\mathbb{R}}_+^{\mathscr{S}} = \{c' \in \overline{\mathbb{R}}_+^{\mathscr{S}} : c' - c \in S\}. \tag{7}$$

We observe, as the notation suggests, that $(c+S) \cap \overline{\mathbb{R}}_+^{\mathscr{S}}$ is the intersection of $\overline{\mathbb{R}}_+^{\mathscr{S}}$ with the parallel of S containing c.

A stoichiometric compatibility class will typically contain a wealth of (strictly) positive compositions. We say that a stoichiometric compatibility class is non-trivial if it contains a member of $\mathbb{R}_+^{\mathscr{S}}$. To see that a stoichiometric compatibility class can be trivial, consider the simple reaction network A + B $\rightleftarrows$ C, and let $\bar{c}$ be the composition defined by $\bar{c}_A = 1$, $\bar{c}_B = 0$, $\bar{c}_C = 0$. Then the stoichiometric compatibility class containing $\bar{c}$ has $\bar{c}$ as its only member.

Definition 3.16

An equilibrium of a kinetic system $\{S, C, R, K\}$ is a composition $c \in \overline{\mathbb{R}}_+^{\mathscr{S}}$ for which $f(c) = 0$. A positive equilibrium of a kinetic system $\{S, C, R, K\}$ is an equilibrium that lies in $\mathbb{R}_+^{\mathscr{S}}$.

In light of Definition 3.8, a kinetic system can admit a positive equilibrium only if its reaction vectors are positively dependent:

Definition 3.17

The reaction vectors for a reaction network $\{S, C, R\}$ are positively dependent if for each reaction y → y' ∈ R there exists a positive number $\kappa_{y \to y'}$ such that

$$\sum_{y \to y' \in \mathscr{R}} \kappa_{y \to y'} (y' - y) = 0.$$

(8)

Remark 3.18

For any weakly reversible network, the reaction vectors are positively dependent [10].

Occasionally we will want to consider changes in the value of the species formation rate function corresponding to small departures from a positive composition c* toward nearby compositions that are stoichiometrically compatible with c*. Thus, for a kinetic system $\{S, C, R, K\}$ with stoichiometric subspace S, with smooth reaction rate functions, and with species formation rate function f : $\overline{\mathbb{R}}_+^{\mathscr{S}} \to S$, we will want to work with the derivative df(c*) : S → S, given by

$$df(c^*)\sigma = \frac{df(c^* + \theta\sigma)}{d\theta}\bigg|_{\theta=0}, \forall \sigma \in S.$$

(9)

In this case, we say that $c* \in \overline{\mathbb{R}}_+^{\mathscr{S}}$ is a degenerate equilibrium if c* is an equilibrium and if, moreover, df(c*) is singular.

CONCORDANT NETWORKS, WEAKLY MONOTONIC KINETICS, AND INJECTIVE KINETIC SYSTEMS

In preparation for the definition of concordance we consider a reaction network $\{S, C, R\}$ with stoichiometric subspace $S \subset \mathbb{R}^S$, and we let $L : \mathbb{R}^R \to S$ be the linear map defined by

$$L\alpha = \sum_{y \to y' \in R} \alpha_{y \to y'} (y' - y).$$

(10)

Note that the real scalar multipliers $\{\alpha_{y \to y'}\}_{y \to y' \in R}$ in (10) are permitted to be positive, negative, or zero.

Definition 4.1

The reaction network $\{S, C, R\}$ is concordant if there do not exist an $\alpha \in$ ker L and a nonzero $\sigma \in S$ having the following properties:

1. For each $y \to y' \in R$ such that $\alpha_{y \to y'} \neq 0$, supp y contains a species s for which sgn $\sigma_s = $ sgn $\alpha_{y \to y'}$.

2. For each $y \to y' \in R$ such that $\alpha_{y \to y'} = 0$, either $\sigma_s = 0$ for all $s \in$ supp y or else supp y contains species s and s' for which sgn $\sigma_s = -$sgn $\sigma_{s'}$, both not zero.

A network that is not concordant is discordant.

Remark 4.2

Concordance is a network property that is discernible by sign-checking in ker L and S. A Windows-based computer program that implements such a concordance test is freely available [1]; it is fast for networks of moderate size. Algorithms underlying the concordance tests implemented in [1] are described in [2].

Examples

While network (1) is a fairly simple "toy" that is concordant, the network depicted in (11) serves as a more intricate and more biologically-motivated concordant example. In Reidl et al. [23], network (11)served to model calcium dynamics in olfactory cilia. The concordance of (11) was ascertained by means of the sign-checking procedure implemented in [1].

$$A \rightleftharpoons B \rightarrow D + B$$

$$C + 4D \rightleftharpoons E \quad \text{(11)}$$

$$B + E \rightarrow F \rightleftharpoons A + E$$

$$D \rightarrow 0$$

A biologically motivated example of a discordant network is shown in (12). This network underlies a model of the bacterial two-component signaling system EnvZ-OmpR [24, 25, 26], which regulates the number and size of membrane pores in response to changes in osmolarity.

$$A \rightleftharpoons B \rightarrow C$$

$$C + D \rightleftharpoons E \rightarrow A + F \quad \text{(12)}$$

$$B + F \rightleftharpoons G \rightarrow B + D$$

Definition 4.3

A kinetic system $\{\mathcal{S}, \mathcal{C}, \mathcal{R}, \mathcal{K}\}$ is injective if, for each pair of distinct stoichiometrically compatible compositions $c^* \in \overline{\mathbb{R}}_+^{\mathcal{S}}$ and $c^{**} \in \overline{\mathbb{R}}_+^{\mathcal{S}}$, at least one of which is positive,

$$\sum_{y \rightarrow y' \in \mathcal{R}} \mathcal{K}_{y \rightarrow y'}(c^{**})(y' - y) \neq \sum_{y \rightarrow y' \in \mathcal{R}} \mathcal{K}_{y \rightarrow y'}(c^*)(y' - y).$$

(13)

Remark 4.4

Clearly, an injective kinetic system cannot admit two distinct stoichiometrically compatible equilibria, at least one of which is positive.

Definition 4.5

A kinetics $\mathcal{K}$ for reaction network $\{\mathcal{S}, \mathcal{C}, \mathcal{R}\}$ is weakly monotonic if, for each pair of compositions c* and c**, the following implications hold for each reaction y → y' ∈ R such that supp y ⊂ supp c* and supp y ⊂ supp c**:

1. $\mathcal{K}_{y \to y'}(c^{**}) > \mathcal{K}_{y \to y'}(c^*) \Rightarrow$ there is a species s $\in$ supp y with c**s>c*s.
2. Ky$\to$y'(c**)=Ky$\to$y'(c*)$\Rightarrow$c**s=c*s for all s $\in$ supp y or else there are species s, s' $\in$ suppy with c**s>c*s and c**s'<c*s'.

We say that the kinetic system $\{\mathcal{S}, \mathcal{C}, \mathcal{R}, \mathcal{K}\}$ is weakly monotonic when its kinetics $\mathcal{K}$ is weakly monotonic.

Remark 4.6

Note that if a kinetic system is weakly monotonic and if c** is a positive composition, then the implications (i) and (ii) will hold for all reactions in the network, even when c* is not strictly positive and even when, for a reaction y $\to$ y', supp y is not contained in supp c*.

Remark 4.7

Clearly, every mass action kinetic system is weakly monotonic. Note that the definition of weak monotonicity makes no assumptions about the continuity or the smoothness of the kinetic rate functions. When, for a weakly monotonic kinetics, a reaction-rate function $\mathcal{K}_{y \to y'}(\cdot)$ is differentiable, there is no requirement that its derivative with respect to c_s, s $\in$ supp y, be strictly positive at all positive compositions.

Proposition 4.8

A weakly monotonic kinetic system $\{\mathcal{S}, \mathcal{C}, \mathcal{R}, \mathcal{K}\}$ is injective whenever its underlying reaction network $\{\mathcal{S}, \mathcal{C}, \mathcal{R}\}$ is concordant. In particular, if the underlying reaction network is concordant, then the kinetic system cannot admit two distinct stoichiometrically compatible equilibria, at least one of which is positive.

Proof

Suppose on the contrary that, for the weakly monotonic kinetic systems $\{\mathcal{S}, \mathcal{C}, \mathcal{R}, \mathcal{K}\}$, there are distinct stoichiometrically compatible compositions c* and c**, with c**$\in$RS+, such that

$$\sum_{y \to y' \in \mathcal{R}} \mathcal{K}_{y \to y'}(c^{**})(y' - y) = \sum_{y \to y' \in \mathcal{R}} \mathcal{K}_{y \to y'}(c^*)(y' - y).$$

(14)

Clearly, then, $\alpha \in \mathbb{R}^{\mathcal{R}}$ defined by

$$\alpha_{y \to y'} := \mathcal{K}_{y \to y'}(c^{**}) - \mathcal{K}_{y \to y'}(c^*), \forall y \to y' \in \mathcal{R}$$

(15)

is a member of ker L. Because c^* and c^{**} are distinct and stoichiometrically compatible, the vector $\sigma := c^{**} - c^*$ is a nonzero member of S. Note that if, for a particular reaction $y \to y'$, we have $\alpha_{y \to y'} > 0$, then weak monotonicity and Remark 4.6 require that, for some species $s \in$ supp y, $\sigma s := c^{**}s - c^*s > 0$. A similar argument can be mounted when we have $\alpha_{y \to y'} < 0$; in this case, there is $s \in$ supp y such that $\sigma s := c^{**}s - c^*s < 0$. Finally, if we have $\alpha_{y \to y'} = 0$, then weak monotonicity requires that the numbers $\{\sigma_s\}_{s \in \text{supp}y}$ are either all zero or else there are two of opposite nonzero sign. The existence of α and σ, so constructed, contradicts the supposition that the network $\{S, C, R\}$ is concordant.

Example

Proposition 4.8 tell us that no weakly monotonic kinetics for the complex but concordant network(11) can give rise to multiple stoichiometrically compatible equilibria, at least one of which is positive. This is consistent with the findings in [23].

The simple Proposition 4.8 — one of the most important of this article — tells us that, for any weakly monotonic kinetic system, injectivity (and therefore the absence of distinct stoichiometrically compatible equilibria, at least one of which is positive) can be precluded merely by establishing concordance of the underlying reaction network, perhaps with the assistance of a concordance-checking computer program such as the one made available in [1]. As we indicate in a subsequent article [3], the Species-Reaction Graph can sometimes also serve to establish network concordance. (See Remark 2.1.)

The converse of Proposition 4.8 is not true. That is, there can be a weakly monotonic kinetic system — in fact, a mass action system — that is injective even when its underlying reaction network is not concordant. On the other hand, the following proposition resembles a converse; proof by construction is provided.

Proposition 4.9

Let $\{S, C, R\}$ be a discordant reaction network. Then there exists for it a weakly monotonic kinetics K such that the resulting kinetic system $\{S, C, R, K\}$ is not injective.

Remark 4.10

(Example). Proposition 4.9 asserts the existence of a weakly monotonic kinetics for the discordant network(12) that gives rise to a non-injective kinetic system. In fact, it can be shown by direct computation that any mass-action system based on (12) possesses at least two stoichiometically compatible equilibria [24] — one positive and one non-positive — thereby violating injectivity.

Taken together, Propositions 4.8 and 4.9 can be summarized in the following theorem—a theorem that tells us that the class of concordant reaction networks is precisely the class of networks that are injective for everyassignment of a weakly monotonic kinetics:

Theorem 4.11

A reaction network has injectivity in all weakly monotonic kinetic systems derived from it if and only if the network is concordant.

We discuss more subtle connections between concordance and injectivity in the next section.

A REMARK ABOUT CONNECTIONS TO EARLIER WORK

In [16, 4, 6] attention was focused on mass action kinetic systems in which the underlying reaction network contains a "degradation reaction" of the form s → 0 for each species s in the network. (Here such networks are called fully open.) In this context it was shown that there are certain fully open networks having the property that all mass action kinetic systems deriving from them are injective — that is, regardless of rate constant values. These were called injective (mass action) networks in [16, 4], and it was shown that a network is injective in this (mass action) sense if its Species-Reaction Graph satisfies the mild conditions stated in Remark 2.1.

With this in mind, it is not unreasonable to suppose that a variant of Proposition 4.9 might be true: A network can have the property that every mass action system deriving from it is injective, regardless of rate constants, only if the network is concordant.

In fact, this is false. We show in (16) an example of a fully open reaction network that is not concordant but for which any mass action system derived from it is injective. Injectivity in the mass action case follows from a theorem in [27][5] that invokes Species-Reaction Graph conditions different from those in Remark 2.1. On the other hand, it can be established that the network is not concordant by a choice of α and σ that satisfy the conditions in Definition 4.1. Thus, the example tells us that "network injectivity" in the mass action case should not be conflated with network concordance.

$$A + B \to C \qquad 2A + B \to D \qquad A \rightleftarrows 0 \rightleftarrows B$$

$$\nwarrow \qquad \searrow$$

$$C \qquad\qquad D$$

$$(16)$$

In a separate article [3], however, we show that when a fully open network's Species-Reaction Graph satisfies the specific conditions stated in Remark 2.1 — conditions that ensured injectivity in the mass action setting considered in [16, 4, 6] and in the far more general NAC kinetics setting considered by Banaji and Craciun [7, 5] — then those same Species-Reaction Graph conditions will ensure that the network itself, divorced from any kinetics at all, is concordant and, in fact, strongly concordant (Section 9). Thus, when the kinetics is weakly monotonic or two-way weakly monotonic (Section 9), injectivity and the absence of multiple positive equilibria are ensured.

In the fully open network setting, then, the Species-Reaction Graph conditions stated in Remark 2.1 will imply injectivity for large classes of kinetics through the very simple Propositions 4.8 and 9.6. While earlier work based on these conditions appealed to Jacobians, special determinant attributes, and the Gale-Nikaido theorem to infer injectivity, the elementary Propositions 4.8 and 9.6 make those appeals unnecessary.Moreover, we shall see in Section 7 how properties of the Species-Reaction Graph will often imply concordance of a network (and, therefore, injectivity of any weakly monotonic kinetic system derived from it) even when the network is not fully open.

PERSISTENCE IN WEAKLY REVERSIBLE CONCORDANT KINETIC SYSTEMS: BOUNDARY BEHAVIOR AND THE EXISTENCE OF POSITIVE EQUILIBRIA

It is our aim in this section to establish two related properties inherent to kinetic systems, not necessarily weakly monotonic, that derive from weakly reversible concordant reaction networks.

These properties are connected to "persistence" questions: Viewed from an ecological perspective, one might be interested to know for a kinetic system whether there will be a positive equilibrium — that is, an equilibrium in which all species coexist. And one might also want to know whether a population that begins at a state in which all species are present can evolve to a different one — in particular to an equilibrium — in which one or more species are extinct (i.e., absent). For kinetic systems in which the underlying network is weakly reversible, questions like these become especially challenging.

As we shall see, however, a weakly reversible concordant reaction network invariably has pleasant properties that make these questions far easier to resolve:

First, we examine behavior of the species formation rate function on the boundary of a nontrivial stoichiometric compatibility class. (Recall that by a nontrivial stoichiometric compatibility class we mean one that contains a strictly positive composition.) In particular, we show that, on the boundary, there can be no equilibria at all. In fact, at every boundary composition $\bar{c}$, the species formation rate vector $f(\bar{c})$ points into the stoichiometric compatibility class in the following sense: There is at least one species s* for which $\bar{c}_{s*} = 0$ while $\dot{c}_{s*} = f_{s*}(\bar{c}) > 0$. (Thus, the inward-pointing boundary vector field thwarts an approach to the boundary along a trajectory that originates at a strictly positive composition.)

Second, we show that if the network is conservative and the kinetic rate functions are continuous then each nontrivial stoichiometric compatibility class contains at least one positive equilibrium. If, in addition, the kinetics is weakly monotonic, then there is precisely one equilibrium in each nontrivial stoichiometric compatibility class, and it is positive. (Of course, in the weakly monotonic case, uniqueness already follows from Proposition 4.8.)

Preliminaries: Reaction Transitive Compositions

We begin with some old reaction network theory ideas, derived from [8][6], that are independent of concordance. The following definition is intended to convey a simple idea: A species set $S^* \subset S$ is reaction-transitive if, whenever all the reactant species of a reaction are present in S^*, then so too are that reaction's product species.

Definition 6.1

For a reaction network $\{S, C, R\}$, a species set $S^* \subset S$ is reaction-transitive if, for all $y \to y' \in R$,

$$\mathrm{supp}\, y \subset \mathscr{S}^* \Rightarrow \mathrm{supp}\, y' \subset \mathscr{S}^*.$$

(17)

We say that a composition $c \in \overline{\mathbb{R}}_+^{\mathscr{S}} +$ is reaction-transitive if supp c is reaction transitive.

Clearly every positive composition is reaction-transitive. It follows from properties presumed of a kinetics in Section 3 that if $\bar{c}$ is a composition on the boundary of $\overline{\mathbb{R}}_+^{\mathscr{S}}$ and if species s is "missing" at composition $\bar{c}$ (i.e., if $\bar{c}_s =$

0), then the formation rate of s at composition $\bar{c}$ is not negative (i.e., $f_s(\bar{c}) \geq$ 0). The following lemma tells us that if $\bar{c}$ is not reaction-transitive, then there must be a missing species which, at composition $\bar{c}$, is produced at a strictly positive rate. (See [10] or Appendix I in [8].)

Lemma 6.2

Consider a kinetic system $\{S, C, R, K\}$ with species formation rate function $f(\cdot)$. If $\bar{c}$ is a composition on the boundary of $\overline{R^S+}$, then $f_s(\bar{c}) \geq 0$ for each species s such that $\bar{c}_s = 0$. Moreover, if $\bar{c}$ is not reaction-transitive, then there is a species s* such that $\bar{c}_{s*} = 0$ and $f_{s*}(\bar{c}) > 0$. Thus, every equilibrium of the kinetic system is reaction transitive.

Boundary Behavior for Weakly Reversible Concordant Networks

With this as background, we are in a position to state a theorem about weakly reversible concordant reaction networks. Note that the theorem makes mention only of network properties; there is no mention of kinetics.

Theorem 6.3

For a weakly reversible concordant reaction network, no non-trivial stoichiometric compatibility class can have on its boundary a composition that is reaction-transitive.

Before proving Theorem 6.3, we observe that the theorem and Lemma 6.2 give us the following corollary:

Corollary 6.4

For a kinetic system that derives from a weakly reversible concordant reaction network, no composition on the boundary of a nontrivial stoichiometric compatibility class is an equilibrium. In fact, at each such boundary composition the species formation rate vector points into the stoichiometric compatibility class in the sense that there is an absent species produced at strictly positive rate.

As might be expected, a species-formation-rate function that, in the sense indicated, points inward on the boundary of a non-trivial stoichiometric compatibility class repels an approach to the boundary by a bounded trajectory that begins at a strictly positive composition. For a formal proof that its ω-limit set is, in fact, disjoint from the boundary when the kinetics is sufficiently smooth see [29].

Remark 6.5

(Concordant deficiency zero networks) Readers acquainted with the deficiency-oriented aspects of chemical reaction network theory — see [8] — will know that a weakly reversible deficiency zero network, taken with mass action kinetics, admits precisely one positive equilibrium within every nontrivial stoichiometric compatibility class; moreover, the sole positive equilibrium is locally asymptotically stable (relative to nearby initial conditions in the same stoichiometric compatibility class). It is not yet known if one can generally assert global asymptotic stability under the same conditions, but if no nontrivial stoichiometric compatibility class has an equilibrium on its boundary, then global asymptotic stability is, in fact, ensured [34].[7] Thus, for concordant weakly reversible deficiency zero networks taken with mass action kinetics, one has global asymptotic stability of equilibria.

Remark 6.6

(About the proof of Theorem 6.3) Although Theorem 6.3 is about network properties alone, divorced from any mention of kinetics, in its proof we will endow the weakly reversible concordant network of the theorem with a mass action kinetics in order to produce a contradiction. To begin, we will presume that there can be a nontrivial stoichiometric compatibility class that has a reaction-transitive composition $\bar{c}$ on its boundary. Then we will endow the network with a mass action kinetics, with rate constants chosen to make the boundary composition $\bar{c}$ an equilibrium.

Now suppose we could take for granted that, for every weakly reversible network and for every assignment of rate constants, the corresponding mass action species-formation rate function admits a positive equilibrium in each nontrivial stoichiometric compatibility class. In that case, we would have a contradiction of Proposition 4.8. In particular, in the nontrivial stoichiometric compatibility class containing $\bar{c}$ we would have, for the mass action system constructed as described, a positive equilibrium that is stoichiometrically compatible with the boundary equilibrium $\bar{c}$.

In fact, an unpublished manuscript by Deng et al. [35] does indeed provide an argument for the existence, in each nontrivial stoichiometric compatibility class, of a positive equilibrium for any mass action system in which the underlying network is weakly reversible. In the absence of a published version, however, we can instead mount a similar but somewhat more complicated argument that relies on the same strategy, this time involving construction of mass action system that is complex balanced at the boundary composition $\bar{c}$. This would permit invocation of known properties of complex balanced mass action systems, in particular the distribution of their positive equilibria.

Because the argument that invokes [35] is more straightforward, we give it here in the main text. The alternative argument, which does not rely on [35], is provided in Appendix B.

Remark 6.7

In preparation for the nearby proof of Theorem 6.3 (and of the proof in the Appendix), it will be helpful if we make the following observation [8]: Consider a weakly reversible network $\{S, C, \mathcal{R}\}$, and suppose that a composition c* is reaction-transitive. It is not difficult to see that, if $y^\dagger$ and $y^{\dagger\dagger}$ are complexes belonging to the same linkage class, then there are two possibilities: Either supp $y^\dagger$ and supp $y^{\dagger\dagger}$ are bothcontained in supp c* or else neither is.

This is to say that the linkage classes are of two distinct kinds: Relative to the reaction-transitive compositionc*, a supported linkage class is one in which the support of every complex is contained in supp c*, while anunsupported linkage class is one in which the support of no complex is contained in supp c*.

If the network is given a kinetics, it follows from properties of a kinetics given in Section 3 that, at the reaction-transitive composition c*, all reactions associated with a supported linkage class will proceed at strictly positive rate, while all reactions associated with an unsupported linkage class will have zero rate. In rougher terms, at composition c*, all reactions associated with a supported linkage class will be "turned on" while those associated with an unsupported linkage class will be "turned off." Moreover, the "turned on" reactions (if there are any) constitute a weakly reversible subnetwork of the original network.

Proof of Theorem 6.3

Suppose that $\{S, C, \mathcal{R}\}$ is a concordant weakly reversible reaction network and that, contrary to what is to be proved, $c* \in \overline{\mathbb{R}}_+^{\mathcal{S}}$ is a reaction-transitive composition on the boundary of a nontrivial stoichiometric compatibility class. Let

$$C* := \{y \in C : \text{supp} y \subset \text{supp} c*\} \text{ and } R* := \{y \to y' \in R : \text{supp} y \subset \text{supp} c^{\mathcal{S}}\}.$$

Provided that $C*$ is not empty, $\{S, C*, \mathcal{R}*\}$ is a weakly reversible subnetwork of the original network. In this case, there are positive numbers $\{\kappa_{y \to y'}\}_{y \to y' \in \mathcal{R}*}$ that satisfy [10]

$$\sum_{y \to y' \in \mathcal{R}*} \kappa_{y \to y'} (y' - y) = 0.$$

$$(18)$$

Now choose rate constants $\{k_{y \to y'}\}_{y \to y' \in \mathcal{R}*}$ for reactions in $\mathcal{R}*$ to satisfy

$$\kappa_{y\to y'}=k_{y\to y'}(c^*)^y, \forall y \to y' \in \mathscr{R}^*.$$

(19)

In any case, for the remaining reactions choose rate constants $\{k_{y\to y'}\}_{y\to y'\in\mathscr{R}\backslash\mathscr{R}^*}$ to be any positive numbers. Note that for any $y \to y' \in \mathcal{R} \setminus \mathcal{R}^*$ we will have $k_{y\to y'}(c^*)^y = 0$ because supp $y \not\subset$ supp c^*. From this, (18), and (19) it follows that

$$\sum_{y\to y'\in\mathscr{R}^*} k_{y\to y'}(c^*)^y(y'-y)+ \sum_{y\to y'\in\mathscr{R}\backslash\mathscr{R}^*} k_{y\to y'}(c^*)^y(y'-y)=0.$$

(20)

This is to say that c* is a (boundary) equilibrium of the mass action system $\{\mathcal{S}, \mathcal{C}, \mathcal{R}, k\}$ so constructed. Because $\{\mathcal{S}, \mathcal{C}, \mathcal{R}\}$ is weakly reversible, [35] asserts, for the same mass action system, the existence of a strictly positive equilibrium c** in the nontrivial stoichiometric compatibility class containing c*. Because the network is concordant, this contradicts the conclusion of Proposition 4.8.

Conservative Weakly Reversible Concordant Networks and the Existence of Positive Equilibria

When, in the following theorem, we say that a kinetics is continuous we mean that its various reaction rate functions are all continuous.

Theorem 6.8

If $\mathcal{K}$ is a continuous kinetics for a conservative reaction network $\{\mathcal{S}, \mathcal{C}, \mathcal{R}\}$, then the kinetic system $\{\mathcal{S}, \mathcal{C}, \mathcal{R}, \mathcal{K}\}$ has an equilibrium within each stoichiometric compatibility class. If the network is weakly reversible and concordant, then within each nontrivial stoichiometric compatibility class there is a positive equilibrium. If, in addition, the kinetics is weakly monotonic, then that positive equilibrium is the only equilibrium in the stoichiometric compatibility class containing it.

We begin with a lemma that is required for application of a fixed point theorem due to Browder [36]. For any kinetic system, the lemma merely makes more firmly geometrical the idea that, on the boundary of a stoichiometric compatibility class, the species formation rate vector does not point outward.

Lemma 6.9

Consider a kinetic system $\{\mathcal{S}, \mathcal{C}, \mathcal{R}, \mathcal{K}\}$ with stoichiometric sub-space S and species formation rate function $f : \overline{\mathbb{R}_+^{\mathcal{S}}}, +\to S$. If $\bar{c}$ is a composition on the boundary of $\overline{\mathbb{R}}_+^{\mathscr{S}} \to S$ then there is a composition $c^\dagger$ in the stoichiometric compatibility class $(\bar{c}+S) \cap \overline{\mathbb{R}}_+^{\mathscr{S}}$ and a positive number λ such that $f(\bar{c}) = \lambda(c^\dagger - \bar{c})$.

Proof

Choose $\theta > 0$ sufficiently small as to ensure that

$$c^-s + \theta fs(c^-) > 0, \forall s \in \mathrm{supp}\, c^-. \tag{21}$$

For $s \notin \mathrm{supp}\, \bar{c}$ we have $\bar{c}_s = 0$ and, by Lemma 6.2, $f_s(\bar{c}) \geq 0$, so

$$\bar{c}_s + \theta f_s(\bar{c}) \geq 0, \forall s \notin \mathrm{supp}\, \bar{c}. \tag{22}$$

Now let $c^\dagger := \bar{c} + \theta f(\bar{c})$. From (21) and (22) it follows that $c^\dagger$ is a member of $\overline{\mathbb{R}}_+^{\mathscr{S}}$. Because $f(\bar{c})$ is a member of S, it follows that $c^\dagger$ is a member of the stoichiometric compatibility class containing $\bar{c}$. Taking $\lambda = 1\theta$, we get the desired result.

Proof of Theorem 6.8

Let $K \subset \overline{\mathbb{R}}_+^{\mathscr{S}} \to S$ be a stoichiometric compatibility class. That is, there is a composition $c\# \in \overline{\mathbb{R}}_+^{\mathscr{S}} \to S$ such that

$$K = (c\# + S) \cap \overline{\mathbb{R}}_+^{\mathscr{S}} \to S.$$

Because the network $\{S, C, \mathcal{R}\}$ is conservative, K is compact [22], and it is easily seen to be convex. Let $g : K \to (c\# + S)$ be defined by $g(c) := f(c) + c$, where $f(\cdot)$ is the species-formation-rate function for the kinetic system $\{S, C, \mathcal{R}, \mathcal{K}\}$. From Lemma 6.9 it is apparent that for each composition $\bar{c}$ on the boundary of K there is a number λ and a composition $c^\dagger \in K$ (both depending perhaps on $\bar{c}$) such that $g(\bar{c}) - \bar{c} = \lambda(c^\dagger - \bar{c})$. In this case, it follows from a theorem of Browder [36] that $g(\cdot)$ has a fixed point in K — that is, a composition c^* $\in K$ such that $g(c^*) = c^*$. But then we must have $f(c^*) = 0$, which is to say that there is an equilibrium in K.

If the network is weakly reversible and concordant and if the stoichiometric compatibility class K is nontrivial, it follows from Corollary 6.4 that c^* is not on the boundary of K; it must be positive. If, in addition, the kinetics is weakly monotonic, the fact that there cannot be another equilibrium in K follows from Proposition 4.8.

CONCORDANCE AND PROPERTIES OF A NETWORK'S FULLY OPEN EXTENSION; EIGENVALUES

By the fully open extension of a reaction network $\{S, C, \mathcal{R}\}$ we mean the network obtained from the original network by adjoining to it the species degradation reactions $\{s \rightarrow 0 : s \in S\}$ if those reactions are not already present. More precisely, the fully open extension of $\{S, C, R\}$ is the network $\{S, \tilde{C}, \tilde{R}\}$, where[8]

$$\tilde{\mathscr{C}} := \mathscr{C} \cup \mathscr{S} \cup \{0\} \text{ and } \tilde{\mathscr{R}} := \mathscr{R} \cup \{s \rightarrow 0 : s \in \mathscr{S}\}. \tag{23}$$

In this section we examine implications of concordance in a reaction network and in its fully open extension. We begin by noting that a network need not be concordant even when its fully open extension is concordant. The simple network $B \leftarrow A \rightarrow C$ provides an example.

Remark 7.1

(Why study a network's fully open extension?). There are two reasons to examine the concordance of a network's fully open extension. The first of these is that concordance of the fully open extension gives partial information about the stability properties of equilibria for kinetic systems derived from the original network, at least when the kinetics satisfies certain modest requirements. This we shall see in Theorem 7.6 and in Section 8.

The second reason is connected to inferences one can make from the original network's Species-Reaction Graph. As we shall see in a subsequent article [3], properties of a network's Species-Reaction Graph will often indicate that the fully open extension of a network is concordant. For this reason, we will want to know when fully-open-network concordance implies concordance of the original network. This is the subject of Theorem 7.4, which also has some stability implications, discussed in Section 8.

In a mass action context, connections were made in [20] between injectivity of a reaction network and of its fully open extension. Here we will be able to generalize those results substantially. By way of preparation, we record the following definition:

Definition 7.2

Consider a reaction network $\{S, C, R\}$ with stoichiometric subspace S.

The network is normal if there are $q \in \mathbb{R}_+^{\mathscr{S}}$ and $\eta \in \mathbb{R}_+^{\mathscr{R}}$ such that the linear transformation $T : S \rightarrow S$ defined by

$$T\sigma := \sum_{y \rightarrow y' \in \mathscr{R}} \eta_{y \rightarrow y'} (y * \sigma)(y' - y) \tag{24}$$

is nonsingular, where "$*$" is the scalar product in $\mathbb{R}^S$ defined by

$$x*x' := \sum_{s \in \mathscr{S}} q_s x_s x'_s.$$

(25)

Remark 7.3

The normality condition is not difficult to satisfy. In fact, it was shown in [20] that every weakly reversible network is normal, regardless of the complexes. Weaker sufficient conditions for network normality are also given in [20].

We are now in a position to state a theorem that has considerable significance when placed alongside another one, provided in [3], that indicates how properties of a network's Species-Reaction Graph can ensure concordance of the network's fully open extension. In that theorem, however, concordance of the original network is left undecided. The following theorem completes the picture when the original network is normal.[9] As we shall see, Theorem 7.4 also has some stability consequences.

Theorem 7.4

A normal network is concordant if its fully open extension is concordant. In particular, a weakly reversible network is concordant if its fully open extension is concordant.

So far in this section all considerations were about the network alone, divorced from a kinetics. Now we will suppose that a network $\{S, C, R\}$ is endowed with a kinetics $\mathcal{K}$ such that the resulting kinetic system $\{S, C, R, \mathcal{K}\}$ admits a positive equilibrium c*. We ask what network concordance — of either the network or of its fully open extension — might tell us about eigenvalues of df(c*) : S → S, where S is the stoichiometric subspace for the network, f(·) is the species formation rate function for the kinetic system, and df(c*) is the derivative of f(·) at c*.

In this case, we will require that the kinetics be "differentiably monotonic" at c*. (It is easy to see that every mass action kinetics for a network is differentiably monotonic at every positive composition.)

Definition 7.5

A kinetics $\mathcal{K}$ for a reaction network $\{S, C, R\}$ is differentiably monotonic at c*∈ $\overline{\mathbb{R}}_+^{\mathscr{S}}$ if, for every reaction y → y' ∈ R, $\mathcal{K}_{y \to y'}(\cdot)$ is differentiable at c* and, moreover, for each species s ∈ S,

$$\frac{\partial \mathcal{K}_{y \to y'}}{\partial c_s}(c^*) \geq 0,$$

(26)

with inequality holding if and only if s ∈ supp y. A differentiably monotonic kinetics is one that is differentiably monotonic at every positive composition.

Theorem 7.6

Let $\{S, C, R, K\}$ be a kinetic system with stoichiometric sub-space S and species formation rate function $f: \mathbb{R}_+^{\mathscr{S}} \to S$. Moreover, suppose that the kinetics is differentiably monotonic at $c* \in \mathbb{R}_+^{\mathscr{S}}$. If the underlying network $\{S, C, R\}$ is concordant then the derivative $df(c^*): S \to S$ is nonsingular (whereupon 0 is not one of its eigenvalues). If the network's fully open extension is concordant then no real eigenvalue of $df(c^*)$ is positive.

Proof of Theorem 7.6 is given in Appendix D.

Corollary 7.7

Let $\{S, C, R, K\}$ be a kinetic system with stoichiometric sub-space S and species formation rate function $f: \mathbb{R}_+^{\mathscr{S}} \to S$. Moreover, suppose that the kinetics is differentiably monotonic at $c* \in \mathbb{R}_+^{\mathscr{S}}$. If the underlying network $\{S, C, R\}$ is normal — in particular, if it is weakly reversible — and if the network's fully open extension is concordant, then every real eigenvalue of $df(c^*)$ is negative.

Proof

If the underlying network is normal then, by virtue of Theorem 7.4, its concordance follows from concordance of the fully open extension. In this case Theorem 7.6 ensures that $df(c^*)$ is nonsingular, so 0 cannot be among its eigenvalues. By the same theorem, concordance of the fully open extension then ensures that all real eigenvalues are negative.

CONSEQUENCES OF DISCORDANCE

Proposition 4.8 told us that a concordant network, taken with any weakly monotonic kinetics, invariably gives rise to an injective kinetic system. In this sense, concordance is a network attribute that enforces injectivity for all weakly monotonic kinetics. Proposition 4.9 went even further: It told us that no discordant network has this same property.

The observations in this section are in the same spirit, this time as counterpoints to Theorem 7.6 and its corollary. Those told us that a normal network with concordant fully open extension, taken with anydifferentiably monotonic kinetics, will invariably have some degree of stability at any positive equilibrium it might admit, in the sense that every real eigenvalue is negative. Here we shall see that no network with discordant fully open extension — in particular, no weakly reversible discordant network — can have this same property (except in the trivial case for which there are no positive equilibria at all).

Because our interest will be in positive equilibria, our focus will be on networks with positively dependent reaction vectors. (Recall from Section 3 that a kinetic system can admit a positive equilibrium only if the reaction vectors for the underlying network are positively dependent.).

Theorem 8.1

Consider a reaction network with positively dependent reaction vectors. If the network is discordant, then there exists for it a differentiably monotonic kinetics such that the resulting kinetic system admits a positive degenerate equilibrium. If the network's fully open extension is discordant then there exists for the original network a differentiably monotonic kinetics such that the resulting kinetic system admits an unstable positive equilibrium — in fact, a positive equilibrium associated with a positive real eigenvalue.

Theorem 8.1 has a number of corollaries. In each corollary, it is left implicit that the unstable positive equilibrium mentioned can be taken to have associated with it a positive real eigenvalue.

Corollary 8.2

For every discordant normal network with positively dependent reaction vectors there exists a differentiably monotonic kinetics such that the resulting kinetic system admits a positive unstable equilibrium.

Proof

It follows from Theorem 7.4 that every normal discordant network has a discordant fully open extension. Corollary 8.2 then follows from Theorem 8.1.

Corollary 8.3

For every weakly reversible discordant network there is a differentiably monotonic kinetics such that the resulting kinetic system admits a positive unstable equilibrium.

Proof

A weakly reversible network is normal and, moreover, has positively dependent reaction vectors. Corollary 8.3 is then a consequence of Corollary 8.2.

In preparation for the next corollary we point out that a concordant weakly reversible network (in fact, a reversible concordant network) can have a discordant fully open extension. Network (27) provides an example, as indicated by computation [1]. (This is a variant of a network considered for different reasons in [14].)

$$A+B \rightleftarrows E \quad C+D \rightleftarrows F \tag{27}$$

$$C \rightleftarrows 2A \rightleftarrows D \rightleftarrows B$$

The following is a consequence of Proposition 4.8, Corollary 6.4, and Theorem 8.1.

Corollary 8.4

For every weakly reversible concordant reaction network with discordant fully open extension there exists for the original network a differentiably monotonic kinetics such that the resulting kinetic system (for the original network) has the following properties: There is an equilibrium that is positive, unique within its stoichiometric compatibility class, and unstable. Moreover, at each point on the boundary of that stoichiometric compatibility class the species- formation-rate vector points inward in the sense that there is an absent species produced at strictly positive rate.

Remark 8.5

Although network concordance, through Proposition 4.8, ensures the uniqueness of a (positive) equilibrium within its stoichiometric compatibility class, network (27) and Corollary 8.4 tell us that concordance by itself cannot also ensure that the sole equilibrium is stable.

We show in [3] that when a normal (in particular, a weakly reversible) network's Species-Reaction Graph satisfies the conditions mentioned in Remark 2.1, concordance of both the network and its fully open extension follow. In this case, the hypothesis of Corollary 8.4 cannot be satisfied.

Nevertheless, even when both a network and its fully open extension are concordant, and even when the kinetics is differentiably monotonic, we still cannot preclude instability of the unique positive equilibrium: Although its associated real eigenvalues must be negative (Corollary 7.7), there might also be complex eigenvalues with positive real part. Such eigenvalues were found to be extant in Reidl et al. [23], where network (11), taken with a weakly monotonic (almost mass action) kinetics, served as a model for calcium oscillations in olfactory cilia.[10] As indicated by computation [1], both network (11) and its fully open extension are concordant.

Thus, it appears that concordance of a network (and of its fully open extension) enforce a degree of dull behavior, at least to the extent that they preclude multiple stoichiometrically compatible equilibria (and positive real eigenvalues), but the capacity for other forms of interesting behavior might nevertheless remain.

Remark 8.6

(Cyclic composition trajectories). Especially when the network in Corollary 8.4 is conservative, in which case stoichiometric compatibility classes are compact, the existence of a (positive) equilibrium that is both unique within its stoichiometric compatibility class and also unstable suggests that, for at least somedifferentiably monotonic kinetics, there might be an attracting periodic orbit.

There are indeed compelling reasons to believe that, for any weakly reversible concordant network with discordant fully open extension, there will exist a differentiably monotonic kinetics such the resulting kinetic system gives rise to a cyclic composition trajectory. In fact, certain considerations point to the presence of a Hopf bifurcation in connection with kinetic variations:

To the weakly reversible network in Corollary 8.4 one can assign a complex balanced mass action kinetics [22, 37, 10]. In this case the resulting mass action system will have the property that the unique equilibrium in each nontrivial stoichiometric compatibility class has associated with it eigenvalues[11] with negative real parts. On the other hand, Theorem 8.1 ensures that, for the same network, there is another differentiably monotonic kinetics that gives rise to a nontrivial stoichiometric compatibility class containing a unique (positive) equilibrium associated with a positive real eigenvalue. Along a suitably parameterized transition from the first differentiably monotonic kinetics to the second, one therefore expects that a pair of complex-conjugate eigenvalues will cross the imaginary axis. (So long as the parameterized kinetics remains differentiably monotonic, it follows from Theorem 7.6 that the crossing cannot be through zero.)

Similar considerations apply even in the case of a weakly reversible concordant network with concordant fully open extension, provided that, for the original network, there is a differentiably monotonic kinetics for which there is a positive equilibrium having eigenvalues with positive real part. However, for a concordant network with concordant fully open extension there is no guarantee, of the kind provided by Corollary 8.4, that such a kinetics will exist.

STRONG CONCORDANCE AND OTHER GENERALIZATIONS

Motivated by a kinetic class considered in work by Banaji and Craciun, we introduce in this section the notion of strong concordance and examine its consequences. Then we consider still other variants of the concordance idea.

Strong Concordance and two-way Weakly Monotonic Kinetics

Some — but certainly not all — of what we have done so far relied on the supposition of weakly monotonic kinetics, in which an increase in the rate of a particular reaction requires an increase in the concentration of at least one of its reactant species. In particular, we argued in Proposition 4.8 that, when the kinetics is weakly monotonic, the species-formation-rate function for a concordant network is invariably injective and that, as a consequence, multiple stoichiometrically compatible equilibria, with at least one positive, are impossible. There was no requirement that there be a degradation reaction for every species.

Although Banaji et al. [17] and Banaji and Craciun [5] do require a complete supply of degradation reactions to get injectivity results, the kinetic requirements they invoke are, in most ways, significantly weaker. We shall be more specific about the meaning of what they call nonautocatalytic (NAC) kinetics in Remark 9.7, but it suffices here to say that, with NAC kinetics, the rate function for a particular reaction is required to have the property that an increase [decrease] in the concentration of one of its reactant [product] species — keeping all other concentrations fixed — cannot result in a decrease of the reaction rate.[12] (Moreover, species that, for the reaction, are neither reactants nor products are not permitted to have an effect on the rate.) When the stoichiometric matrix is "strongly sign determined" — i.e., every one of its square submatrices has a certain property — Banaji and Craciun are also able to assert injectivity and the absence of multiple steady states, but, again, only in the fully open setting. (In the fully open setting, the issue of stoichiometric compatibility becomes moot.)

With weakly monotonic kinetics, then, the rate of a particular reaction can be influenced (positively) only by an increase in the concentration of a reactant species. The "influences" that NAC kinetics admits are broader: The rate of a reaction can also be influenced (positively) by a decrease in the concentration of aproduct species. In rough terms, NAC kinetics admits "two-way influences": It embraces product inhibition, while weakly monotonic kinetics does not.

For the most part, it is the purpose of this section to show that, without much difficulty, Proposition 4.8 can be generalized to subsume such two-way influences in the kinetics, provided that network concordance is replaced by strong concordance. We will assert that, when the kinetics is two-way weakly monotonic, the species-formation-rate function for a strongly concordant network is invariably injective and that, as a consequence, multiple stoichiometrically compatible equilibria, with at least one positive, are impossible. As was the case with Proposition 4.8 (and in contrast to the otherwise broad Banaji and Craciun results) there will be no requirement of a degradation reaction for every species.

We begin by making precise what we mean by a two-way weakly monotonic kinetics. Hereafter, following Banaji and Craciun, we restrict our attention to networks having the property that no reaction has a species common to both its reactant and product complexes. In practical terms, this is a very small price to pay for the loss of ambiguity in speaking of a reactant or product species for a particular reaction.

The following definition is intended to describe a kinetics in which an increase in the rate of a particular reaction requires that there be either an increase in the concentration of a reactant species or a decrease in the concentration of a product species. And for the rate to remain unchanged, it is necessary that there be no change in the concentrations of any reactant species or else that there be appropriately opposing changes in concentrations of the reactant and/or product species.

Definition 9.1

A kinetics $\mathcal{K}$ for a reaction network $\{\mathcal{S}, \mathcal{C}, \mathcal{R}\}$ is two-way weakly monotonic if, for each pair of compositions c^* and c^{**}, the following implications hold for each reaction $y \rightarrow y' \in \mathcal{R}$ such that supp $y \subset$ supp c^* and supp $y \subset$ supp c^{**}:

1. $\mathcal{K}_{y \rightarrow y'}(c^{**}) > \mathcal{K}_{y \rightarrow y'}(c^*) \Rightarrow$ there is a species s such that $\text{sgn}(c^{**}s - c^*s) = \text{sgn}(y - y')s \neq 0$.

2. $K_{y \rightarrow y'}(c^{**}) = K_{y \rightarrow y'}(c^*) \Rightarrow c^{**}s = c^*s$ for all $s \in$ supp y, or else there are species s, s' with $\text{sgn}(c^{**}s - c^*s) = \text{sgn}(y - y')s \neq 0$ and $\text{sgn}(c^{**}s' - c^*s') = -\text{sgn}(y - y')s' \neq 0$.

Remark 9.2

Definition 9.1 provides for a kinetic class that subsumes the class of weakly monotonic kinetics (Definition 4.5). It is intended to permit, but not require, the concentration of a reaction's product species to influence its rate. The terminology two-way weakly monotonic should not be construed to suggest that there is a complete anti-symmetry with respect to the influence of reactant and product species: Note that — as with mass action kinetics — the rate of a reaction can remain unchanged even when the concentrations of all of the reaction's product species increases, the concentrations of all other species remaining fixed. The same is not true of the reaction's reactant species. An increase in the concentrations of all of its reactant species, the others remaining fixed, must result in an increase in the reaction rate. Again, any kinetics that is weakly monotonic (e.g., mass action kinetics) is also two-way weakly monotonic.

Remark 9.3

In contrast to the definition of NAC kinetics [17, 5], there is no requirement here that the rate functions in a two-way weakly monotonic kinetics be differentiable or even continuous. Although Definition 9.1 is intended to capture the general spirit of NAC kinetics, there are also subtle, but consequential, differences that we shall discuss in Remark 9.7.

Because the class of two-way weakly monotonic kinetics is broader than the class of weakly monotonic kinetics, analogous theory will require focus on a narrower class of reaction networks — in particular the class of networks we call strongly concordant. In the following definition, the map L is as in eq. (10).

Definition 9.4

A reaction network $\{S, C, R\}$ with stoichiometric subspace S is strongly concordant if there do not exist $\alpha \in \ker L$ and a non-zero $\sigma \in S$ having the following properties:

1. For each $y \to y'$ such that $\alpha_{y \to y'} > 0$ there exists a species s for which sgn $\sigma_s = \text{sgn} \ (y - y')_s \neq 0$.

2. For each $y \to y'$ such that $\alpha_{y \to y'} < 0$ there exists a species s for which sgn $\sigma_s = -\text{sgn} \ (y - y')_s \neq 0$.

3. For each $y \to y'$ such that $\alpha_{y \to y'} = 0$, either (a) $\sigma_s = 0$ for all $s \in$ supp y, or (b) there exist species s, s' for which sgn $\sigma_s = \text{sgn} \ (y - y')_s \neq 0$ and sgn $\sigma_{s'} = -\text{sgn} \ (y - y')_{s'} \neq 0$.

Note that a network that is strongly concordant is also concordant. Network (28) is concordant but not strongly concordant.

A+B→P

B+C→Q   (28)

C→2A

Remark 9.5

(Ascertaining strong concordance). We show in [3] that strong concordance of a fully open network can be asserted when its Species-Reaction Graph satisfies the conditions stated in Remark 2.1. More generally, strong concordance of a network, not necessarily fully open, can be determined, either positively or negatively, by a sign-checking procedure implemented in [1].

The following proposition generalizes Proposition 4.8.

Proposition 9.6

(Generalization of Proposition 4.8). A two-way weakly monotonic kinetic system $\{S, C, R, K\}$ is injective whenever its underlying reaction network $\{S, C, R\}$ is strongly concordant. In particular, if the underlying re- action network is strongly concordant, then the kinetic system cannot admit two distinct stoichiometrically compatible equilibria, at least one of which is positive.

The proof of the proposition closely parallels the fairly simple proof of Proposition 4.8. In fact, Proposition 9.6 is merely a special case of a more general proposition that we state and prove in §9.2, one that allows for very general species-influences in the kinetics.

Before we do that, however, we want to emphasize once again that — in contrast to the work of Banaji and Cracun — Proposition 9.6 does not require the fully open setting. That they were compelled to require a full supply of degradation reactions in order to ensure injectivity of the species-formation-rate function is due, in part, to difficulties inherent in the NAC kinetic class. Although NAC kinetics is close in spirit to what we call two-way weakly monotonic kinetics, there are subtle distinctions between the two that, for NAC kinetics in the true chemical reactions, makes a result resembling Proposition 9.6 impossible, at least in the absence of additional special reactions (e.g., degradation reactions) conforming to stronger kinetic requirements. This we explain in the following remark.

Remark 9.7

(Difficulties instrinsic to NAC kinetics). In the sense of [17, 5], a (differentiable) kinetics K for a reaction network $\{S, C, R\}$ is non-autocatalytic (NAC) if for each $c* \in \mathbb{R}_+^{S}$, for each reaction $y \to y' \in R$, and for each species $s \in S$, the following conditions hold:

1. $\partial K y \to y' \partial c s(c*)(y's - ys) \leq 0,$

2. $y's-ys=0 \Rightarrow \partial Ky \rightarrow y' \partial cs(c*)=0$.

For any reaction network $\{\mathcal{S}, \mathcal{C}, \mathcal{R}\}$, it is not difficult to imagine an NAC kinetic system $\{\mathcal{S}, \mathcal{C}, \mathcal{R}, \mathcal{K}\}$ in which, for some open set $\Omega \in \mathbb{R}_+^{\mathcal{S}}$, each rate function $\mathcal{K}_{y \rightarrow y'}(\cdot)$ is constant on Ω. In this case, the species-formation-rate function is also constant on Ω, so the kinetic system is not injective. Thus, there can be no assertion resembling Proposition 9.6, in which every reaction network within a certain class, when taken with every NAC kinetics, invariably gives rise to an injective kinetic system.

In the work of Banaji and Craciun injectivity results from the supposition that the "true" chemical reactions — those that conform to NAC kinetics — are supplemented by degradation reactions that conform to still stronger kinetic requirements.

The following theorem is a variant of Theorem 7.4. It will primarily find use in the following way: We show in [3] that, when a reaction network's Species-Reaction Graph satisfies the conditions of Remark 2.1, then the network's fully open extension is strongly concordant. Thus, when the network is normal, those same Species-Reaction Graph conditions will ensure strong concordance of the original network.

Theorem 9.8

A normal network is strongly concordant if its fully open extension is strongly concordant. In particular, a weakly reversible network is strongly concordant if its fully open extension is strongly concordant.[13]

Proof of the theorem is very similar to the proof of Theorem 7.4. See Appendix C.

Theorem 7.6 also has a generalization. In preparation for it we need the following definition:

Definition 9.9

A kinetics $\mathcal{K}$ for a reaction network $\{\mathcal{S}, \mathcal{C}, \mathcal{R}\}$ is differentiably two-way monotonic at $c* \in \mathbb{R}_+^{\mathcal{S}}$ if, for every reaction $y \rightarrow y' \in \mathcal{R}$, $\mathcal{K}_{y \rightarrow y'}(\cdot)$ is differentiable at $c*$ and, moreover,

1. $\partial Ky \rightarrow y' \partial cs(c*) > 0$ $\forall s \in suppy$,
2. $\partial Ky \rightarrow y' \partial cs(c*) \leq 0$ $\forall s \in suppy'$,
3. $\partial Ky \rightarrow y' \partial cs(c*) = 0$ $\forall s \in \mathcal{S} \backslash (suppy \cup suppy')$.

Proofs of the following theorem and its corollary proceed very much like the proofs of Theorem 7.6 and its corollary.

Theorem 9.10

Let $\{S,\, C,\, R,\, K\}$ be a kinetic system with stoichiometric subspace S and species formation rate function $f: \mathbb{R}_+^{S} \to S$. Moreover, suppose that the kinetics is differentiably two-way monotonic at $c* \in \mathbb{R}_+^{S}+$. If the underlying network $\{S, C, R\}$ is strongly concordant then the derivative $df(c*) : S \to S$ is nonsingular (whereupon 0 is not one of its eigenvalues). If the network's fully open extension is strongly concordant then no real eigenvalue of $df(c*)$ is positive.

Corollary 9.11

Let $\{S,\, C,\, R,\, K\}$ be a kinetic system with stoichiometric subspace S and species formation rate function $f: \mathbb{R}_+^{S} \to S$. Moreover, suppose that the kinetics is differentiably two-way monotonic at $c* \in \mathbb{R}_+^{S}$. If the underlying network $\{S, C, R\}$ is normal — in particular, if it is weakly reversible — and if the network's fully open extension is strongly concordant, then every real eigenvalue of $df(c*)$ is negative.

Still more General Species Influences on Reaction Rates

In our definition of two-way weakly monotonic kinetics, we generalized weakly monotonic kinetics to permit (but not require) for a particular reaction an inhibitory effect exerted by the reaction's product species. That is, a particular kinetics within the two-way weakly monotonic class was permitted to have the property that the rate of a particular reaction $y \to y'$ might increase whenever the concentration of $s \in \operatorname{supp} y'$ decreases, the concentrations of all other species remaining fixed. The companion notion of strong concordance — an attribute of network structure alone, divorced from kinetic considerations — worked hand-in-hand with the broader kinetic class to give rise to the injectivity result given in Proposition 9.6.

It is not substantially more difficult to consider, in an analogous way, kinetic classes with far wider ranges of permissible species influences. Indeed, for a particular reaction we might want to permit inducer or inhibitor effects on the reaction rate by species that, for the reaction, are neither reactants nor products. (We shall continue to require that, for each reaction, an increase in the concentration of one its reactant species, with all other species concentrations held constant, results in an increase of the reaction's rate.)

We begin by indicating formally what we mean by an influence specification for a network. This amounts to indicating, for each reaction, which species are deemed to be possible inducers and which are deemed to be possible inhibitors (and which species are required to have no effect whatsoever).

Definition 9.12

An influence specification $\mathcal{I}$ for a reaction network $\{\mathcal{S}, \mathcal{C}, \mathcal{R}\}$ is an assignment to each reaction $y \rightarrow y'$ of a function $\mathcal{I}_{y \rightarrow y'} : \mathcal{S} \rightarrow \{1, 0, -1\}$ such that

$$\mathcal{I}_{y \rightarrow y'}(s) = 1, \forall s \in \text{supp} y. \tag{29}$$

If $\mathcal{I}_{y \rightarrow y'}(s) = 1$ [resp., -1] then species s is an inducer [inhibitor] of reaction $y \rightarrow y'$.

Recall that, for the two-way monotonic kinetic class, we permitted, but did not require, the product species to be inhibitors for a particular kinetics within the class. In the following definition, our intent is to specify, for a broad kinetic class, which species are permitted to be inhibitors or inducers — without requiring them to be one or the other — for a particular kinetics within the class. (The exception is that, in the sense of Definition 9.12, we require that, for each reaction, its reactant species are inducers.)

Definition 9.13

A kinetics $\mathcal{K}$ for a reaction network $\{\mathcal{S}, \mathcal{C}, \mathcal{R}\}$ is weakly monotonic with respect to influence specification $\mathcal{I}$ if, for every pair of compositions c* and c**, the following implications hold for each reaction $y \rightarrow y' \in \mathcal{R}$ such that supp $y \subset$ supp c* and supp $y \subset$ supp c** :

1. $\mathcal{K}_{y \rightarrow y'}(c^{**}) > \mathcal{K}_{y \rightarrow y'}(c^*) \Rightarrow$ there is a species s such that sgn(c**s−c*s)=J$y \rightarrow$y'(s)≠0.

2. $\mathcal{K}_{y \rightarrow y'}(c^{**}) = \mathcal{K}_{y \rightarrow y'}(c^*) \Rightarrow$ either (a) c**s=c*s for all s ∈ supp y or (b) there are species s, s' with sgn(c**s−c*s)=J$y \rightarrow$y'(s)≠0 and sgn(c**s'−c*s')=−J$y \rightarrow$y'(s')≠0.

Remark 9.14

Note that if a kinetics $\mathcal{K}$ is weakly monotonic with respect to influence specification $\mathcal{I}$ and if c** is a positive composition then implications (i) and (ii) in Definition 9.13 will hold for all reactions in the network, even when c* is not strictly positive and even when, for a reaction $y \rightarrow y'$, supp y is not contained in supp c*.

In the following definition, the map L is again as in eq. (10).

Definition 9.15

A reaction network $\{S, C, R\}$ with stoichiometric subspace S is concordant with respect to influence specification $\mathcal{I}$ if there do not exist $\alpha \in \ker L$ and a non-zero $\sigma \in S$ having the following properties:

1. For each $y \to y'$ such that $\alpha_{y \to y'} > 0$ there exists a species s for which sgn $\sigma_s = \mathcal{I}_{y \to y'}(s) \neq 0$.

2. For each $y \to y'$ such that $\alpha_{y \to y'} < 0$ there exists a species s for which sgn $\sigma_s = -\mathcal{I}_{y \to y'}(s) \neq 0$.

3. For each $y \to y'$ such that $\alpha_{y \to y'} = 0$, either (a) $\sigma_s = 0$ for all $s \in$ supp y or (b) there are species s, s' for which sgn $\sigma_s = \mathcal{I}_{y \to y'}(s) \neq 0$ and sgn $\sigma_{s'} = -\mathcal{I}_{y \to y'}(s') \neq 0$.

Remark 9.16

It is expected that [1] will be expanded to ascertain not only concordance and strong concordance but also the more general form of concordance indicated in Definition 9.15.

Remark 9.17

The definitions of the weakly monotonic kinetic class (Definition 4.5) and of ordinary network concordance (Definition 4.1) emerge as special cases of Definitions 9.13 and 9.15 applied with the influence specification given, for each reaction, by $\mathcal{I}_{y \to y'} =$ sgn y. Similarly, the definitions of the two-way monotonic class and of strong concordance emerge from the influence specification given, for each reaction, by $\mathcal{I}_{y \to y'} =$ sgn $(y - y')$.

The following proposition generalizes Proposition 9.6. In light of Remark 9.17, its proof will serve also as a proof of Proposition 9.6.

Proposition 9.18

(Generalization of Proposition 9.6). A kinetic system $\{S, C, R, K\}$ is injective whenever there exists an influence specification $\mathcal{I}$ such that:

1. The kinetics K is weakly monotonic with respect to $\mathcal{I}$.
2. The underlying network $\{S, C, R\}$ is concordant with respect to $\mathcal{I}$.

Proof

Let $\{S, C, R, K\}$ be a kinetic system and suppose that $\mathcal{I}$ is an influence specification that satisfies conditions (i) and (ii) in the proposition statement. Suppose also that, contrary to what is to be proved, there are distinct stoichiometrically compatible compositions c* and c**, with $c^{**} \in \mathbb{R}^{\mathcal{S}}_{+}$, such that

$$\sum_{y \to y' \in \mathcal{R}} \mathcal{K}_{y \to y'}(c^{**})(y' - y) = \sum_{y \to y' \in \mathcal{R}} \mathcal{K}_{y \to y'}(c^{*})(y' - y).$$

(30)

Then $\alpha \in \mathbb{R}^{\mathcal{R}}$, defined by

$$\alpha_{y \to y'} := \mathcal{K}_{y \to y'}(c^{**}) - \mathcal{K}_{y \to y'}(c^{*}), \quad \forall y \to y' \in \mathcal{R},$$

(31)

is a member of ker L. The vector $\sigma := c^{**} - c^{*}$ is clearly a member of the stoichiometric subspace, S.

Note that, if for a particular reaction $y \to y'$ we have $\alpha_{y \to y'} > 0$, then weak monotonicity with respect to $\mathcal{I}$ implies that, for some species s, sgnσs=sgn(c**s−c*s)=Jy→y'(s)≠0. Similarly, when $\alpha_{y \to y'} < 0$, there is a species s such that sgnσs=sgn(c**s−c*s)=−sgn(c*s−c**s)=−Jy→y'(s)≠0. Finally, if $\alpha_{y \to y'} = 0$, then σs=c**s−c*s=0 for all s ∈ supp y, or else there exist distinct species s, s' such that sgnσs=sgn(c**s−c*s)=Jy→y'(s)≠0 and sgnσs'=sgn(c**s'−c*s')=−Jy→y'(s')≠0.

The existence of α and σ so constructed contradicts the supposition that the network $\{\mathcal{S}, \mathcal{C}, \mathcal{R}\}$ is concordant with respect to influence specification $\mathcal{I}$.

It is not difficult to see that other propositions and theorems in this paper can also be generalized in various ways to accommodate the broader notions of concordance and kinetic monotonicity, both relative to a common influence specification.

Remark 9.19

(Kinetic classes with mandatory inducer-inhibitor influences). For a reaction network $\{\mathcal{S}, \mathcal{C}, \mathcal{R}\}$, consider the class of all kinetics that are weakly monotonic with respect to a certain influence specification, $\mathcal{I}$. Moreover, let $y \to y'$ be a fixed reaction, and suppose that species s is such that $\mathcal{I}_{y \to y'}(s) < 0$. (This choice of sign is made only for the sake of concretion.)

Now if $\mathcal{K}$ is a fixed kinetics in the kinetic class under consideration, then, with respect to $\mathcal{K}$, species s might have no effect at all on the rate of reaction $y \to y'$. With another kinetics $\mathcal{K}'$ in the same class, s might indeed be a nontrivial inhibitor of $y \to y'$. That $\mathcal{I}_{y \to y'}(s)$ is negative merely indicates that the kinetic class under consideration is sufficiently wide as to admit some kinetics, such as $\mathcal{K}'$, in which s is an inhibitor of $y \to y'$.

In this way, mass action kinetics (for which there are no inhibitors) can be viewed to sit within a larger class of kinetics, certain members of which might include inhibition. Theorems can then be stated for broad classes of kinetics that include mass action kinetics as a special subclass.

We could just as well have chosen to consider kinetic classes within which every kinetics is required to include, for example, a nontrivial inhibition of species s on reaction $y \rightarrow y'$. Specification of such compulsory influences can be accomplished in the following way: Definition 9.12 would be expanded to include, for each reaction $y \rightarrow y'$, a specified species set $\mathcal{M}_{y \rightarrow y'} \subset \mathcal{S}$, with supp $y \subset \mathcal{M}_{y \rightarrow y'}$, for which the influences given by $\mathcal{I}_{y \rightarrow y'}$ are mandatory.

In item (ii)(a) of Definition 9.13, supp y would then be replaced by $\mathcal{M}_{y \rightarrow y'}$, and the same replacement would be made in item (iii)(a) of Definition 9.15. With very minor adjustments in the proof, Proposition 9.18 would emerge as before.

DISCUSSION: REACTION NETWORK DESCRIPTIONS OF ENZYME-DRIVEN BIOCHEMISTRY

The cell requires, among other things, biochemical mechanisms that can underlie a variety of switches and oscillators dedicated to specific purposes. Each of these devices will generally execute its function against a mostly stable but nevertheless complex biochemical background, one that must maintain roughly fixed concentrations of critical metabolic components and roughly fixed rates of crucial reactions in the presence of intermittent disturbances that the cell is likely to experience. Thus, it appears that certain cellular reaction networks are required to behave stably against a variety of perturbations, while others are required to admit richer behavior, for example rapid switching between two widely separated stable steady states in response to small changes in a signal.

One purpose of this article is to draw distinctions, based on structural features, between reaction networks that might, for example, provide the basis for a sensitive biochemical switch and those more stably-behaved reaction networks that admit only mundane behavior, at least when the kinetics resides within a very broad class. Such structural differences can be very subtle, so it is not surprising that the idea of concordancerequires sufficient mathematical precision to differentiate between highly similar networks that might differ only in very fine detail. Recall that no concordant network taken with any weakly monotonic kinetics can give rise to two distinct stoichiometrically-compatible positive equilibria that might be required for a bistable switch.

Like the work of Banaji and Craciun [5], the work here gives results for very broad classes of kinetics, not just kinetics of the mass action kind. Thus, these results have the capacity to deliver information about biochemical reaction networks in which the kinetics might be, for example, of the generalized Michaelis-Menten type — that is, kinetics derived from more detailed elementary-step (mass action) enzyme-kinetic mechanisms via well-accepted approximation procedures [38].

However, when one invokes such approximations, there is a need for caution in how the theory is used, be it theory presented here or in [5]. In particular, one needs to be concerned with distinctions between, on one hand, information that the theory gives for a coarse-grained biochemical network with approximate kinetics and, on the other hand, information the theory gives for the original detailed mechanistic (mass action) network that is being approximated.

It is instructive to consider a toy example. Suppose that an enzyme E serves as a catalyst for the simple overall reaction $S1 + S2 \rightarrow P$ whereby two small-molecule substrates, S1 and S2, combine to form the small-molecule product P. Suppose also that the reaction takes place in a well-stirred cell to which substrates are supplied through an entry port at fixed rates and from which the substrates and product are removed through an effluent port at rates proportional to their instantaneous concentrations within the cell. Imagine too that the large-molecular-weight enzyme is entrapped within the cell by means of membranes at the entry and effluent ports. In this case, transport of substrates and product can be described by pseudo-reactions of the form $S_1 \rightleftarrows 0$, $S_2 \rightleftarrows 0$, and $P \rightarrow 0$, which we hereafter presume to be taken with mass action kinetics. (These transport reactions can instead be regarded to model the effects of constant-rate synthesis of S_1 and S_2 and first-order degradation of S_1, S_2, and P.)

We begin by supposing that, at the level of elementary (mass action) reactions, the operative enzyme-catalytic mechanism involves what is sometimes called ordered binding of substrates [38]. In particular, S1 binds first to enzyme E to form ES_1, and only then does S_2 bind to ES_1 to form ES_1S_2. Thereafter, the product P, once formed, unbinds from E. In this case, the reaction network that gives rise to the appropriate mass action differential equations for the toy cell is shown in (32).

$$S1 + E \quad\rightleftharpoons\quad ES1$$

$$S2 + ES1 \rightleftharpoons \quad ES1S2 \quad\rightarrow\quad P + E$$

$$S1 \;\rightleftharpoons\quad 0 \quad\leftrightarrows\; S2$$

$$\uparrow$$

$$P \tag{32}$$

By means of software provided in [1] we can readily determine that network (32) is concordant (and, in fact, strongly concordant). Thus, all the properties that accrue to concordant networks accrue to (32). In particular, because mass action kinetics is weakly monotonic, regardless of rate constant values, the corresponding mass action differential equations cannot admit two distinct stoichiometrically compatible steady states, at least one of which is positive (Proposition 4.8). In fact, by means of [1] we can determine that the fully open extension of network (32) is also concordant. Thus, we can assert, by means of Theorem 7.6, that any real eigenvalues associated with a positive equilibrium of the mass action differential equations for (32) are invariably negative.[14]

Now suppose instead that, at the level of elementary (mass action) reactions, enzyme catalysis proceeds by the so-called random [38] or unordered mechanistic scenario, whereby S1 and S2 are required to bind independently at nearby sites on the enzyme before they join to form P. In this case, the governing mass action differential equations derive from the somewhat different network displayed in (33).

$$S1 + E \rightleftharpoons ES1 \qquad\qquad S2 + E \rightleftharpoons ES2$$

$$S2 + ES1 \rightleftharpoons \quad ES1S2 \quad\leftrightarrows S1 + ES2$$

$$\downarrow$$

$$P + E$$

$$S1 \;\rightleftharpoons\quad 0 \quad\leftrightarrows\; S2$$

$$\uparrow$$

$$P \tag{33}$$

From software provided in [1] we can determine quickly that network (33) is not concordant. In fact, we can also determine from [1] that, in contrast to the situation for network (32), there are parameter values such that the mass action differential equations corresponding to network (33) give rise to two distinct stoichiometrically-compatible positive steady states [6].

These highly similar examples tell us that, in consideration of complex enzyme driven reaction networks written at the level of elementary reactions, the fine mechanistic details of enzyme catalysis can make a difference — not only in the concordance or discordance of the resulting reaction network but also in the capacity for various kinds of qualitative behavior. As the examples demonstrate, this is already true even if there is only a single overall coarse-grained reaction, in our case $S_1 + S_2 \rightarrow P$.

What, then, are we to make of approximate coarse-grained network descriptions in which fine-grained mechanistic details are obscured?

Suppose, for example, that in our toy system we assume that the overall reaction $S1 + S2 \rightarrow P$ has associated with it an approximate kinetic rate function $\mathcal{K}_{S1+S2\rightarrow P}(\cdot, \cdot)$ such that $\mathcal{K}_{S1+S2\rightarrow P}(c_{S1}, c_{S2})$ gives the molar occurrence rate per unit volume when the local concentrations of $S1$ and $S2$ are c_{S1} and c_{S2}. Such an overall reaction rate function would presumably derive from one or another detailed mechanistic description — for example, those indicated in (32) or (33) — by means of a rapid equilibrium or pseudo-steady-state approximation procedure of the kind described in [38]. In this case, the coarse-grained differential equations for our toy system would correspond to the coarse-grained reaction network (34).

$$S1 + S2 \quad \rightarrow \quad P$$

$$S1 \; \rightleftarrows \quad 0 \quad \leftrightharpoons \; S2$$

$$\uparrow$$

$$P \tag{34}$$

The simple network (34) is concordant, as is readily confirmed by [1]. Then, so long as the coarse-grained reaction rate function $\mathcal{K}_{S1+S2\rightarrow P}(\cdot, \cdot)$ conforms to the seemingly natural requirement of weak monotonicity (Definition 4.5), Proposition 4.8 tells us that the differential equations derived from network (34) cannot admit two positive stoichiometrically-compatible steady states.

But if, for certain parameter values, the mass action differential equations corresponding to the more fully-articulated random binding mechanistic network (33) do indeed admit two positive stoichiometrically-compatible steady states, and if the classical kinetic-approximation procedures described in [38] are in fact highly reliable, how are we to account for the denial by Theorem 4.11 of two such positive steady states in consideration of the coarse-grained concordant network (34)?

The answer resides in the implicit but faulty presumption that the derived coarse-grained rate function $\mathcal{K}_{S1+S2\rightarrow P}(\cdot, \cdot)$ should invariably be weakly monotonic — in particular, that an increase in the rate at which P is produced will invariably require an increase in the concentration of at least one substrate. Indeed, calculations in [38] indicate that, for certain values of the mass action rate constants in the random-binding enzyme-catalysis mechanism included in network (33), the coarse-grained rate function $\mathcal{K}_{S1+S2 \to P}$ $(\cdot, \cdot)$ derived from standard pseudo-steady-state approximations can increase in value even when no substrate concentration increases, at least in certain concentration ranges. (See, in particular, p. 659 of [38].) In such cases, Theorem 4.11 stands silent when applied to the coarse-grained network (34), just as it stands silent when applied to the discordant (mass action) network (33).

This is not to say that use of Proposition 4.8 or of the broader Proposition 9.6 should be avoided when coarse-grained networks and kinetics are employed. Rather, users of the theory should retain some sensitivity to the idea that the seemingly natural presumption of weak monotonicity in the kinetics might be inappropriate in certain concentration ranges when, in certain underlying mass action enzyme-catalysis mechanisms, rate constants take certain combinations of values. In the passage from finer-grained to approximate coarser-grained reaction network descriptions, such possibilities should kept in mind.

In any case, it should be remembered that The Chemical Reaction Network Toolbox [1] provides means to assess concordance properties of both fine-grained and coarse-grained network descriptions of a particular biochemistry. When the kinetics is mass action, the Toolbox provides additional tests that are even more incisive.

HIGHLIGHTS

We describe a large class of chemical reaction networks, those endowed with a subtle structural property called concordance. We show that the class of concordant networks coincides precisely with the class of networks which, when taken with any weakly monotonic kinetics, invariably give rise to kinetic systems that are injective --- a quality that, among other things, precludes the possibility of switch-like transitions between distinct positive steady states. We also provide persistence characteristics of concordant networks, instability implications of discordance, and consequences of stronger variants of concordance. Some of our results are in the spirit of recent ones by Banaji and Craciun, but here we do not require that every species suffer a degradation reaction. This is especially important in studying biochemical networks, for which it is rare to have all species degrade.

ACKNOWLEDGMENTS

We are grateful to Uri Alon for his encouragement and support and to Avi Mayo for his help. We especially thank Haixia Ji and Daniel Knight for expanding The Chemical Reaction Network Toolbox [1] to implement concordance tests.

FOOTNOTES

[1]GS was supported by an advanced ERC grant.

[2]MF was supported by NSF grant EF-1038394 and NIH grant 1R01GM086881-01.

[3]Algorithms underlying [1] are described in [2].

[4]A positive composition is one in which all species concentrations are strictly positive.

[5]A similar theorem appeared earlier in [15, 14]. In the case of (16), injectivity follows from the fact that the Species-Reaction Graph has just one cycle, and it is a c-cycle — that is, a cycle whose edge set is the union of c-pairs.

[6]See in particular Propositions 5.3.1–5.3.2, Remark 6.1.E, and Appendices I and II in [8]. The idea behind Def. 6.1 was important not only in [8] but also in more recent papers, where the formulation is phrased differently. A species set is called semi-locking in [28] or a siphon in [29] if its complement in the full set of species is reaction-transitive. See also recent articles by Gnacadja [30, 31, 32], which, among other things, contain a review of persistence results by A. I. Vol'pert [33] that first appeared in the 1970s Russian literature. In [30] a reaction-transitive species set is called reach-closed.

[7]For a discussion of this and extensions to certain weakly reversible deficiency zero networks that do admit boundary equilibria in nontrivial stoichiometric compatibility classes see [28].

[8]Note that in the definition of $\tilde{C}$ we are viewing S as a set in $\mathbb{R}^s$ — that is, as a surrogate for the standard basis of $\mathbb{R}^s$. See Section 3.

[9]In Theorem 7.4 and Corollary 7.7 normality can, in fact, be replaced by a milder condition, weak normality. See [3].

[10]We are grateful to Sayanti Banerjee for calling this paper to our attention.

[11]We are referring here to eigenvalues associated with eigenvectors in the stoichiometric subspace.

[12]Without much loss of generality, Banaji and Craciun presume that no reaction has a species common to both its reactant and product complexes.

[13]Here again, in Theorem 9.8 and Corollary 9.11, normality can be replaced by a milder condition, weak normality. See [3].

[14]We are referring here to eigenvalues corresponding to eigenvectors in the stoichiometric subspace.

[15]We have not precluded the possibility that a degradation reaction, say $\bar{s} \rightarrow 0$, might be a member of $\mathcal{R}$, the set of reactions in the original network. In this case the term $\alpha_{\bar{s} \rightarrow 0}(-\bar{s})$ would appear twice in (D.14), once in the first sum and once in the second. Nevertheless, it is easy to confirm that the sentence remains true as written.

REFERENCES

1. Ji H, Ellison P, Knight D, Feinberg M. The chemical reaction network toolbox, version 2.1. 2011Available at http://www.chbmeng.ohio-state.edu/~feinberg/crntwin/.

2. Ji H. Ph.D. thesis, Department of Mathematics. The Ohio State University; 2011. Uniqueness of Equilibria for Complex Chemical Reaction Networks.

3. Shinar G, Feinberg M. Concordant reaction networks and the species-reaction graph, arXiv:1203.6560v2 [q-bio.MN] 2012 Available at http://arxiv.org/pdf/1203.6560v2.pdf Submitted to Mathematical Biosciences.

4. Craciun G, Feinberg M. Multiple equilibria in complex chemical reaction networks. II. the species-reaction graph. SIAM Journal on Applied Mathematics. 2006;66:1321–1338.

5. Banaji M, Craciun G. Graph-theoretic criteria for injectivity and unique equilibria in general chemical reaction systems. Advances in Applied Mathematics. 2010;44:168–184. [PMC free article] [PubMed]

6. Craciun G, Tang Y, Feinberg M. Understanding bistability in complex enzyme-driven reaction networks.Proceedings of the National Academy of Sciences. 2006;103:8697–8702. [PMC free article] [PubMed]

7. Banaji M, Craciun G. Graph-theoretic approaches to injectivity and multiple equilibria in systems of interacting elements. Communications in Mathematical Sciences. 2009;7:867–900.

8. Feinberg M. Chemical reaction network structure and the stability of complex isothermal reactors i. the deficiency zero and deficiency one theorems. Chemical Engineering Science. 1987;42:2229–2268.

9. Feinberg M. Chemical reaction network structure and the stability of complex isothermal reactors ii. multiple steady states for networks of deficiency one. Chemical Engineering Science. 1988;43:1–25.

10. Feinberg M. Written version of lectures given at the Mathematical Research Center. Madison, WI: University of Wisconsin; Lectures on chemical reaction networks, 1979. Available athttp://www.chbmeng. ohio-state.edu/~feinberg/LecturesOnReactionNetworks.

11. Feinberg M. The existence and uniqueness of steady states for a class of chemical reaction networks.Archive for Rational Mechanics and Analysis. 1995;132:311–370.

12. Feinberg M. Multiple steady states for chemical reaction networks of deficiency one. Archive for Rational Mechanics and Analysis. 1995;132:371–406.

13. Ellison P, Feinberg M. How catalytic mechanisms reveal themselves in multiple steady-state data: I. basic principles. Journal of Molecular Catalysis. A, Chemical. 2000;154:155–167.

14. Schlosser PM, Feinberg M. A theory of multiple steady states in isothermal homogeneous CFSTRs with many reactions. Chemical Engineering Science. 1994;49:1749–1767.

15. Schlosser PM. Ph.D. thesis. University of Rochester; 1988. A Graphical Determination of the Possibility of Multiple Steady States in Complex Isothermal CFSTRs.

16. Craciun G, Feinberg M. Multiple equilibria in complex chemical reaction networks. I. the injectivity property. SIAM Journal on Applied Mathematics. 2005;65:1526–1546.

17. Banaji M, Donnell P, Baigent S. P-matrix properties, injectivity, and stability in chemical reaction systems. SIAM Journal on Applied Mathematics. 2007;67:1523.

18. Gale D, Nikaido H. The jacobian matrix and global univalence of mappings. Mathematische Annalen.1965;159:81–93.

19. Craciun G, Feinberg M. Multiple equilibria in complex chemical reaction networks: extensions to entrapped species models. IEE Proc. Syst. Biol. 2006;153:179–186. [PubMed]

20. Craciun G, Feinberg M. Multiple equilibria in complex chemical reaction networks: Semiopen mass action systems. SIAM Journal on Applied Mathematics. 2010;70:1859–1877.

21. Biggs N. Algebraic Graph Theory. second edition. Cambridge University Press; 1993.

22. Horn F, Jackson R. General mass action kinetics. Archive for Rational Mechanics and Analysis.1972;47:81–116.

23. Reidl J, Borowski P, Sensse A, Starke J, Zapotocky M, Eiswirth M. Model of calcium oscillations due to negative feedback in olfactory cilia. Biophysical Journal. 2006;90:1147–1155. [PMC free article] [PubMed]

24. Shinar G, Milo R, Rodriguez-Martinez M, Alon U. Input output robustness in simple bacterial signaling systems. Proceedings of the National Academy of Sciences. 2007;104:19931–19935. [PMC free article][PubMed]

25. Shinar G, Feinberg M. Structural sources of robustness in biochemical reaction networks. Science.2010;327:1389–1391. [PubMed]

26. Shinar G, Feinberg M. Design principles for robust biochemical reaction networks: What works, what cannot work, and what might almost work. Mathematical Biosciences. 2011;231:39–48. [PMC free article] [PubMed]

27. Craciun G. Ph.D. thesis. The Ohio State University; 2002. Systems of Nonlinear Differential Equations Deriving from Complex Chemical Reaction Networks.

28. Anderson DF. Global asymptotic stability for a class of nonlinear chemical equations. SIAM Journal on Applied Mathematics. 2008;68:1464.

29. Angeli D, Leenheer PD, Sontag ED. Persistence results for chemical reaction networks with time-dependent kinetics and no global conservation laws. SIAM Journal on Applied Mathematics. 2011;71:128.

30. Gnacadja G. Reachability, persistence, and constructive chemical reaction networks (part I): reachability approach to the persistence of chemical reaction networks. Journal of Mathematical Chemistry.2011;49:2117–2136.

31. Gnacadja G. Reachability, persistence, and constructive chemical reaction networks (part II): a formalism for species composition in chemical reaction network theory and application to persistence. Journal of Mathematical Chemistry. 2011;49:2137–2157.

32. Gnacadja G. Reachability, persistence, and constructive chemical reaction networks (part III): a mathematical formalism for binary enzymatic networks and application to persistence. Journal of Mathematical Chemistry. 2011;49:2158–2176.

33. Vol'pert A, Hudjaev S. Analysis in Classes of Discontinuous Functions and Equations of Mathematical Physics. Springer; 1985.

34. Siegel D, MacLean D. Global stability of complex balanced mechanisms. Journal of Mathematical Chemistry. 2000;27:89–110.

35. Deng J, Feinberg M, Jones CKRT, Nachman A. On the steady states of weakly reversible chemical reaction networks. 2011 ArXiv:1111.2386v2 [q-bio.QM] available at http://arxiv.org/pdf/1111.2386.pdf.

36. Browder FE. A new generalization of the Schauder fixed point theorem. Mathematische Annalen.1967;174:285–290.

37. Horn F. Necessary and suffcient conditions for complex balancing in chemical kinetics. Archive for Rational Mechanics and Analysis. 1972;49:172–186.

38. Segel IH. Enzyme Kinetics: Behavior and Analysis of Rapid Equilibrium and Steady-State Enzyme Systems. Wiley-Interscience; 1993.

39. Kato T. A Short Introduction to Perturbation Theory for Linear Operators. Springer-Verlag; 1982.

CITATION

CHAPTER 1

Görlich D, Dittrich P (2013) Molecular Codes in Biological and Chemical Reaction Networks. PLoS ONE 8(1): e54694. doi:10.1371/journal.pone.0054694

CHAPTER 2

Latino DARS, Aires-de-Sousa J (2014) Automatic NMR-Based Identification of Chemical Reaction Types in Mixtures of Co-Occurring Reactions. PLoS ONE 9(2): e88499. doi:10.1371/journal.pone.0088499

CHAPTER 3

Hongsheng Chen, Zhong Zheng, Zhiwei Chen, and Xiaotao T. Bi, A Lattice Gas Automata Model for the Coupled Heat Transfer and Chemical Reaction of Gas Flow Around and Through a Porous Circular Cylinder, doi:10.3390/e18010002

CHAPTER 4

Dnyaneshwar B. Rasale and Apurba K. Das, Chemical Reactions Directed Peptide Self-Assembly, doi:10.3390/ijms160510797

CHAPTER 5

Hassan H. Abdallah, Janez Mavri, Matej Repič, Vannajan Sanghiran Lee, and Habibah A. Wahab, Chemical Reaction of Soybean Flavonoids with DNA: A Computational Study Using the Implicit Solvent Model, doi:10.3390/ijms13021269

CHAPTER 6

Ramakrishnan N, Bhalla US (2008) Memory Switches in Chemical Reaction Space. PLoS Comput Biol 4(7): e1000122. doi:10.1371/journal.pcbi.1000122.

CHAPTER 7

Ernesto Suárez, Distinguishability in Entropy Calculations: Chemical Reactions, Conformational and Residual Entropy, doi:10.3390/e13081533

CHAPTER 8

Guy Shinar, and Martin Feinberg, Concordant Chemical Reaction Networks and the Species-Reaction Graph, doi: 10.1016/j.mbs.2012.08.002

CHAPTER 9

Wim Hordijk, Stuart A. Kauffman, and Mike Steel, Required Levels of Catalysis for Emergence of Autocatalytic Sets in Models of Chemical Reaction Systems, doi:10.3390/ijms12053085

CHAPTER 10

Katalin M. Hangos,. Engineering Model Reduction and Entropy-based Lyapunov Functions in Chemical Reaction Kinetics, doi:10.3390/e12040772

CHAPTER 11

Shinar G, Feinberg M. Concordant Chemical Reaction Networks. Mathematical biosciences. 2012;240(2):92-113. doi:10.1016/j.mbs.2012.05.004.

INDEX